T0281295

Microwave Dielectric Spectroscopy of Ferroelectrics and Related Materials

Ferroelectricity and Related Phenomena

Editor: **George W. Taylor**, Princeton Resources, Princeton, New Jersey, USA.
Associate Editor: **Lev A. Shuvalov**, Institute of Crystallography, Russian Academy of Sciences, Moscow, Russia

This book is part of a series. The publisher will accept continuation orders which may be cancelled at any time and which provide for automatic billing and shipping of each title in the series upon publication. Please write for details.

Microwave Dielectric Spectroscopy of Ferroelectrics and Related Materials

Jonas Grigas
Vilnius University
Lithuania

CRC Press
Taylor & Francis Group
Boca Raton London New York

CRC Press is an imprint of the
Taylor & Francis Group, an **informa** business

CRC Press
Taylor & Francis Group
6000 Broken Sound Parkway NW, Suite 300
Boca Raton, FL 33487-2742

First issued in paperback 2020

ISBN-13: 978-0-367-45594-1 (pbk)
ISBN-13: 978-2-88449-190-7 (hbk)

Visit the Taylor & Francis Web site at
http://www.taylorandfrancis.com

and the CRC Press Web site at
http://www.crcpress.com

British Library Cataloguing in Publication Data

Grigas, Jonas
  Microwave Dielectric Spectroscopy of
  Ferroelectrics and Related Materials.—
  (Ferroelectricity & Related Phenomena;
  Vol. 9)
  I. Title   II. Series
  537.2448

CONTENTS

SERIES EDITORS' PREFACE

The behavior of dielectric properties is one of the most important characteristics of ferroelectrics, and for a long time was even considered as a definition of ferroelectricity. The first ferroelectrics and a majority of the ones discovered in following years became known due especially to their dielectric properties. If we also take into account that unique dielectric properties of many ferroelectrics are widely used in modern technology, then the role of dielectric spectroscopy in fundamental and applied investigations of ferroelectrics is obviously important.

The frequency region of dielectric spectroscopy of ferroelectrics now stretches from hundredths of hertz to thousands of gigahertz into the infrared region. Dielectric dispersion at infralow, low, and medial frequencies permits one to judge in particular the interaction of the domains with defects and impurities. At high frequencies the main role in dielectric dispersion belongs to resonance and relaxation behavior of domains and domain walls. In the technologically complicated microwave region, dielectric spectroscopy permits one to obtain very important information about the lattice dynamic of ferroelectrics, especially in the neighborhood of structural phase transitions (soft modes, central peaks, and such). Thanks to the efforts of many scientists, dielectric spectroscopy of ferroelectrics and related materials in the entire frequency region has been developing effectively within the last decades. However, spectroscopy within the 1 m to 0.1 mm region fell behind slightly until relatively recently. Now this lag has been eliminated. One of the significant contributions to this elimination belongs to Professor Jonas Grigas in the Department of Radiophysics of the Physical Faulty of Vilnius University, Lithuania, and his pupils.

In this monograph, Professor Grigas summarizes and generalizes results of his 25-year work in the field of microwave dielectric spectroscopy of ferroelectrics and related materials.

Knowing well the difficulties of dielectric measurements in the microwave region and how to avoid artifacts, the author pays attention to his own approach to the methodology, and only later considers the results of his experimental investigations, which form the main part of the book.

The reader will find many crystals investigated in the microwave region only, or first, by Professor Grigas, including low-dimensional and ferroelectric semiconductors, quasi-one-dimensional H-bonded ferroelectrics, other order-disorder ferroelectrics, and protonic conductors.

All results of microwave dielectric investigations are considered in close connection with the atomic structure of the crystals.

New methodology, original fundamental experimental results, and profound analysis make this book not only indispensable for everybody who works—or wants to work—in the field of microwave dielectric spectroscopy or uses dielectrics in this frequency region, but also make it very useful for all scientists and graduate students whose interests are connected to the physics of ferroelectrics and related materials, physics of structural phase transition, and physics of superionic conductors.

Lev A. Shuvalov
George W. Taylor

PREFACE

Felix qui potuit rerum cognoscere causas.
—Virgil

Recent technological advances have not only affected the size of electronic elements but also greatly extended the useful frequency range.
These advances to a great extent depend on the discovery of suitable
materials useful for applications in the development of new radar
systems, satellite broadcasting, and multichannel terrestrial and cosmic
communications via the use of microwaves. These applications at
present are in great demand and are going to increase multifold in the
near future. Information transfer by digital means in communications is
controlled by high-speed computers requiring time scales of pulses
falling within the microwave range.

Ferroelectrics and related materials have become of increasing
importance for the utilization of microwave techniques. The parameters
of these materials, due to phase transitions, are extreme and nonlinear.
A number of active and passive devices, because of their intrinsic
high-speed response time, can be realized by electrical control of the
dielectric properties of these materials at frequencies up to approximately 1000 gigahertz.

Dynamic processes in solids proceed in such a way that the characteristic frequencies lie in the microwave region. A soft mode frequency in
a number of ferroelectrics on approaching the Curie temperature drops
into the microwave range. A frequency of relaxational soft modes lies
entirely in the microwave range. Spectra of the fundamental ferroelectric
dispersion and loss give information on the microscopical dynamical
mechanism of phase transitions and show the frequency regions where
these materials are useful for applications. Therefore, there is a critical
need to have reliable broad-band methods of microwave dielectric
spectroscopy of ferroelectrics and related materials. The term *related
materials* is applied to materials in which structural phase transitions are
possible.

The choice of methods for the microwave measurements depends
on the shape of a sample, value of permittivity and loss, accuracy desired,
and so on. Von Hippel made revolutionary contributions in the field of
microwave dielectric measurements. His technique has been used in
many laboratories all over the world. Various modifications of this
technique have been suggested. It is not my intention to review all the

techniques in this book. However, the book will cover only some new broad-band methods suitable for obtaining spectra of dielectric permittivity and loss of ferroelectrics and related materials in the meter to submillimeter wavelength range. I intend to show the importance of microwave dielectric spectroscopy in the field of material sciences, and leave the reader to judge the extent to which this objective is achieved.

Although I must accept sole responsibility for the opinions expressed in this book, it is my pleasure to acknowledge invaluable help from my colleagues in the Department of Radiophysics of Vilnius University in Lithuania. Among them I would like to mention my former students: Drs. Jūras Banys, Rimas Beliackas,* Saulius Lapinskas, Rolandas Mizaras, and Ričardas Sobiestianskas; their contributions can be measured by the number of references in these pages. I am, in particular, indebted to Drs. Algirdas Brilingas and Vidas Kalesinskas, who helped develop and run computer-controlled dielectric spectrometers in our laboratory; and to Dr. Vytautas Samulionis for cooperation in ultrasonic studies of most of the materials discussed in this book. I am grateful to Professors V. Shugurov and V. Ivaška, who helped me to solve some electrodynamical problems. Also, I have benefitted from cooperation and discussions with scientists from universities abroad, namely: R. Davidovich, A. Levanyuk, A. Volkov, and N. Mozgova (Russia); R. Levitsky, J. Poplavko, and I. Stasyuk (Ukraina); Z. Czapla (Poland); K. Gesi, Y. Ishibashi, W. Kinase, E. Nakamura, K. Toyoda, Y. Uesu, K. Wakino, and T. Yagi (Japan); A. M. Glazer and A. K. Jonscher (United Kingdom); G. E. Kugel and M. D. Fontana (France); J. Petzelt and J. Fousek (Czeck Republic); K. R. Rao (India); L. J. Gauckler (Switzerland); J. Toulouse (United States); and others. I deeply acknowledge the invaluable help of Professor Bruno Jaselskis for his improvement of the English.

I am indebted to Professor Lev A. Shuvalov and Dr. George W. Taylor for their encouragement to write this book.

Finally, financial support by the Foundation for Promotion of Material Science and Technology of Japan during the writing of the manuscript is gratefully acknowledged.

It is hoped that our investigations prove to be useful for researchers, engineers, and graduate students.

---

*Deceased

1 INTRODUCTION

1.1. MICROWAVES

The term *microwaves* indicates wavelength in the micron ranges. In this book, microwaves mean those wavelengths measured from 1 m to 0.1 mm, which correspond to the frequency range of 0.3 to 3000 gigahertz. The microwave frequency range is usually grouped into different wave bands: (1) the decimeter wave band for microwave frequencies from 0.3 to 3 GHz, (2) the centimeter wave band for microwave frequencies from 3 to 30 GHz, and (3) the millimeter wave band for microwave frequencies from 30 to 300 GHz. There is also a subset of microwave wavelengths measured from 1 to 0.1 mm (the submillimeter wave band) for the 300 to 3000 GHz frequency range. Microwaves are bounded on the long wavelength side by radio (meter) waves and on the short wavelength side by infrared (IR) waves. The frequency bands commonly used in electronics industries and academic institutes are shown in Tables 1.1 and 1.2.

Since wavelengths at microwave frequencies are of the same order of magnitude as the dimensions of circuit elements, the time of propagation of electrical signals from one part of the circuit to the other is comparable to the period of oscillating currents and charges. Thus, in the microwave region one deals with electric and magnetic fields rather than voltage and current.

The advantages of microwaves over other regions of the electromagnetic spectrum arise from the short wavelength monochromatic radiation and from the width of the spectrum available for use. The short

1

Table 1.1. IEEE frequency bands

Designation	Frequency	Wavelength
Voice frequency	30–3000 Hz	1–0.1 Mm
Very low frequency	3–30 kHz	100–10 km
Low frequency	30–300 kHz	10–1 km
Medium frequency	0.3–3 MHz	1–0.1 km
High frequency	3–30 MHz	100–10 m
Very high frequency	30–300 MHz	10–1 m
Ultrahigh frequency	0.3–3 GHz	10–1 dcm
Superhigh frequency	3–30 GHz	10–1 cm
Extreme high frequency	30–300 GHz	10–1 mm
Decimillimeter	300–3000 GHz	1–0.1 mm

Table 1.2. Microwave bands

Designation	Frequency (GHz)	Wavelength
P band	0.23–1.0	13.0–3.0 dcm
L band	1.0–2.0	30.0–15 cm
S band	2.0–4.0	15.0–7.5 cm
C band	4.0–8.0	7.5–3.75 cm
X band	8.0–12.5	3.75–2.4 cm
Ku band	12.5–18.0	2.4–1.67 cm
K band	18.0–26.5	1.67–1.13 cm
Ka band	26.5–40.0	1.13–0.75 cm
Millimeter	40.0–300	0.75–1.0 mm
Submillimeter	300–3000	1.0–0.1 mm

wavelength monochromatic radiation results in high directivity and resolving power of microwave antennas. Infrared sources suffer from significant attenuation when propagating through the atmosphere, except in certain narrow "windows." The microwaves give the best overall system performance, having properties of high directivity and resolving power and low atmospheric attenuation. Thus, the most important applications of microwaves are in communications and radar. In the realm of communications, microwave relay systems for telephone and television are the largest markets, with satellite communications growing rapidly. Numerous special applications such as point-to-point communications have replaced the use of telephone lines in heavily accessible metropolitan regions. Direct broadcast satellites open an immensely competitive microwave market.

Radar provided the major incentive for the development of microwave technology because only microwaves could provide the required resolution with antennas of a reasonable size. Microwaves find extensive application in electronic warfare, from specialized broadband receivers to high-power jamming transmitters.

The most familiar consumer application of microwaves is the microwave oven which uses a minimum of sophisticated circuitry. And yet, due to the high dielectric loss in water (and, therefore, high absorption of microwaves) the technique for heating foods and other water containing products has been revolutionized. Also, for the same reason, microwaves are used in medical equipment.

Microwaves have also flourished in basic research and science applications such as radio astronomy, elementary particle acceleration, nuclear fusion, spectroscopy, and materials research, which has led to the development of new solid-state active and passive microwave devices.

Several types of vacuum tubes are used for microwave power generation. Conventional yet sophisticated vacuum tubes, such as triodes and tetrodes, are still used as signal sources below 10 GHz. The most important state-of-the-art microwave tubes are the linear-beam (O-type) and the crossed-field (M-type), named after the French TPOM (tubes à propagation des ondes à champs magnetique) devices. In the linear beam, electric and magnetic fields are parallel and the electron flow forms a beam traversing the length of the tube. Resonant klystron amplifier, the two-cavity klystron oscillator, and the reflex klystron oscillator belong to the O-type tubes. The latter is capable of supplying an output power from 0.5 watts to a fraction of milliwatt in a frequency range of 1 to 100 GHz. The helix traveling-wave tube, coupled-cavity traveling-wave tube, forward-wave amplifier, and backward-wave amplifier and oscillator are nonresonant wide-band O-type tubes. The backward-wave oscillator, for instance, is a low-power, monochromatic tunable generator, which has a milliwatts output covering a frequency range of 300 MHz to 900 GHz.

In M-type devices, the dc magnetic field is perpendicular to the dc electric field and plays a direct role in the electron-field interaction process. The magnetron is the oldest powerful oscillator of this family, operating up to 100 GHz. Magnetrons are found in many thousands of radars and microwave ovens around the world. Meanwhile, radar systems have become more sophisticated in their extraction of data from radar returns. New radars require frequency and phase coherence between transmitted and received signals. This is impossible to achieve with the magnetron. Other M-type devices are the forward-wave crossed-field amplifier and oscillator (dematron) and the backward-wave crossed-field amplifier (amplitron) and oscillator (carcinotron).

In recent years, many fast-wave relativistic electron microwave devices, such as gyrotrons, ubitrons, and peniotrons, have been developed. They generate a power of megawatts in the millimeter wave region. The laser of free electrons can cover the frequency range from microwaves to visible-light.

Many solid-state devices can be operated in the microwave region. When a p-n junction is doped so heavily on both sides that the field becomes sufficiently high for quantum-mechanical tunneling, the diode generates microwaves. When a p-n junction is operated in avalanche breakdown, an IMPATT diode can, under proper conditions, generate microwaves up to several hundreds of gigahertz. The transferred-electron device has been extensively used as microwave oscillators and amplifiers, covering the 1 to 100 GHz frequency range.

The reflex klystrons, the backward-wave oscillators and the solid-state devices are the predominant choice for microwave spectroscopy and research of the linear parameters of materials.

Two types of measurements can be performed with microwaves: time domain and frequency domain. Time domain reflectometry is used to display the reflected response of systems over a frequency range from dc to about 10 GHz (see Section 3.1.3) and to investigate parameters of discontinuities in a microwave network. Two types of network analyzers are used in frequency domain measurements: the scalar network analyzer, which measures the magnitude of the ratio of the microwave signals in a network, and the vector network analyzer, which measures both the magnitude and phase of the microwave in the network (Section 3.3.6).

1.2. DIELECTRIC SPECTROSCOPY

A complete study of ferroelectric or structural phase transitions requires the contribution of two types of experimental techniques: one type sensitive to long range cooperative phenomena, the other to changes in the local order. Techniques in the first category tend to have large interaction volume and high sensitivity to symmetry changes of crystals. One such technique is dielectric spectroscopy.

The aim of dielectric spectroscopy is to determine the complex permittivity of various excitations related to polarization. The frequency range of maximum interest from the point of view of phase transition dynamics is 10^7 to 10^{12} Hz. In this frequency region the relaxational dynamics are related to various mechanisms of disorder which manifest in solids. A soft mode frequency on approaching Curie temperature in ferroelectrics also drops to this frequency region. However, possibilities of scattering and infrared reflectivity techniques are mostly limited by the wave number of $10 \, cm^{-1}$ or $300 \, GHz$ frequency.

Dielectric spectroscopy has a principal advantage over scattering methods: dielectric spectroscopy enables one to determine independently both real and imaginary parts of the response function (i.e., complex permittivity), whereas the scattering techniques permit one to determine only its imaginary part. Therefore, dielectric spectroscopy may be more informative than scattering methods and may allow for better distinction between various theoretical models used to describe phase transitions. Another advantage is the possibility to compare the soft mode contribution with the static value of permittivity and to directly examine the central-peak type excitations, if such exist.

Dielectric spectroscopy is a fairly restrictive method in relation to the selection rules. The soft mode can be dielectrically active in the first-order absorption spectrum in both paraelectric and ferroelectric phases only for a proper ferroelectric phase transition. For improper ferroelectric or nonferroelectric phase transitions the dielectric anomalies are of known and very characteristic shape and enable one to determine the presence and type of various phase transitions in solids.

Many authors have examined microwave behavior of a number of ferroelectrics and related materials. However, no complete and consistent description of the ferroelectric dispersion is available, because of restrictions of their frequency range or lack of accuracy of their measurements.

Achievements in microwave techniques during the last decades, and the new approach to the microwave dielectric spectroscopy enable one to study quantitatively ferroelectric dispersion and other excitations in solids. The strongly temperature-dependent dielectric dispersion, caused by the motion of those atoms or groups of atoms responsible for ferroelectricity, is called *ferroelectric dispersion*.

This book attempts to survey new approaches to microwave dielectric spectroscopy of ferroelectrics, semiconductors and superionics. It deliberately avoids incursions into the fundamentals of ferroelectricity or solid state physics that have been covered by a number of recent comprehensive monographs. Similarly, no attempt has been made to give a systematic description of the behavior of all ferroelectrics at microwaves. Also, nonlinear properties of these materials are not considered, beyond all manner of doubt, interesting and important for applications.

Another limitation of this work springs from the present state of incompleteness of the subject and the fact that this survey was undertaken during an initial state of intensive development of microwave dielectric spectroscopy. Infrared and scattering techniques are well developed and generally available, however, practical microwave dielectric spectrometers are not yet commercially available.

In Chapter 2, the dielectric response of solids at high and microwave frequencies, and the characteristic behavior of permittivity for various excitations are surveyed. The starting point of the original discussion is Chapter 3. It deals with electrodynamic problems and methods, which can be used in the microwave dielectric spectroscopy of solids of various shapes and electrical parameters of samples in the meter to submillimeter wavelength range. The results of the solved electrodynamic problems can also be used in the development of microwave spectrometers and for calculations of various active and passive waveguide devices. The following chapters review the most interesting results revealed by microwave dielectric spectroscopy. Chapter 4 describes a new family of crystals: low-dimensional semiconductive antimony and bismuth chalcogenides (and halogenides). Dielectric spectroscopy revealed anomalous dielectric properties, soft and "friable" lattices, and various phase transitions. This led to the conclusion that these ionic-covalent semiconductors, containing highly polarizable chalcogenide or halogenide ions, are favorable for phase transitions and ferroelectricity. A number of these crystals are either novel ferroelectrics or incipient ones.

Chapter 5 is devoted to ferroelectric dynamics of three different families of primarily displacive semiconductive ferroelectrics, such as SbSI and others. The soft mode and phase transition problem of SbSI have been discussed for three decades because the soft mode frequency near Curie temperature lies beyond the limits of conventional spectroscopical techniques. These examples of ferroelectrics show that in the vicinity of the Curie temperature the frequency of the soft mode can be extremely low. Due to large anharmonicity, it can be strongly overdamped or split into several components and cause broad absorption and dielectric dispersion in the 10^{10} to 10^{12} Hz frequency range. The soft mode behavior in these ferroelectrics shows that displacive-like and order-disorder like behaviors in anharmonic ferroelectrics are in fact regimes dominant at different temperatures. The crossover of these mechanisms occurs on approaching Curie temperature. Taking into account microwave spectroscopic results, the model of phase transition in SbSI-type crystals is proposed. The examples of ferroelectric dynamics of incommensurate crystals with amplitudon and phason excitations at microwaves are also presented.

In Chapter 6, examples of complete dielectric spectra of ferroelectric dispersion in quasi-one-dimensional H-bonded ferroelectrics of the KDP-family are presented. Chapter 6 also discusses in cluster approximation dynamical models of the phase transitions, treating them as the result of crystal instability in a proton subsystem, which give a quantitative description of both dynamical and static properties. The ferroelectric dispersion at microwaves, and mechanism of phase transitions in other new or less studied order-disorder ferroelectrics (where one or more than one subsystem is involved in the ferroelectric dynamics) is discussed in Chapter 7. Finally, Chapter 8, which was written in collaboration with Dr. Saulius Lapinskas, is devoted to microwave dielectric spectroscopy and dynamic properties of protonic conductors. A dynamic model of superionic phase transition in protonic conductors is also proposed.

2 DIELECTRIC DISPERSION

The theory of the dielectric dispersion of dielectrics has been discussed in many textbooks and reviews (see, e.g., Jonscher 1983; Kneubühl 1989, and references therein). Recalled here are some definitions and theoretical models used by experimenters to interpret measured results, with emphasis on the microwave spectral range.

2.1. COMPLEX PERMITTIVITY: Kramers–Kronig Relations

Dielectric dispersion is the frequency-dependent relation between electric displacement D or the dielectric polarization P and the electric field E, and can be described by complex permittivity or susceptibility. Assuming harmonically oscillating fields of circular frequency ω, the electric field, electric displacement, and dielectric polarization are given as

$$E = E_0 \exp(i\omega t),$$
$$D = \varepsilon_0 \varepsilon(\omega) E_0 \exp(i\omega t), \quad (2.1)$$
$$P = \varepsilon_0 \chi(\omega) E_0 \exp(i\omega t),$$

where relative permittivities $\varepsilon(\omega)$ and susceptibilities $\chi(\omega)$ are found. Relative permittivities are related to susceptibilities by

$$\varepsilon(\omega) = 1 + \chi(\omega). \quad (2.2)$$

9

In general, the polarization vector, due to the moments of various types of polarization, may be unable to follow the change of the field. As a result $\varepsilon(\omega)$ and $\chi(\omega)$ become complex:

$$\varepsilon^*(\omega) = \varepsilon'(\omega) - i\varepsilon''(\omega),$$

$$\chi^*(\omega) = \chi'(\omega) - i\chi''(\omega),$$

(2.3)

with

$$\varepsilon'(\omega) = 1 + \chi'(\omega) \quad \text{and} \quad \varepsilon''(\omega) = \chi''(\omega).$$

There are three main types of polarization: dipole, ionic and electronic. Dipole polarization is observed in the frequency range from low to microwave frequencies. Ionic polarization is detected in short microwave and infrared ranges. In a wide frequency range, various kinds of dipoles and phonon modes cause several dispersion regions, which produce the dielectric spectrum of the material.

The real part of the permittivity or susceptibility is the measure of polarization of a medium. It also determines the short-time average of the energy density of the oscillating electric field. The imaginary part defines the short average power absorbed by the unit volume of the medium, and it is the measure of energy loss. It is related to electrical conductivity σ for the given frequency by

$$\varepsilon'' = \frac{\sigma}{\varepsilon_0 \omega}.$$

The real and imaginary parts of the permittivity are functionally related by Kramers–Kronig relations:

$$\varepsilon'(\omega) - \varepsilon_\infty = \frac{2}{\pi} P \int_0^\infty \frac{\omega' \varepsilon''(\omega')}{(\omega')^2 - \omega^2} d\omega',$$

(2.4)

$$\varepsilon'(\omega) = \frac{2}{\pi} P \int_0^\infty \frac{[\varepsilon'(\omega') - \varepsilon_\infty]\omega}{(\omega')^2 - \omega^2} d\omega',$$

(2.5)

where P indicates Cauchy's principal part of the integral, ω' is an integration variable, and ε_∞ is a contribution of the polarization, which adapts instantaneously to changes of the electric field of a given frequency. Equations 2.4 and 2.5 show that any excitation giving rise to an electromagnetic field absorption peak, gives the contribution $\varepsilon'(\omega) - \varepsilon_\infty$ to the static permittivity, and any dielectric dispersion is followed by an absorption peak. These relations in optical spectroscopy are used to calculate $\Delta\varepsilon$ from the measurements of absorption. They have also been adapted to the fast Fourier algorithm of digital computers. Kramers–Kronig relations are applicable to all linear dissipation mediums. A simple way of deriving the Kramers–Kronig equations was proposed by Frölich (1958).

2.2. DEBYE-TYPE DISPERSION

In an external electric field, a molecule with a permanent dipole moment undergoes orientation. In an alternating field, the orientation depends on the frequency and causes a dielectric dispersion. The classical theory of the dielectric dispersion for polar liquids, which occurs mainly through reorientation of the noninteracting permanent dipoles, was developed by Debye (1954). In the Debye model, the relaxing polarization P_r is assumed to decay after the step removal of the electric field according to the following equation:

$$\frac{dP_r}{dt} = -\frac{1}{\tau}[P_r(t) - P_e(t)], \tag{2.6}$$

where τ corresponds to the average relaxation time, and the equilibrium polarization is given by

$$P_r(t) = \varepsilon_0\chi(0)E(t), \tag{2.7}$$

which vanishes after switching off the electric field. The normalized macroscopic relaxation function $\Phi(t)$ describes the decay of the relaxing polarization and is expressed as

$$\Phi(t) = \exp\left(-\frac{t}{\tau}\right). \tag{2.8}$$

Thus, the Debye model postulates a simple exponential decay of the relaxing polarization $P_r(t)$.

Approximations of $\Phi(t)$ are

$$\Phi(t \to 0) = 1 - (t/t),$$
$$\Phi(t \to \infty) = \exp(-t/\tau). \tag{2.9}$$

The relation between the complex permittivity and the function $\Phi(t)$ was first derived by Kubo (1957):

$$\frac{\varepsilon^*(\omega) - \varepsilon_\infty}{\varepsilon(0) - \varepsilon_\infty} = \int_0^\infty -\frac{d\Phi(t)}{dt}\exp(-i\omega t)dt. \tag{2.10}$$

The characteristic permittivity $\varepsilon(\omega)$ of the Debye model can be derived from $\Phi(t)$ with Kubo's relation (Equation 2.10). The result is

$$\varepsilon^*(\omega) = \varepsilon_\infty + \frac{\varepsilon(0) - \varepsilon_\infty}{1 + i\omega\tau} = \varepsilon_\infty + \frac{S_r}{1 + i\omega\tau}, \tag{2.11}$$

where $S_r = \varepsilon(0) - \varepsilon_\infty$ is the strength of the relaxator. Equation 2.11 is known as the *Debye equation*. The real part of the permittivity decreases to ε_∞ with increasing frequency, while the imaginary part (and, consequently, the absorption) exhibits a maximum at the frequency $\omega = 1/\tau$, but vanish at zero and infinite frequencies.

The permittivity at high frequencies using Equation 2.11 can be approximated by

$$\frac{\varepsilon^*(\omega \rightarrow \infty) - \varepsilon_\infty}{S_r} \cong (\omega\tau)^{-2} - i(\omega\tau)^{-1}. \tag{2.12}$$

This demonstrates that for Debye relaxation the imaginary part of permittivity decreases as $1/\omega$ at high frequencies, which contradicts the Kubo's linear-response theory.

For analysis of experimental results it is convenient to plot the imaginary part $\varepsilon''(\omega)$ of the permittivity versus the real part $\varepsilon'(\omega)$ as a function of frequency. Elimination of $\omega\tau$ in the Debye equation (2.11) gives an equation representing a semicircular arc with the symmetry axis going through the point $[\varepsilon(0) + \varepsilon_\infty]/2$. This plot was introduced by Cole and Cole (1941).

However, a number of dielectrics exhibit deviations from the simple Debye model, and the Cole–Cole plot is inside a semicircle, defined by Equation 2.11. Thus, Cole and Cole (1941) introduced the empirical equation

$$\varepsilon^*(\omega) = \varepsilon_\infty + \frac{S_r}{1 + (i\omega\tau)^{1-\alpha}}, \tag{2.13}$$

which differs from Equation 2.11 by the power exponent α. The values of the parameter α are contained in the range $0 \leq \alpha \leq 1$; α increases with increasing internal degrees of freedom of dipoles.

Some dielectrics give an asymmetric arc. Davidson and Cole (1951) derived an empirical formula that describes the asymmetric contour of permittivity as

$$\varepsilon^*(\omega) = \varepsilon_\infty + \frac{S_r}{(1 + i\omega\tau)^\beta}, \tag{2.14}$$

where β is contained within the limits $0 < \beta \leq 1$. As β decreases, the arc becomes progressively more asymmetric.

Debye's theory and the two-minimum potential barrier model, which is often adopted in the case of ferroelectrics, give a single relaxation time. More complex cases give more than one relaxation time. The resultant dispersion diagram can be a superposition of Debye's curves for a single relaxation time. The relaxation time in proper ferroelectrics, due to a critical slowing down of polarization fluctuations, is strongly temperature dependent (i.e., $\tau = \tau_0/(T - T_c)$). In a dispersion region for frequencies $\omega > 1/\tau$ at the T_c temperature, a minimum permittivity is obtained as opposed to a static maximum.

2.3. RELAXATION AT HIGH FREQUENCIES

Debye theory usually represents a good approximation at low frequencies but deviates from the basic Kubo formalism at high frequencies. It does not fulfill the sum rule for the conductivity spectrum and leads to too high of an ε'' value at $\omega \gg 1/\tau$. As a consequence, many attempts have been made (see Kneubühl 1989) to improve phenomenological relaxation formalism, which agrees with the Debye relaxation at low frequencies and fulfills the restrictions of Kubo's theory at high frequencies. (In this chapter the terms *low* and *high* frequency are applied only in relation to a given polarization mechanism). The most simple is the sech-approximation.

In the sech-approximation, the relaxation function (Equation 2.8) is given by

$$\Phi_s(t) = \text{sech}(t/\tau), \tag{2.15}$$

which for short and long times, is in agreement with Kubo's theory

$$\Phi_s(t \to 0) \cong 1 - \frac{1}{2}\left(\frac{t}{\tau}\right)^2,$$

$$\Phi_s(t \to \infty) \cong 2\exp\left(-\frac{t}{\tau}\right). \tag{2.16}$$

For this approximation, the short-time relaxation time is $\tau_s = \tau$, while the average relaxation time is

$$\langle \tau_s \rangle = \frac{\pi}{2}\tau. \tag{2.17}$$

The complex permittivity of the sech-approximation, derived from Kubo's theory for low frequencies, gives

$$\varepsilon(\omega \to 0) - \varepsilon_\infty \cong 1 - 2G(\omega\tau)^2 - \frac{\pi}{2}\omega\tau, \qquad (2.18)$$

where $G = 0.915965$ is the Catalan's constant. At high frequencies, the complex permittivity is given by

$$\varepsilon^*(\omega \to \infty) - \varepsilon_\infty \cong -(\omega\tau)^{-2} - i\pi\omega\tau \exp\left(-\frac{\pi}{2}\omega\tau \right). \qquad (2.19)$$

The sech-approximation deviates from the Debye model (Figure 2.1) at high as well as medium frequencies because in this approximation the

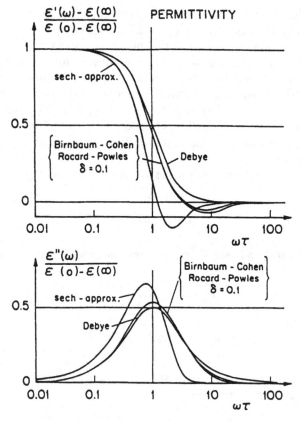

Figure 2.1. Dispersion of complex permittivities of various models. For designations and definitions see text. From Kneubühl (1989).

short-time relaxation time equals the long-time relaxation time. In real dielectrics these times may be different.

These two relaxation times can be distinguished by introducing (Birnbaum and Cohen 1970) a different relaxation function instead of Equation 2.15:

$$\Phi(t) = \exp\{\delta - [\delta^2 + (t/\tau)^2]^{1/2}\}, \tag{2.20}$$

where the difference between these two relaxation times is described by the factor δ, which is of the order 10^{-2} to 10^{-3}. In this approximation

$$\Phi(t \to 0) \cong \left[1 - \frac{1}{2}\left(\frac{t}{\tau_s}\right)^2\right],$$

$$\Phi(t \to \infty) \cong \exp[\delta - (t/\tau)], \tag{2.21}$$

$$\tau_s = \delta^{1/2}\tau.$$

Using this approximation the complex permittivity at low frequencies is given by

$$\frac{\varepsilon^*(\omega \to 0) - \varepsilon_\infty}{S_r} \cong 1 - (\omega\tau)^2(1 + \delta) - i(\omega\tau)\delta\exp(\delta)J_1(\delta), \tag{2.22}$$

and at high frequencies by

$$\frac{\varepsilon^*(\omega \to \infty) - \varepsilon_\infty}{S_r} \cong -\frac{1}{(\omega\tau_s)^2} - i\left(\frac{\pi\delta}{2\omega\tau}\right)^{1/2}\exp[\delta(1 - \omega\tau)], \tag{2.23}$$

where J_1 is the modified Bessel function of the second kind. The average relaxation time is

$$\langle\tau\rangle(\delta \to 0) = \tau. \tag{2.24}$$

The Birnbaum–Cohen relaxation for $\delta \leq 0.1$ is similar to the Debye relaxation for medium and low frequencies $\omega \leq 1/\tau$ (Figure 2.1).

As a compromise between the stringent conditions of Kubo's theory and the mathematical complications of the Birnbaum–Cohen relaxation, Rocard and Powles (see Powles 1948) have introduced the simple relaxation function

$$\Phi(t) = \frac{\exp(-|t|/\tau) - \delta\exp(-|t|/\delta\tau)}{1 - \delta}. \tag{2.25}$$

For short-time relaxation times this becomes

$$\Phi(t \to 0) \cong 1 - \frac{1}{2}\left(\frac{t}{\tau_s}\right)^2,$$

where $\tau_s = \delta^{1/2}\tau$. And for long-time relaxation times:

$$\Phi(t \to \infty) \cong (1 - \delta)^{-1} \exp(-t/\tau).$$

For small δ this approximation corresponds to the Debye relaxation function. The average relaxation time is

$$\langle \tau \rangle = (1 + \delta)\tau$$

with τ given by Equation 2.24. The complex permittivity of Rocard–Powles relaxation is given by

$$\varepsilon^*(\omega) = \varepsilon_\infty + \frac{S_r}{(1 + i\omega\tau)(1 + i\delta\omega\tau)}. \qquad (2.26)$$

At low frequencies and for small δ the difference between this and Birnbaum–Cohen relaxation is negligible. However, for high frequencies the imaginary part of the complex permittivity as related by the Rocard–Powles relaxation differs from the imaginary part of permittivity as given by the Birnbaum–Cohen relaxation due to the violation of Kubo's formalism. All the relaxations are compared in Figure 2.1.

Finally, the simple Rocard–Powles relaxation in most cases can replace the proper Birnbaum–Cohen relaxation at high frequencies. The Debye and related relaxation models are usually suited for the dielectric dispersion and absorption analysis at microwaves and lower frequencies. However, they are not feasible at infrared or even submillimeter ranges. Figure 2.1 demonstrates that the relaxational dispersion at high frequencies yields a negative value of dielectric permittivity, which usually is supposed to be the characteristic feature of oscillatory type dielectric dispersion.

2.4. OSCILLATORY-TYPE DISPERSION

The Debye relaxation and its modifications, described in previous sections, are adequate for relatively slow orientational polarization due to permanent electric dipoles. In most cases in the short microwave and infrared regions the dielectric dispersion and absorption mechanism are dealt with on the atomic scale. H.A. Lorentz's classical damped harmonic oscillator (DHO) model describes the complete dielectric function of dielectrically active phonon modes:

$$\varepsilon^*(\omega) = \varepsilon_\infty + \frac{S\omega_T^2}{\omega_T^2 - \omega^2 - i\gamma\omega}, \qquad (2.27)$$

where, as in Equation 2.11, $S = \varepsilon(0) - \varepsilon_\infty$ is the oscillator strength, γ is the damping (line width), ω_T is the frequency where $\varepsilon''(\omega)$ is peaked (i.e., the transversal optical mode frequency $\omega_{T0} = \omega_T$ (see Section 4.4.3)), as long as γ is not too large compared to ω_T. The line width of the soft mode in ferroelectrics is much broader than that of ordinary phonon modes, especially near Curie temperature, T_c, where the soft mode becomes overdamped. Though any spectrum can be decomposed into a sum of classical oscillator terms, the interpretation of a soft mode spectrum is rather problematic. It is usually believed that DHO is reduced at $\omega \ll \omega_T$ to Debye relaxator by the replacement of $\tau = \gamma/\omega_T^2$ in the limit of high relative damping (i.e., $\Gamma^{-1} = \omega_T/\gamma \ll 1$). However, this correspondence is only a formal one without physical meaning.

There are cases which cannot be described by Equation 2.27. In general, a system of oscillators in solids does not necessarily relax towards a thermodynamic equilibrium but relaxes towards a time-dependent quasi-equilibrium. In such cases, average positions of oscillators cannot follow the variation of the external force, and it may be considered to relax towards an instantaneous equilibrium position. In contrast to the case of the DHO model, which is valid for any short time interval, Van Vleck–Weisskopf and Fröhlich (see Takagi 1979) have derived a generalized form of resonance permittivity (GVWF) for an oscillatory model:

$$\varepsilon^*(\omega) = \varepsilon_\infty + \frac{S(\omega_T^2 + \gamma\gamma' - i\gamma'\omega)}{\omega_T^2 + \gamma\gamma' - i(\gamma + \gamma')\omega - \omega^2}. \qquad (2.28)$$

The appearance of the γ'-term is due to a coarseness of time interval where the evolution of the system is neglected.

If $\omega_T = 0$, the GVWF oscillator changes smoothly into Equation 2.11 of the Debye relaxator. In the case of $\gamma' \to 0$, the GVWF oscillator follows Equation 2.27 of DHO. When $\gamma' = \gamma$, the GVWF oscillator becomes the so-called Van Vleck–Weisskopf–Fröhlich (VWF) oscillator.

A closer inspection of Cole–Cole plots for DHO and VWF models shows that with the increase of Γ, Cole–Cole plots of the VWF oscillator approach smoothly to a semicircle given by Debye (Equation 2.11). In contrast, those of DHO (Equation 2.27) become distorted in the high frequency part as given by Equations 2.19, 2.23 or 2.26. This reflects the fact that the real part of permittivity for DHO for any value of Γ becomes negative for $\omega/\omega_T > 1.0$, and that of the Debye model is always positive for any frequency. Thus, the dielectric spectra of DHO and the relaxators, described by Equations 2.19, 2.23 and 2.26, are similar.

The temporal behavior of the decay of the DHO and VWF oscillators shows that in case of $\Gamma \geq 2$ they are not strictly "oscillators", as the relaxational motion goes on without the oscillation in the phase plane.

By solving Equation 2.27 in the form

$$\varepsilon^*(\omega) = 0 \qquad (2.29)$$

we find two roots given by

$$\omega_L = \pm \left[\frac{\omega_T^2 \varepsilon(0)}{\varepsilon_\infty} - \frac{1}{4}\gamma^2 \right]^{1/2} - i\left(\frac{1}{2}\gamma\right). \qquad (2.30)$$

We define ω_L as the longitudinal optic mode frequency. It is a complex one. Equation 2.29 states the exact definition of the longitudinal optic mode frequency. To derive the Lyddane–Sachs–Teller (LST) relation (see Equation 4.15) from Equation 2.27 we merely use the absolute value of ω_L squared:

$$|\omega_L|^2 = \frac{\omega_T \varepsilon(0)}{\omega_\infty}. \qquad (2.31)$$

This is the LST relation for the classical damped oscillator, where the frequencies are the poles and zeros of the permittivity.

The location of the poles and zeros of $\varepsilon^*(\omega)$ for the DHO and VWF oscillators also shows distinct differences between the DHO and VWF models. Poles of VWF reach the imaginary axis only in the limiting case as $\Gamma \to \infty$, while two poles of DHO reach the imaginary frequency axis at $\Gamma = 2.0$. When $\Gamma \to \infty$ one of the poles moves in the complex frequency plane towards the origin. For DHO, in case of $\Gamma \geq 2$, the frequencies are given by

$$\omega_{12} = -i\left[\frac{\gamma}{2} \pm \left(\frac{\gamma^2}{4} - \omega_T^2 \right)^{1/2} \right]. \qquad (2.32)$$

In this case, only the absolute value of ω_1 given by Equation 2.29 has a physical meaning. Meanwhile, ω_T and γ lose their original meanings, separately.

For the Debye dispersion we also can write the LST relation. The right-hand side of Equation 2.11 can be transformed into

$$\varepsilon^*(\omega) = \varepsilon_\infty \left(\frac{\omega_L - \omega}{-i\omega_T - \omega} \right), \qquad (2.33)$$

where

$$\omega_T = \frac{1}{\tau}, \quad \omega_L = -\frac{i(\varepsilon_\infty + S_r)}{\varepsilon_\infty \tau}. \qquad (2.34)$$

The definitions given by Equation 2.34 parallel those for the DHO; ω_L is the complex frequency where the zero of $\varepsilon^*(\omega)$ occurs and ω_T is the absolute value of the frequency where the pole occurs. By writing the dielectric spectrum to exhibit explicitly, the poles and zeros, the LST relation for the relaxator is obtained:

$$\frac{\varepsilon(0)}{\varepsilon_\infty} = \frac{|\omega_L|}{\omega_T}. \qquad (2.35)$$

The pole and zero frequencies enter the LST relation linearly rather than quadratically for the Debye spectrum.

The LST relation is exact as long as we include all modes in the sense of accounting for all peaks in the spectrum of ε''. Attempts have been made, from the spectrum in a limited frequency range, to state that the LST relation has "broken down". However, the LST relation can be derived (Barker 1975) from the general principles of statistical mechanics and breaks down only when statistical mechanics or casual behavior break down.

The polarization fluctuations of transverse and longitudinal character at long wavelengths are given by

$$\langle P^2 \rangle_\omega = \frac{kT}{\hbar\omega} \text{Im}[\varepsilon^*(\omega)], \qquad (2.36)$$

$$\langle P^2 \rangle_\omega = \frac{kT}{\hbar\omega} \text{Im}\left[-\frac{1}{\varepsilon^*(\omega)} \right], \qquad (2.37)$$

where k is Boltzman's constant. The weighing factor $1/\omega$ can cause a pronounced peak in dielectric spectra having sufficient weight at a lower frequency limit. This is manifested as a central peak as, for instance, in Raman spectra. However, the occurrence of a central peak does not imply Debye-type dispersion. Any dielectrically active excitation with a suitable low- frequency form can show a central peak when Equations 2.36 and 2.37 are evaluated.

Free carriers. From Equation 2.3 it follows that for real frequencies, ε'' must be an odd function of frequency. In the above sections, materials that have $\varepsilon'' \sim \omega$ at low frequencies were considered. In semiconductors for instance, due to free carriers, ε'' does not vanish as $\omega \to 0$ but depends on frequency as $\varepsilon'' \sim 1/\omega$ (see Jonscher 1983). The dielectric spectrum for quasi-free carriers is given by

$$\varepsilon^*(\omega) = \varepsilon_\infty + \frac{\omega_p^2}{-\omega^2 - i\omega\omega_c}, \tag{2.38}$$

where ω_c is the carrier collision frequency and ω_p is the plasma frequency (see Equation 3.102). For this type of spectrum, the permittivity at low frequencies does not reach a limiting static value. The imaginary part also increases without limit. In highly conductive materials the real part can level off even at a negative value. There are two poles of this spectrum that lie on the imaginary frequency axis. The longitudinal fluctuation spectrum has a peak near the zero of Equation 2.38 and is called the plasma frequency. The microwave dielectric spectroscopy of such materials is discussed in Section 3.3.7.

Frequency-dependent damping arises when the electric field couples to an optic mode, which, in turn, is coupled to an optically inactive mode or group of modes (Barker 1975). Consider the dielectric function

$$\varepsilon^*(\omega) = \varepsilon_\infty + \frac{S\omega_T^2}{\omega_T^2 - \dfrac{\alpha\omega_T^2}{1 - i\omega\tau} - \omega^2}, \tag{2.39}$$

where the designations are the same as in Equations 2.11 and 2.27. Equation 2.39 satisfies the general requirements of statistical mechanics. The second term in the denominator, which has the frequency-dependence characteristic of Debye dispersion, can be considered as a correction to the restoring force of the first term. In the high-frequency limit it approaches zero, but at lower frequencies it has significant real and imaginary parts that affect the total denominator. If these manifestations are caused by strong interactions of the primary oscillator of strength S with other phonons, the coupling constant α governs this strength of interaction, which contributes to the damping. There are three poles and three zeros that can result from Equation 2.39. The LST relation must include all of them. They give a resonance near ω_T and a broad peak at lower frequencies. The latter one corresponds to driving all the coupled phonons via the effective charge of the principle mode. Essentially, S is spread over both peaks, and the parameter α controls the fraction of S which appears in each. When Equation 2.39 describes a soft mode behavior and when the spectra are examined only by infrared or Raman spectroscopy in the region of high-frequency peak, it is quite easy to miss the true soft-mode behavior (see Section 5.2.1). The divergence of $\varepsilon(0)$ near a ferroelectric phase transition temperature may consist entirely of the lower frequency pole on the imaginary axis moving towards the origin (i.e., the frequency peak in the spectrum may not soften).

Coupled modes. The experimental findings show that in a number of proper ferroelectrics, the temperature dependence of $\varepsilon(0)$ predicted by the LST relation is different from what is measured, and a considerable part of the contribution to $\varepsilon(0)$ comes not from the modes active in infrared or Raman spectra but from excitations at lower frequencies. This stimulated exploration of the models that involve the coupling of the soft mode to another plausible lower frequency oscillator (central mode after Petzelt et al. 1987) or to low-frequency localized impurity (disorder) mode (Burns 1976). Some of these later investigations have not been confirmed (Chapter 5); some were very fruitful.

The coupling of the soft mode to the disorder mode also yields to dielectric dispersion in microwave or lower frequency ranges. In the case of a considerable contribution of other phonons to the permittivity, this coupling results in large enhancement of the static dielectric permittivity (see Equations 4.19 and 4.20 in Section 4.4.3) for a relatively small number of impurities.

Aforementioned manifestations of various excitations, which lie beyond the limits of optical spectroscopy, and the variety of empirical models to describe them show the importance of microwave dielectric spectroscopy of ferroelectrics and related materials for solid state physics.

3 MICROWAVE DIELECTRIC SPECTROSCOPY

3.1. STATUS OF MICROWAVE DIELECTRIC MEASUREMENTS OF MATERIALS

3.1.1. Introduction

The dielectric response of materials provides information about the orientational adjustment of dipoles and the translational adjustment of mobile charges present in a dielectric medium in response to an applied electric field. Microwave dielectric spectroscopy of ferroelectrics enables the independent determination of the dielectric permittivity and loss in the fundamental ferroelectric dispersion region, as well as the parameters of the soft modes related to phase transitions.

Besides scientific purposes, microwave dielectric measurements are of increasing importance in telecommunications related applications and the design of microwave circuit components. These applications would include imaging radars, guidance systems, surveillance and secure communications. The magnetic properties are also of crucial importance. Dielectric and magnetic parameters fully characterize the manner in which electromagnetic waves propagate within the medium. The difficulties of making measurements on a wide range of materials over a wide frequency (and temperature) range have led to the development of various direct and indirect methods.

At microwave frequencies, the direct single-frequency methods were enriched in recent years with the more convenient broad-band

Frequency-Domain Dielectric Spectroscopy (FDDS) (Brilingas et al. 1986; 1987; Grigas et al. 1990), Time-Domain Spectroscopy (TDS) (Gestblom and Jonsson 1980) and Dispersive Fourier Transform Spectroscopy (DFTS) (Afsar and Button 1981). TDS now operates over 5 decades with an upper limit in the 20 GHz region. FDDS operates from low frequencies to 1000 GHz, while DFTS is used from ultraviolet down to 60 GHz (Birch 1987). Significant contributions to DFTS have come from NPL, Teddington, UK (Afsar and Chantry 1977; Birch 1981; 1987; Birch and Clarke 1982), and from MIT, Cambridge, USA (Afsar et al. 1986).

In spite of the development of the broad-band spectrometers, the discrete frequency and narrow-band methods have not had their day. Many applications require a degree of sensitivity that is not achievable by the broad-band techniques. Engineering applications are usually narrow-band, and discrete frequency measurements are sufficient.

Computer-controlled spectrometers are now the norm in dielectric spectroscopy. Computers allow the computation of electromagnetic fields in entirely new measurement geometries and the use of numerical analyses in the direct measurement process. The use of such computer-centered spectrometers is now one of the most fruitful factors in generating entirely new approaches to microwave dielectric spectroscopy. Each investigator employs a method adequate for the size and the shape of a sample. The most important problem now is the rigorous mathematical solution of the microwave interaction with the samples of various geometries.

Earlier general accounts by von Hippel (1954), Brandt (1963), Bussey (1967), Fellers (1967), Petrov (1972), and Chamberlain and Chantry (1973), which are still recommended, and more recent surveys by Lynch (1974), Birch and Clarke (1982), Afsar and Button (1981, 1985), Afsar (1984, 1985), and Birch (1987) give a common view of microwave dielectric and magnetic measurements. Biological dielectric measurements (Pething 1979; Grant et al. 1978) are also now of great importance. Modern applications of ferroelectrics (Vendik 1979) and dielectrics (Laverghetta 1984) in microwave engineering stimulate microwave dielectric investigations.

Although there is now complete overlap and coverage of the radio-frequency to the infrared band, the discussion of the experimental methods based on coaxial, waveguide and free-space techniques is still divided. Conventional microwave resonant measurements will not be discussed despite their high sensitivity. They are not convenient for microwave dielectric spectrometry in relation to ferroelectrics, especially in the fundamental dispersion region.

3.1.2. Frequency-Domain Measurements

In the lower megahertz region, the classical approach to the measurement of dielectric parameters of materials treats the sample as a lossy capacitance to be measured in a bridge circuit or else a resonated one. For lossy materials, nonresonant bridges are more appropriate and operate up to ultrahigh frequencies (UHF) (Lynch 1977; Pratt and Smith 1982). The resonance substitution method has also remained in use because of its ability to measure very low dielectric loss (Ogawa and Kakimoto 1978). Various mathematical corrections are used to overcome the effects of fringing fields in two-terminal cells (see Section 3.2.1).

In the UHF range, coaxial lines with a small sample (as a lumped admittance terminating line) (Rzepecka and Stuchly 1975), the analog of the two-terminal cell, and the TEM re-entral cavity (Kaczkowski and Milewski 1980) are used for wide-band measurements of complex permittivity.

Biological substances exhibit a very wide range of complex dielectric permittivity (Stuchly and Stuchly 1980a). Numerous geometries using the coaxial line reflection method for measurements of dielectric parameters of biological substances at radio and microwave frequencies were reviewed in depth by Stuchly and Stuchly (1980b). Since 1980 the open-ended coaxial line has become popular for nondestructive dielectric measurements of biological substances *in vivo* (Gaida and Stuchly 1983) when the specimen does not have to be contained in a cell but is assumed to be sufficiently large so that effects of the back surface and sides can be ignored.

Microstrips (Pannel and Jervis 1981; Kent and Kohler 1984) and other modern transmission lines (Maj and Modelski 1984) can be used for broad-band measurements of dielectric parameters of materials below 10 GHz.

The noncontact method, using two dielectric resonators that support the $TE_{01\delta}$ mode, was developed by Nishikawa et al. (1990) for nondestructive measurements of complex permittivity of an isolated small area in a dielectric substrate in the 1 to 12 GHz frequency range.

Reflection and transmission measurements are mostly used in waveguides. They have become far more convenient to use with the advent of automatic network analyzers and six-port reflectometers. Van Loon and Finsy (1975) have described a waveguide technique for liquids which operate from 5 to 150 GHz.

A combined reflection-transmission approach using modern network analyzers facilitates the measurements of dielectric parameters of samples placed in a transmission line between the two ports of the analyzer (Ligthart 1983).

3.1.3. Time-Domain Measurements

With the development of fast oscilloscopes and tunnel-diode step generators, the traveling-wave time-domain spectroscopy appeared. The time-domain spectrometer, operating in the 10^{-4} to 10^6 Hz frequency range, was described by Hyde (1970). By 1973 many of the basic features had become established (Van Germet 1973). Microwave time-domain spectrometers, using numerous coaxial sample cell geometries, have been developed as discussed in survey by Yu et al. (1979). Both reflection (Rzepecka and Stuchly 1975) and transmission (Gestblom and Noreland 1977) methods of TDS are now commonly used to 10 GHz or above.

The combined development of the theoretical basis for viewing dielectric behavior as a purely time-domain phenomenon rather than a Fourier transform of a frequency response (Cole 1976), the critical comparisons of the different methods with respect to time-domain spectroscopy (Gabriel et al. 1984), and microprocessor-controlled (Parisien and Stuchly 1979) or computer-controlled (Gestblom and Jonsson 1980; Boned and Peyrelasse 1982) time-domain spectrometers, has had a significant influence upon the continuing popularity of these time-domain techniques for the dielectric spectroscopy of various materials in the long wavelength range of microwaves.

New femtosecond time-resolved spectroscopy through impulsive stimulated Brillouin and Raman scattering permits observation in real time of acoustic and optic phonon oscillations of the amplitude 10^{-5} to 10^{-3} Å respectively (Dougherty et al. 1991). Thus direct observation of the frequency and damping of the soft mode and relaxational mode in ferroelectrics is possible.

3.1.4. Microwave Open Resonators and Free-Space Methods

Open resonators are the quasi-optical microwave analogs of the Fabri–Perot resonator, which has been developed for dielectric measurements of low-loss materials in the millimeter-wave region 30 to 200 GHz. The comprehensive reviews by Clarke and Rosenberg (1982) and Cullen (1983) give all details on this technique and its development.

The most sensitive open resonators are of the bi-concave type (Figure 3.1). The electromagnetic field in the resonator is of the form of a standing TEM-wave Gaussian beam. Yu and Cullen (1982) developed the Gaussian-beam theory for dielectric measurements with open res-

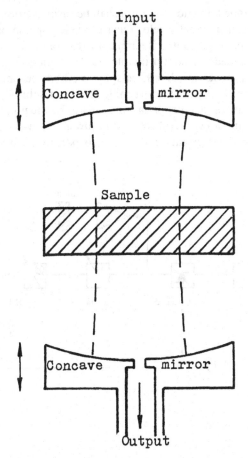

Figure 3.1. Open-resonator geometry. From Afsar et al. (1986).

onators. The resonant fields have the Gaussian-like maximum amplitude on the axis and fall away moving away from it. Sample diameters down to six wavelengths are needed for measurements at the "waist" of the resonator. Open resonators are as sensitive as closed cavities and can be of value over the 2 to 300 GHz frequency range, provided large diameter samples are available. The remarkably high Q-factor of about 10^5 of the resonators allows the measurement of low loss very accurately. Open resonators have many other applications (Clarke and Rosenberg 1982) including measurements of magnetic parameters of materials.

For large-aperture samples free-space microwave methods possess an advantage over guided-wave methods because the cross-sectional shape is not important, and the sample can be easily introduced into the beam of the electromagnetic field. The sample is of plane-parallel form and a measurement geometry is similar to that of the open-resonator, but all the basic measurement principles are the same as in guided-wave measurements (see Section 3.3). Free-space methods become more convenient in some cases in the short millimeter and submillimeter wave range when quasi-optical propagation tends to be the norm.

Both transmission and reflection measurements are possible (Cook and Rosenberg 1979) (Figure 3.2). The incorporation of the free-space

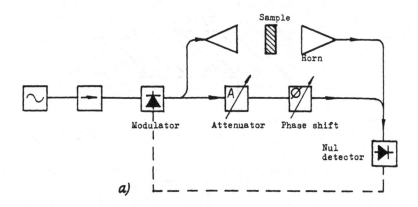

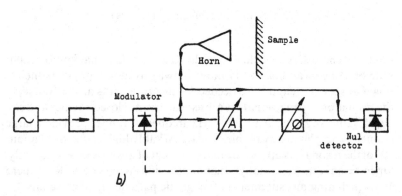

Figure 3.2. Free-space transmission (a) and reflection (b) bridge. From Afsar et al. (1986).

range into one arm of a bridge improves accuracy and, with the additional application of corrections for multiple reflections in the sample, one is able to measure dielectric parameters of materials.

Among the direct monochromatic methods one can mention the spectrometer (Jones et al. 1984), based on a Mach–Zehnder configuration, using a 245 GHz optically pumped laser as a radiation source. The spectrometer allows both refractive and absorption measurements. By incorporating backward-wave oscillators into the Mach–Zehnder configuration (Volkov et al. 1990), a submillimeter wave spectrometer has been developed which operates in the 150 to 900 GHz frequency range (see Section 3.4). Rachford and Forestor (1983) have applied the backward-wave oscillators spectrometer for transmission and reflection measurements between 75 and 110 GHz in order to find the complex permittivity and permeability of materials.

Dispersive Fourier transform spectroscopy (see Section 3.1.1) is the direct broad-band method, which yields the full spectral variation of the optical constants of materials from a measurement of either its complex transmission or reflection coefficient. It covers a frequency range from 3 mm wavelength to about 15 μm where the phonon spectra of solids usually lie. The source is a high-pressure mercury-vapor lamp. The accuracy of measurements critically depends on the degree of opacity, plane-parallelism of the specimen, and its thickness. Using DFTS, Birch (1983) determined optical constants of some commercial microwave materials in the frequency range between 90 and 1200 GHz. It is useful for the wide frequency band characterization of low-loss polymers, promising millimeter-wave ceramics, and optical glasses.

3.1.5. Stochastic Cavity Methods

A stochastic or untuned cavity is an oversized cavity, which, as a result of the action of the mode stirrer, maintains throughout its volume an isotropic electromagnetic field (Figure 3.3). In the 1980s such cavities have been used for the spectroscopy of solid and liquid dielectric materials. Unlike the conventional resonant cavities, the response to an excited electromagnetic field is independent of frequency. Therefore, such cavities are wide-band devices that can be applied from the microwave up to light frequencies. Various aspects of the application of such cavities have been covered in reviews by Birch and Clarke (1982) and Kremer et al. (1984).

The advantage of such cavities for the near-millimeter wavelength studies comes from the independence of measurements on the specimen geometry. The measurement parameters are sensitive only to absorption

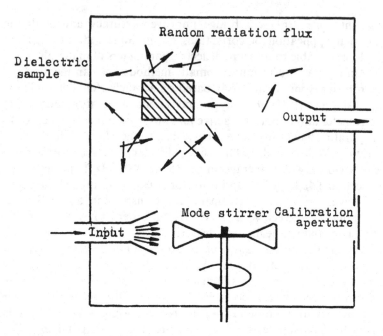

Figure 3.3. "Stirred-mode" cavity: the mode stirrer rotates rapidly to render the internal fields homogeneous and isotropic on a time-average basis. From Afsar et al. (1986).

loss, not to reflection or scattering loss (i.e., for low-permittivity materials they depend only on the volume of the specimen). Using this technique, absorption measurements have been reported by Birch et al. (1983) for a number of low-loss materials and powders up to 1000 GHz. This technique does not apply exclusively to the study of isotropic powders. Izatt and Kremer (1981) have shown that by using lamellar specimens it is possible to determine both optical parameters (i.e., from studies of several lamellae of different thicknesses, both the refractive index and the absorption can be determined).

In toto, only a few of the numerous available methods can be successfully used for dielectric spectroscopy of anisotropic ferroelectric and related single crystals each of individual size and shape, especially in a vicinity of phase transitions temperature and fundamental ferroelectric dispersion region, when both permittivity and loss vary by several orders. In the next sections some particular cases of various excitations in ferroelectric, superionic, semiconductor and other crystals, and dielectrics in general will be discussed, which can be useful to any investigator.

3.2. METER AND DECIMETER WAVE RANGE

The coaxial technique is the most convenient for the dielectric spectroscopy of solids in this wavelength range. In this section we consider at greater length a method intended for the dielectric spectroscopy of small specimens of crystals. The specimen usually is placed at the end of the coaxial line between the inner conductor and the short piston (Figure 3.4) and forms a capacitor. Such a configuration allows us to easily place the capacitor into a temperature-control device or to apply a bias voltage. The direction and homogeneity of the electric field in the capacitor allow investigation of both isotropic and anisotropic materials in the weak or strong electromagnetic fields. In case of anisotropic materials one can cut the sample along the desirable crystallographic axis. It is possible also to apply a hydrostatic pressure.

Coaxial lines are broad-band lines. From the low-frequency end they can be used at any frequency. From the superhigh-frequency end the condition of propagation of the main T-wave limits application of the coaxial lines. This condition is given by

$$\lambda > \pi(r_3 + r_4),$$

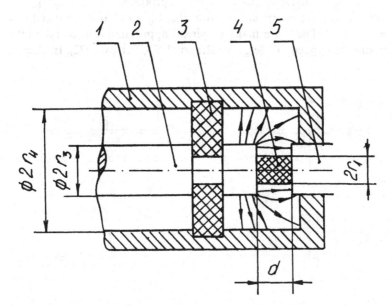

Figure 3.4. Dielectric specimen in a coaxial line: (1) outward conductor, (2) inner conductor, (3) dielectric washer, (4) specimen, (5) short piston.

where r_3 and r_4 are radii of the inner and outward conductors of the coaxial line. The limiting wavelength is about $\lambda_1 = 1\,\text{cm}$ or frequency $v_1 = 30\,\text{GHz}$.

3.2.1. Measuring Capacitor

Measuring capacitors are used for the determination of dielectric parameters of solids in a frequency range below $10^{10}\,\text{Hz}$. We recall the standard expressions for the determination of dielectric parameters. The capacitor consists of two metallic rectangular or disk electrodes and a dielectric sample between them (Figure 3.5a). The shapes of capacitors depend on the frequency range, measuring device and other circumstances. The area of the capacitor S must be much larger than the distance between the electrodes d (i.e., $S \gg d$). Due to fringing field effects, the capacity of the capacitor without a dielectric is larger than C_0 given by

$$C_0 = \frac{S}{36\pi d}. \tag{3.1}$$

The additional spurious capacitor C_s, produced by the fringing field effects, is connected in shunt with the C_0 so that all capacity is $C = \varepsilon C_0 + C_s$. There are many empirical approximations to take into account the capacity C_s (e.g., see Brandt 1963; Petrov 1972). In case of

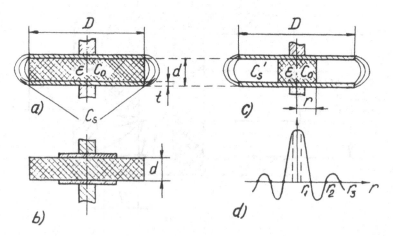

Figure 3.5. Disk capacitors (a–c) filled up by dielectric and distribution of high-frequency electric field in the capacitor (d).

disk electrodes, one can obtain the exact analytical solution:

$$C_s = \frac{D}{72\pi}\left(\ln\frac{8\pi D}{d} - 1\right). \tag{3.2}$$

Scattered field is concentrated in air, therefore, C_s does not depend on the value of permittivity. The error of calculation of the permittivity ε, neglecting the fringing field effects (i.e., the capacity C_s), is given by $\Delta\varepsilon/\varepsilon = C_0/C_s$.

When $D/d \simeq 10$, the error is $\Delta\varepsilon \simeq 20\%/\varepsilon$ (i.e., $\Delta\varepsilon \simeq 1\%$ for $\varepsilon = 20$ and 0.01% for $\varepsilon = 2000$). For ferroelectrics with the permittivity of $\varepsilon > 1000$, the fringing field effects are small, even for $D/d \simeq 1$.

In connection with the application of circular disk capacitors for microwave integrated circuits, new methods of calculation of such capacitors recently were developed on the basis of the numerical solution of the integral equation (Pavilonis et al. 1978). It was found that C_s depends also on the thickness of the electrodes t. When $D = d$ and $t = D/2$, $C/C_0 = 5$, where C is the calculated capacity. Even the hole of the radius r_h in the center of the plates does not change the capacity if $r_h/D \leq 0.08$.

In some cases, the sample is only partly covered by the electrodes (Figure 3.5b). Scattering field is concentrated in the sample and the capacity C_s (von Hippel 1954) for the electrodes of arbitrary form, with the perimeter P, is given by the empirical formula

$$C_s = 10^{-3}P[1.97\varepsilon + 6\log(P/d) - 8.87].$$

Due to redistribution of the electric field, the capacity C_s changes after placing the sample between the plates, especially if $\varepsilon \geq 20$. It is difficult to take it into account in every particular case. This change strongly decreases when the sample is placed in the center of the disk capacitor (Figure 3.5c).

Capacity of this capacitor is given by

$$C = \frac{\varepsilon r^2}{36d} + C_s'. \tag{3.3}$$

Dimensions of the capacitor in all formulas are in millimeters and the capacity is in picofarads. When $r \ll D$, the electric field is homogeneous in the capacitor. In case of thin or needle-shaped cylindrical crystals with $r/d \leq 1$ and $D/d > 10$, dielectric permittivity can be obtained from the measurements of the capacity C_1 with the sample and C_2 without it,

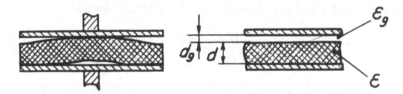

Figure 3.6. Unparallel sample in the capacitor produces the gap.

according to

$$\varepsilon = 1 + \frac{36d}{r^2}(C_1 - C_2). \tag{3.4}$$

The sample should cling closely to the electrodes. Even a small gap d_g (Figure 3.6) causes the equivalent series capacitor C_g, decreasing the measured capacity C_m, given by

$$C_m = \frac{C}{1 + \dfrac{C}{C_g}} = \frac{C}{1 + \dfrac{\varepsilon d_g}{\varepsilon_g d}}, \tag{3.5}$$

where C is the true capacity.

Neglecting the gap between the sample and the hard plates of the capacitor, the error is given by

$$\frac{\Delta\varepsilon}{\varepsilon} = \frac{\varepsilon d_g}{\varepsilon_g d}, \tag{3.6}$$

which is significant in case of the high permittivity samples.

In a similar way, one can find the error of the tangent of the loss angle, $\tan\delta$. When $\varepsilon = 1000$, $d = 1$ mm and $\varepsilon_g = 1$ (free space), $\Delta\varepsilon = 100\%$ even for $d_g = 1\ \mu$m. To avoid these errors, the samples should be covered by gold or silver (using vacuum evaporation) or by silver paste. The resistance of the covered electrodes R_e should be small in comparison with the resistance of the sample, R_s, in series representation because the measured value of $\tan\delta$ is given by

$$\tan\delta = (\omega R_s + R_e)C. \tag{3.7}$$

When determining the dielectric parameters of the ferroelectrics with a high dielectric permittivity, the defective surface layers, which usually appear during the preparation of a sample, should also be taken into account through a capacity, existing in series with the bulk. Even a defective thickness of the surface layer of $0.26\ \mu$m considerably changes the dielectric parameters of ferroelectrics with a high dielectric permittivity at microwaves (Grigas et al. 1988a).

As dielectric permittivity is complex, $\varepsilon^* = \varepsilon' - i\varepsilon''$, the capacity is also complex, $C^* = C' - iC''$. The values of permittivity are given by

$$\varepsilon' = 1 + \frac{36\pi d(C' - C_0)}{S}, \quad \varepsilon'' = \frac{36\pi d C''}{S}, \tag{3.8}$$

where $C' = C_0 + C_s$, and C_0 is given by Equation 3.1.

Afore-cited equations are valid for a quasistatic capacitor in which capacitance is independent of frequency and the electric field is homogeneous in the sample, just as it is when the dimensions of the capacitors are much smaller in comparison to a wavelength λ of exciting electric field. However, with the increase of frequency the electric field in the disk capacitor, in contrast to the static case, depends on r and is given by

$$E = A J_0 [k(\varepsilon\mu)^{1/2} r],$$

where A is the constant, depending on the dimensions of the capacitor, $k = 2\pi/\lambda$ is the wave number, r is the distance from the center of a capacitor along the radius, J_0 is the Bessel function of the first kind of order zero, and μ is the magnetic permeability. The electric field between the electrodes of the capacitor is nonhomogeneous (Figure 3.5c). Finding the zeros of the Bessel function, the radii $r_1, r_2, r_3, \ldots$, where the field between the electrodes of a capacitor is equal to zero, can be obtained. The radius r_1 is given by

$$r_1 = \frac{2.405\lambda}{2\pi(\varepsilon\mu)^{1/2}}. \tag{3.9}$$

Taking the radius of the sample $r \leq 0.1 r_1$, one finds the conditions of the quasi-stationary electric field distribution in the capacitor:

$$r \leq \frac{0.24\lambda}{2\pi(\varepsilon\mu)^{1/2}} = \frac{3.8 \times 10^{-2}\lambda}{(\varepsilon\mu)^{1/2}}, \tag{3.10}$$

$$d < \frac{\lambda}{2(\varepsilon\mu)^{1/2}}.$$

In the microwave range, when the permittivity of the sample is high its radius should be very small. For example, when $\lambda = 10\,\text{cm}$ and $\varepsilon = 1000$, the radius of the sample should be $r \leq 0.12\,\text{mm}$.

Hence, Equation 3.10 sets the conditions of quasi-stationarity. When $r > r_1$, the capacity loses its meaning (i.e., the permittivity cannot be calculated from the formulas of the static capacitor). In this case, a dynamic capacitor, which takes into account inhomogeneous distribution of the electric field in the sample, should be used.

3.2.2. Dynamic Coaxial Capacitor

Consider now the capacitor, which is formed by the cylindrical sample, placed at the end of the coaxial line between the inner conductor and the short piston (Figure 3.4). The line is excited by the harmonic electromagnetic field of frequency ω, and the main monochromatic T-wave propagates along the line without variation along the coordinates z and φ. The wave has only the components of the electromagnetic field E_z and H_φ. In the cylindrical system of coordinates, with the center at the axis of the line, the components E_z and H_φ are given by

$$E_z = -ik^2\mu(\varepsilon\mu)^{1/2}AJ_0[\alpha(\varepsilon\mu)^{1/2}],$$
$$H_\varphi = -k^2\varepsilon\mu AJ_0'[\alpha(\varepsilon\mu)^{1/2}], \tag{3.11}$$

where $\alpha = kr$, and J_0' is the derivative of the Bessel function J_0. As mentioned before, the components E_z and H_φ depend on r. Therefore, from Equation 3.10, the radius of the ferroelectric sample of high permittivity should be very small. It is a complicated task not only to make such a sample but especially to fix it in the capacitor. Moreover, single crystals of ferroelectrics in the majority of cases are brittle. A number of perfect needle-shaped single crystals (see Chapter 4) can be grown of radii ranging from a few microns to a few millimeters. The crystals, especially those possessing high piezoelectric moduli and high coefficients of electromechanical coupling, should be mechanically free in the capacitor (Figure 3.4). This is due to the fact that the mechanical squeeze considerably distorts the measured results, especially in the vicinity of ferroelectric and ferroelastic phase transitions.

In such cases, the cylindrical sample of radius r_1 is placed into a dielectric washer-shaped lossless sample holder of the small permittivity ε_3 and small dielectric loss (e.g., teflon with the metallized butt-ends). The external diameter of the sample holder should be equal to the diameter of the inner conductor of the coaxial line. The sample should be a little shorter than the sample holder, and the butt ends covered by gold (silver) paste or liquid metal (e.g., In–Ga). The sample holder with the sample is squeezed between the inner conductor of the coaxial and the short, keeping the sample mechanically free.

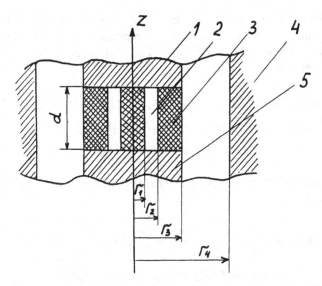

Figure 3.7. Dynamic coaxial capacitor: (1) specimen, (2) air clearance, (3) dielectric sample holder, (4) external and (5) internal conductors of coaxial line.

Calculation of the capacity of this dynamic coaxial capacitor shown in Figure 3.7 takes into account nonhomogeneous distribution of electric field.

Inside the crystal ($r \leq r_1$) the components of the electromagnetic field (Equations 3.11) take the form

$$E_z = ik^2 \mu_1 A_1 J_0[\alpha(\varepsilon_1 \mu_1)^{1/2}],$$
$$H_\varphi = k^2 (\varepsilon_1 \mu_1)^{1/2} A_1 J'_0[\alpha(\varepsilon_1 \mu_1)^{1/2}],$$

(3.12)

where $A_1 = -A(\varepsilon_1 \mu_1)^{1/2}$, ε_1 is the dielectric permittivity, and μ_1 is the magnetic permeability of the sample.

The components of the electromagnetic field in the free space clearance between the crystal and crystal holder are given by

$$E_z = ik^2 \mu^2 \{ B_2 J_0[\alpha(\varepsilon_2 \mu_2)^{1/2}] + C_2 N_0[\alpha(\varepsilon_2 \mu_2)^{1/2}] \},$$
$$H_\varphi = k^2 (\varepsilon_2 \mu_2)^{1/2} \{ B_2 J'_0[\alpha(\varepsilon_2 \mu_2)^{1/2}] + C_2 N'_0[\alpha(\varepsilon_2 \mu_2)^{1/2}] \},$$

(3.13)

where ε_2 is the dielectric permittivity, μ_2 is the magnetic permeability of free space, B_2 and C_2 are constants, N_0 is the Neiman function, and N'_0 is its derivative.

Coefficients B_2 and C_2 can be found from the continuity of the electric and magnetic field tangential components at the interface between the sample and air condition. For this purpose the following system of equations is used:

$$\mu_1 A_1 J_0[\alpha(\varepsilon_1\mu_1)^{1/2}] = \mu_2\{B_2 J_0[\alpha(\varepsilon_2\mu_2)^{1/2}] + C_2 N_0[\alpha(\varepsilon_2\mu_2)^{1/2}]\},$$

$$(\varepsilon_1\mu_1)^{1/2} A_1 J_0'[\alpha(\varepsilon_1\mu_1)^{1/2}] = (\varepsilon_2\mu_2)^{1/2}\{B_2 J_0'[\alpha(\varepsilon_2\mu_2)^{1/2}]$$
$$+ C_2 N_0'[\alpha(\varepsilon_2\mu_2)^{1/2}]\}. \tag{3.14}$$

By solving this system of equations (3.14) with respect to B_2 and C_2, one obtains

$$B_2 = A_1 \frac{\pi\alpha}{2\mu_2}\{\mu_1(\varepsilon_2\mu_2)^{1/2} J_0[\alpha(\varepsilon_1\mu_1)^{1/2}]N_0'[\alpha(\varepsilon_2\mu_2)^{1/2}]$$

$$- \mu_2(\varepsilon_1\mu_1)^{1/2} J_0'[\alpha(\varepsilon_1\mu_1)^{1/2}]N_0[\alpha(\varepsilon_2\mu_2)^{1/2}]\},$$
$$\tag{3.15}$$
$$C_2 = A_1 \frac{\pi\alpha}{2\mu_2}\{\mu_2(\varepsilon_1\mu_1)^{1/2} J_0[\alpha(\varepsilon_2\mu_2)^{1/2}]J_0'[\alpha(\varepsilon_1\mu_1)^{1/2}]$$

$$- \mu_1(\varepsilon_2\mu_2)^{1/2} J_0'[\alpha(\varepsilon_2\mu_2)^{1/2}]J_0[\alpha(\varepsilon_1\mu_1)^{1/2}]\}.$$

Now the components of the electromagnetic field in th-e next medium $r_2 \le r \le r_3$ (i.e., in the sample holder) may be written in the form

$$E_z = ik^2\mu_3\{B_3 J_0[\alpha(\varepsilon_3\mu_3)^{1/2}] + C_3 N_0[\alpha(\varepsilon_3\mu_3)^{1/2}]\},$$
$$H_\varphi = k^2(\varepsilon_3\mu_3)^{1/2}\{B_3 J_0[\alpha(\varepsilon_3\mu_3)^{1/2}] + C_3 N_0[\alpha(\varepsilon_3\mu_3)^{1/2}]\}. \tag{3.16}$$

The coefficients B_3 and C_3 can be found from boundary conditions. The set of equations are obtained in the similar way as in Equations 3.14:

$$\mu_2\{B_2 J_0[\alpha(\varepsilon_2\mu_2)^{1/2}] + C_2 N_0[\alpha(\varepsilon_2\mu_2)^{1/2}]\}$$
$$= \mu_3\{B_3 J_0[\alpha(\varepsilon_3\mu_3)^{1/2}] + C_3 N_0[\alpha(\varepsilon_3\mu_3)^{1/2}]\},$$
$$\tag{3.17}$$
$$(\varepsilon_2\mu_2)^{1/2}\{B_2 J_0'[\alpha(\varepsilon_2\mu_2)^{1/2}] + C_2 N_0'[\alpha(\varepsilon_2\mu_2)^{1/2}]\}$$
$$= (\varepsilon_3\mu_3)^{1/2}\{B_3 J_0'[\alpha(\varepsilon_3\mu_3)^{1/2}] + C_3 N_0'[\alpha(\varepsilon_3\mu_3)^{1/2}]\}.$$

By solving Equations 3.17 with regard to B_3 and C_3, one obtains

$$B_3 = \frac{\pi\alpha}{2\mu_3}\{\mu_2(\varepsilon_3\mu_3)^{1/2}\{B_2J_0[\alpha(\varepsilon_2\mu_2)^{1/2}] + C_2N_0[\alpha(\varepsilon_2\mu_2)^{1/2}]\}N'_0[\alpha(\varepsilon_3\mu_3)^{1/2}]$$

$$- \mu_3(\varepsilon_2\mu_2)^{1/2}\{B_2J'_0[\alpha(\varepsilon_2\mu_2)^{1/2}]$$

$$+ C_2N'_0[\alpha(\varepsilon_2\mu_2)^{1/2}]\}N'_0[\alpha(\varepsilon_3\mu_3)^{1/2}]\}, \tag{3.18}$$

$$C_3 = \frac{\pi\alpha}{2\mu_3}\{\mu_3(\varepsilon_2\mu_2)^{1/2}J_0[\alpha(\varepsilon_3\mu_3)^{1/2}][B_2J'_0[\alpha(\varepsilon_2\mu_2)^{1/2}]$$

$$+ C_2N'_0[\alpha(\varepsilon_2\mu_2)^{1/2}]] - \mu_2(\varepsilon_3\mu_3)^{1/2}J'_0[\alpha(\varepsilon_3\mu_3)^{1/2}][B_2J'_0[\alpha(\varepsilon_2\mu_2)^{1/2}]$$

$$+ C_2N_0[\alpha(\varepsilon_2\mu_2)^{1/2}]]\}.$$

For standard coaxial lines of wave impedance 50 Ohm, taking $\mu_1 = \mu_2 = \mu_3 = 1$, $\varepsilon_2 = 1$, and $\varepsilon_3 = 2.2$ (for teflon), the argument of the Bessel function $z = \alpha(\varepsilon_2\mu_2)^{1/2} \ll 1$, and with the error less than 3% one may write

$$J_0(z) \approx 1, \quad J_1(z) \approx \frac{z}{2},$$

$$N_1(z) \approx -\frac{2}{\pi z}, \quad N_0(z) \approx \frac{2(\ln z + C_e)}{\pi}, \tag{3.19}$$

where $C_e = 0.577216$ is Euler's constant. By substituting given values into Equations 3.15 and 3.18, the coefficients may be written in the form

$$B_2 = A_1\left\{J_0[k(\varepsilon_1)^{1/2}r_1] + k(\varepsilon_1)^{1/2}r_1J_1[k(\varepsilon_1)^{1/2}r_1]\left[\ln\frac{kr_1}{2} + C_e\right]\right\},$$

$$C_2 = C_3 = A_1\frac{\pi kr_1}{2}\left\{-(\varepsilon_1)^{1/2}J_1[k(\varepsilon_1)^{1/2}r_1] + \frac{kr_1}{2}J_0[k(\varepsilon_1)^{1/2}r_1]\right\},$$

$$B_3 = A_1\left\{J_0[k(\varepsilon_1)^{1/2}r_1] + k(\varepsilon_1)^{1/2}r_1J_1[k(\varepsilon_1)^{1/2}r_1]\left(\ln\frac{kr_1}{2} + C_e\right)\right.$$

$$\left. - kr_1\ln(\varepsilon_3)^{1/2}\left[\left(\frac{kr_1}{2}\right)J_0(k(\varepsilon_1)^{1/2}r_1 - (\varepsilon_1)^{1/2}J_1(k(\varepsilon_1)^{1/2}r_1)\right]\right\}. \tag{3.20}$$

The impedance at the cross section $r = r_3$ of the capacitor is

$$Z(r_3) = \frac{U(r_3)}{I(r_3)}.$$

The voltage between the electrodes of the capacitor is

$$U(r_3) = E_z(r_3)d$$

and the current is

$$I(r_3) = 2\pi r_3 H_\varphi(r_3).$$

Therefore, the impedance is

$$Z(r_3) = \frac{d}{2\pi r_3} \frac{E_z(r_3)}{H_\varphi(r_3)} = \frac{1}{i\omega C} \tag{3.21}$$

and the capacity is then given by

$$C = \frac{2\pi r_3(\varepsilon_3)^{1/2}}{\omega d} \times \frac{B_3 J_1[k(\varepsilon_3)^{1/2}r_3] + C_3 N_1[k(\varepsilon_3)^{1/2}r_3]}{B_3 J_0[k(\varepsilon_3)^{1/2}r_3] + C_3 N_0[k(\varepsilon_3)^{1/2}r_3]}. \tag{3.22}$$

By substituting Equation 3.20 in Equation 3.22 the capacity of a dynamic coaxial capacitor is obtained:

$$C = \frac{\dfrac{r_3(\varepsilon_3)^{1/2}}{36kd}\left\{ J_1(\beta_3)\left[J_0(\beta_1) + \beta_1 J_1(\beta_1)\left(\ln\left(\dfrac{\alpha}{2}\right) + C_e\right) + \beta_1 J_1(\beta_1) - \right.\right.}{J_0(\beta_3)\left[J_0(\beta_1) + \beta_1 J_1(\beta_1)\left[\ln\left(\dfrac{\alpha}{2}\right) + C_e\right] + \beta_1 J_1(\beta_1) - \right.}$$

$$\frac{\left. -\left(\dfrac{\alpha}{2}\right)^2 J_0(\beta_1)\ln(\varepsilon_3)^{1/2}\right] + N_1(\beta_3)\left[\pi\left(\dfrac{\alpha}{2}\right)^2 J_0(\beta_1) - \dfrac{\pi}{2}\beta_1 J_1(\beta_1)\right]\right\}}{\left. -\left(\dfrac{\alpha}{2}\right)^2 J_0(\beta_1)\ln(\varepsilon_3)^{1/2}\right] + N_0(\beta_3)\left[\pi\left(\dfrac{\alpha}{2}\right)^2 J_0(\beta_1) - \beta_1 J_1(\beta_1)\right]}, \tag{3.23}$$

where $\beta_1 = k(\varepsilon_1)^{1/2}r_1$ and $\beta_3 = k(\varepsilon_3)^{1/2}r_3$. When the dimensions of the capacitor are in millimeters, the capacity is in picofarads. To take into account loss, one replaces the permittivity of a sample ε_1 by the complex value $\varepsilon' - i\varepsilon''$. Then the capacity in Equations 3.22 and 3.23 also becomes complex.

    The general formula (3.23) of the dynamic capacitor can be simplified in certain cases. At frequencies below 1 GHz, when the sample is not fragile and permittivity ε_1 is low enough, there is no need to use the sample holder. In such a case, for $r_1 = r_3$, Equation 3.23 transforms into the simple expression

$$C_1 = \frac{\beta_1 J_1(\beta_1)}{18k^2 dJ_0(\beta_1)}. \qquad (3.24)$$

If $\beta_1 \ll 1$ (see Equation 3.10 of the quasi-stationarity), then $J_1(\beta_1) \approx \beta_1/2, J_0(\beta_1) \approx 1$, and Equation 3.24 coincides with Equation 3.3 for the static capacitor. The static approximation for calculation of a capacity with an accuracy of 3% is valid up to $\beta_1 = 0.474$.

When the radius of a sample $r_1 < r_3$ (Figure 3.7), one should take into account the capacity C_s of fringing field effects. In that case, Equation 3.23 transforms into

$$C_2 = \frac{AC_1 + C_s}{1 - BC_1}, \qquad (3.25)$$

where

$$A = \frac{r_3}{r_1} \frac{J_0(\alpha_1)N_1(\alpha_3) - J_1(\alpha_3)N_0(\alpha_1)}{J_0(\alpha_3)N_1(\alpha_1) - J_1(\alpha_1)N_0(\alpha_3)},$$

$$B = \frac{18kd}{r_1} \frac{J_0(\alpha_3)N_0(\alpha_1) - J_0(\alpha_1)N_0(\alpha_3)}{J_0(\alpha_3)N_1(\alpha_1) - J_1(\alpha_1)N_0(\alpha_3)}, \qquad (3.26)$$

and

$$C_s = \frac{r_3}{18kd} \frac{J_1(\alpha_3)N_1(\alpha_1) - J_1(\alpha_1)N_1(\alpha_3)}{J_0(\alpha_3)N_1(\alpha_1) - J_1(\alpha_1)N_0(\alpha_3)},$$

where C_1 is given by Equation 3.24, $\alpha_1 = kr_1$, $\alpha_3 = kr_3$, and $r_3 = a$ (Figure 3.4). Coefficient $A \simeq 1$, while B strongly decreases with the increase of r_1. The capacity C_s can be approximated by the static capacity (Petrov 1972)

$$C_s = \frac{r_3^2 - r_1^2}{36d}.$$

The static capacity (Equation 3.6) depends linearly on dielectric permittivity. This dependence for the dynamic capacity (Equations 3.22 to 3.25) is nonlinear. Figure 3.8 illustrates this dependence calculated by Equation 3.23. With the increase of permittivity (or radius r_1 of a sample) the series resonance occurs. The capacitor becomes equivalent to the short-circuited end of the coaxial line. Above the resonance the reactance becomes inductive (i.e., the C' becomes negative). From Equation

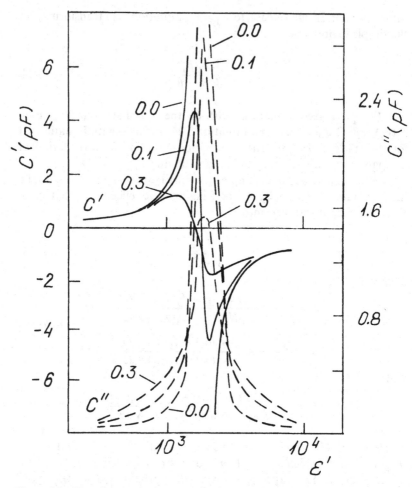

Figure 3.8. Complex capacity vs. dielectric permittivity of a sample of $r = 0.2$ mm at a frequency of 3 GHz. Solid lines represent C', dotted lines C''. Numbers are values of $\tan \delta$.

3.24 it follows that at every root of the Bessel functions J_0 and J_1 resonances and antiresonances will occur. The condition of resonance in Equation 3.25 is $C_1 = 1/B$. This resonance can be used for sensitive investigations of dielectric loss of ferroelectrics and related materials in coaxial lines within the limits of $\tan \delta$ values 3×10^{-4} to 10^{-1}. At resonance the reactance of the impedance given by Equation 3.21 is zero. Therefore, the impedance gives the value of $\tan \delta$.

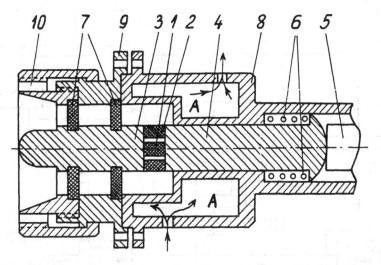

Figure 3.9. Measuring coaxial capacitor: (1) sample, (2) sample holder, (3) inner conductor, (4) piston, (5) micrometrical screw, (6) spring, (7) washers holders, (8) frame of the capacitor, and (9) and (10) are fishes.

At frequencies below 1 GHz the shape of the sample is nonessential, while above this frequency the sample should be cylindrical. Such samples in some cases can be made by an ultrasonic vibrator.

Hence, the general formula of the dynamic coaxial capacitor (Equation 3.23) and its approximations (Equations 3.24 and 3.25) are obtained, taking into account the inhomogeneous distribution of the electric field in the capacitor at microwave frequencies. One of the coaxial capacitors is shown in Figure 3.9. The capacitor is easy to place into a heater or cooler. The space A ensures the thermostatic adjustment of the sample.

3.2.3. Coaxial Dielectric Spectrometer

The spectra of the complex dielectric permittivity $\varepsilon^*(v, T)$ in the meter, decimeter and long centimeter wave range up to 10–20 GHz can be obtained (Brilingas et al. 1986) with a coaxial dielectric spectrometer setup (Figure 3.10). These spectra are obtained from the results of the measurements of the complex reflection coefficient $R^*(v, T)$ of the T-wave in the coaxial line loaded with the sample in the measuring capacitor (Figure 3.9). The impedance of the capacitor (Equation 3.21) is related to the amplitude reflection coefficient $R^* = R \exp e^{i\varphi}$ and the

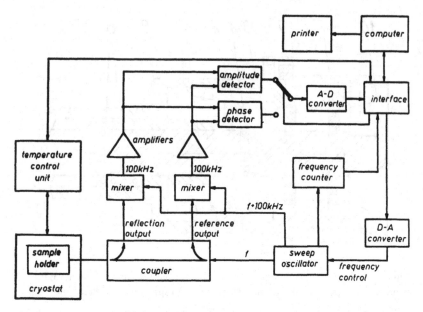

Figure 3.10. Diagram of the coaxial dielectric spectrometer setup. From Brilingas et al. (1986).

capacity of the sample according to

$$R^*(v) = \frac{1 - Z_0 C_0 i\omega\varepsilon^*(\omega)}{1 + Z_0 C_0 i\omega\varepsilon^*(\omega)}, \tag{3.27}$$

where $\omega = 2\pi v$, Z_0 is the characteristic impedance of the coaxial line, and C_0 is defined as in Equation 3.1. From Equation 3.27 the real and imaginary parts of the complex dielectric permittivity are given by

$$\varepsilon'(v) = \frac{2R(v)\sin\varphi(v)}{\omega C_0 Z_0 [R^2(v) + 2R(v)\cos\varphi(v) + 1]},$$

$$\varepsilon''(v) = \frac{1 - R^2(v)}{\omega C_0 Z_0 [R^2(v) + 2R(v)\cos\varphi(v) + 1]}.$$

The parasitic capacity and C_0 can be determined experimentally. If the dielectric loss are not high and the conditions of quasi-stationarity (Equation 3.10) are satisfied, the values of dielectric permittivity are determined according to Equation 3.8. If the conditions (Equation 3.10) are not satisfied, the dielectric parameters are determined according to the equations of the dynamic capacitor (Equation 3.23 or Equations 3.24 and 3.25).

The complex reflection coefficient, $R^*(v, T)$, is obtained by amplifying the incident and reflected signals by a frequency converter and detected by synchronous amplitude and phase detectors. At the output of the phase detector, the voltage is proportional to the difference of the phases of both the signals:

$$U_\varphi = k_\varphi(U_r - U_i) = k_\varphi\varphi, \qquad (3.28)$$

while at the output of the amplitude detector, the voltage is proportional to the ratio of the amplitude of the signals:

$$U_A = k_A \frac{A_r}{A_i} = k_A R, \qquad (3.29)$$

where k_φ and k_A are the characteristics of the detectors.

The measured amplitude reflection coefficient is given by

$$R_m^* = \frac{U_A}{k_A} \exp i\left(\frac{U_\varphi}{k_\varphi}\right). \qquad (3.30)$$

Inhomogeneities of the line and distortions in a high frequency part of the spectrometer (influence of which increases with the increase of frequency) can be taken into account using a computer and digital processing of information by an analysis of the six-port between the capacitor and the output planes of the directional couples (Figure 3.11).

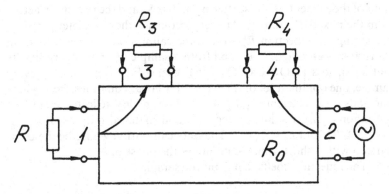

Figure 3.11. Diagram of the reflectometer setup: (1) sample input, (2) oscillator input, outputs of the mixers of measuring (3) and reference (4) channels.

The linear eight-port can be described by a scattering matrix $\|S_{ri}\|$ of the complex coefficients $S_{ri} = b_r/a_i$, relating a reflected signal b_r from the input to the signal a_i at input i. The indicator of the setup measures the reflection factor R_m (i.e., the ratio of outcoming signals from the measuring and referenced output $R_m = b_3/b_4$).

For an ideal reflectometer setup ($R_2 = R_3 = R_4 = 0$):

$$R_m = \frac{S_{12}S_{31}}{S_{42}} R = kR, \tag{3.31}$$

where the coefficient k can be found by calibrating the spectrometer using a shot with $R = -1$. In general one should solve the set of linear equations

$$\begin{Vmatrix} a_1 \\ a_2 \\ a_3 \\ a_4 \end{Vmatrix} \times \|S\| = \begin{Vmatrix} b_1 \\ b_2 \\ b_3 \\ b_4 \end{Vmatrix} \tag{3.32}$$

respect to b_1, b_2, b_3, and b_4. Taking into account the fact that $a_1 = R_s b_1$, $a_3 = R_3 b_3$ and $a_4 = R_4 b_4$, one finds the relation between the measured reflection factor R_m and the reflection coefficient R:

$$R_m^* = \frac{k_1 R^* + k_2}{k_3 R^* + 1}. \tag{3.33}$$

The coefficients k_1, k_2 and k_3 are composed from products and sums of the elements of the scattering matrix S_{ri} and the reflection factors from the mixers R_3 and R_4. At every frequency they are determined by measuring the reflection R_m^i from three calibration samples (e.g., from short and open-circuit lines, and from a sample of known permittivity and small loss (TiO_2, $CaTiO_3$, etc.). Using Equation 3.33 for every sample, one obtains a set of three complex linear equations, from which one finds the coefficients k_1, k_2 and k_3. For an ideal reflectometer setup (no inhomogeneities, ideal directivity and matched outputs) $R = -1$, $k_1 = -1/R^*$, and $k_2 = k_3 = 0$. The only calibration would be the comparison with a short, which determines the phase of k_1.

The reflection coefficient from the sample is

$$R^* = \frac{R_m^* - k_1}{k_2 - k_3 R_m^*},$$

where R_m^* is given by Equation 3.30. The computer receives the signals (Equations 3.28 and 3.29) from the detectors, controls the frequency and temperature of the sample, and successively calculates $\varepsilon'(v, T)$ and $\varepsilon''(v, T)$ spectra. By using precise coaxial impedance analyzers and the dynamic coaxial measuring capacitor (Equation 3.23), one can make coaxial dielectric spectrometers for the frequency range up to 20–30 GHz.

3.3. CENTIMETER AND MILLIMETER WAVE RANGE

High dielectric loss in the range of fundamental ferroelectric dispersion prohibits the use of any resonance method. Therefore, in this section, only the broad-band waveguide methods are discussed, based on the measurements of wave parameters which allow acquisition of dielectric spectra in the centimeter and millimeter wave range of ferroelectrics and related materials possessing different values of permittivity and loss. From the historic measurements of von Hippel (1954), the waveguide measurements have undergone phenomenal changes. The use of computers for data processing, the network analyzers for reflection and transmission measurements, and also the solution of formidable electrodynamic problems make it possible to look anew at the waveguide technique for dielectric spectroscopy. The main methods used, depending on the shape and electrical parameters of the samples, will be considered.

3.3.1. Method of Filling Cross-Section of Waveguide

Recall the most general case of the plane-parallel dielectric plate filling the cross section of the rectangular waveguide. The dielectric plate is characterized by a complex dielectric permittivity ε^* and complex magnetic permeability μ^*. As the TE_{10}-mode is assumed incident on the plate from the medium 1, as shown in Figure 3.12, the wave is partly reflected from the medium 2 and partly transmitted to the medium 3. The complex amplitude reflection and transmission coefficients, R^* and T^*, of TE_{10}-mode, travelling in the x-direction, and their relation with the material parameters ε^* and μ^* of the dielectric plate will be obtained.

The wave equations for all three mediums can be written. For medium 1, characterized by the free space permittivity ε_0 and permeability μ_0, the wave equations are given by

$$\dot{E}_y = \omega\mu_0 E_0 e^{-i\gamma_0 x} + \omega\mu_0 R^* E_0 e^{i\gamma_0 x},$$
$$\dot{H}_x = \gamma_0 H_0 e^{-i\gamma_0 x} - \gamma_0 H_0 R^* e^{i\gamma_0 x}. \tag{3.34}$$

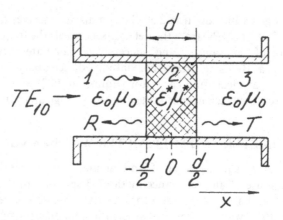

Figure 3.12. Plane-parallel plate in rectangular waveguide.

For medium 2, the wave equations are

$$\dot{E}_y = \omega\mu^* A E_0 e^{-i\gamma x} + \omega\mu^* B E_0 e^{i\gamma x},$$
$$\dot{H}_x = \gamma A H_0 e^{-i\gamma x} - \gamma B H_0 e^{i\gamma x}.$$

$$(3.35)$$

And for medium 3:

$$\dot{E}_y = \omega\mu_0 T^* E_0 e^{-i\gamma_0 x},$$
$$\dot{H}_x = \gamma_0 T^* H_0 e^{-i\gamma_0 x}.$$

$$(3.36)$$

The constants A and B are related to the dielectric and magnetic parameters of medium 2. The propagation constants in an empty and uniformly filled dielectric plate waveguide are γ_0 and γ, respectively, and are defined as

$$\gamma_0 = \left[\omega^2 \varepsilon_0 \mu_0 - \left(\frac{\pi}{a}\right)^2 \right]^{1/2},$$

$$(3.37)$$

$$\gamma = \left[\omega^2 \varepsilon_0 \mu_0 \varepsilon^* \mu^* - \left(\frac{\pi}{a}\right)^2 \right]^{1/2},$$

$$(3.38)$$

where a is the width of the wide wall of the waveguide.

The boundary conditions are such that the components of field vectors E and H, tangential to the boundary of two mediums, are continuous. For the boundary of mediums 1 and 2, $x = -d/2$. From Equations 3.34 and 3.35:

$$\mu_0 e^{i\gamma_0 d/2} + \mu_0 R^* e^{-i\gamma_0 d/2} = \mu^* A e^{i\gamma d/2} + \mu^* B e^{-i\gamma d/2}, \qquad (3.39)$$

and

$$\mu_0 e^{i\gamma_0 d/2} - R^* \gamma_0^{-i\gamma_0 d/2} = \gamma A e^{i\gamma d/2} - \gamma B e^{-i\gamma d/2}. \qquad (3.40)$$

For mediums 2 and 3 at $x = d/2$ from Equations 3.34 and 3.35:

$$\mu^* A e^{-i\gamma d/2} + \mu^* \beta e^{i\gamma d/2} = \mu_0 T^* e^{-i\gamma_0 d/2}, \qquad (3.41)$$

and

$$\gamma A e^{-i\gamma d/2} - \gamma B e^{i\gamma d/2} = \gamma_0 T^* e^{-i\gamma_0 d/2}. \qquad (3.42)$$

Solution of the sets of Equations 3.39 through 3.42, with respect to R^*, T^*, A and B from the coefficients of those equations, are summarized in Table 3.1, from which follows:

$$\Delta = e^{-i\gamma_0 d}[\gamma_0 \gamma \mu^* \mu_0 e^{i\gamma d} - \gamma^2 \mu_0^2 e^{-i\gamma d} + \gamma_0 \gamma \mu^* \mu_0 e^{-i\gamma d} + \gamma^2 \mu_0^2 e^{-i\gamma d}$$

$$+ \mu^{*2} \gamma_0^2 e^{-i\gamma d} + \gamma_0 \mu_0 \gamma \mu^* e^{-i\gamma d} - \mu^{*2} \gamma_0^2 e^{-i\gamma d} + \mu_0 \mu^* \gamma \gamma_0 e^{i\gamma d}].$$

By canceling similar terms:

$$\Delta = e^{-i\gamma_0 d}[2\gamma\gamma_0 \mu^* \mu_0(e^{i\gamma d} + e^{-i\gamma d}) + (\gamma^2 \mu_0^2 + \gamma_0^2 \mu^{*2})e^{i\gamma d} - e^{-i\gamma d}].$$

The determinant ΔR is composed of the coefficients of the unknown, T^*, A, B and of the free term (see Table 3.1). It is given by

$$\Delta R = -\gamma_0 \gamma \mu_0 \mu^* e^{i\gamma d} + \mu_0^2 \gamma^2 e^{-i\gamma d} - \gamma_0 \gamma \mu_0 \mu^* e^{-i\gamma d} - \mu_0^2 \gamma^2 e^{i\gamma d}$$

$$+ \mu^{*2} \gamma_0^2 e^{i\gamma d} + \mu_0 \gamma_0 \gamma \mu^* e^{-i\gamma d} - \mu^{*2} \gamma_0^2 e^{-i\gamma d} + \mu_0 \gamma_0 \gamma \mu^* e^{i\gamma d}$$

$$= (\gamma_0^2 \mu^{*2} - \gamma \mu_0^2)(e^{i\gamma d} - e^{-i\gamma d}).$$

Table 3.1. Coefficients of the set of Equations (3.39) to (3.42)

R^*	T^*	A	B	Free term	No of Eqs.
$\mu_0 e^{-i\gamma_0 d/2}$	0	$-\mu^* e^{-i\gamma d/2}$	$-\mu^* e^{-i\gamma d/2}$	$-\mu_0 e^{i\gamma_0 d/2}$	3.39
$-\gamma_0 e^{-i\gamma_0 d/2}$	0	$-\gamma e^{-i\gamma d/2}$	$\gamma e^{-i\gamma d/2}$	$-\gamma_0 e^{i\gamma_0 d/2}$	3.40
0	$-\mu_0 e^{-i\gamma_0 d/2}$	$\mu^* e^{-i\gamma d/2}$	$\mu^* e^{-i\gamma d/2}$	0	3.41
0	$-\gamma_0 e^{-i\gamma_0 d/2}$	$\gamma e^{-i\gamma d/2}$	$-\gamma e^{-i\gamma d/2}$	0	3.42

The determinants ΔT, ΔA and ΔB are calculated in a similar way from Table 3.1. Then the complex reflection and transmission coefficients are given by

$$R^* = \frac{\Delta R}{\Delta} \quad \text{and} \quad T^* = \frac{\Delta T}{T^*},$$

or

$$R^* = \frac{(\gamma_0^2 \mu^{*2} - \gamma^2 \mu_0^2)(e^{\gamma d} - e^{-i\gamma d})}{e^{-i\gamma d}[2\mu_0 \mu^* \gamma_0 \gamma(e^{i\gamma d} + e^{-i\gamma d}) + (\gamma^2 \mu_0^2 + \gamma_0^2 \mu^{*2})(e^{i\gamma d} - e^{-i\gamma d})]} \quad (3.43)$$

and

$$T^* = \frac{4\mu_0 \mu^* \gamma_0 \gamma}{e^{-i\gamma d}[2\mu_0 \mu^* \gamma_0 \gamma(e^{i\gamma d} + e^{-i\gamma d}) + (\gamma^2 \mu_0^2 + \gamma_0^2 \mu^{*2})(e^{i\gamma d} - e^{i\gamma d})]}. \quad (3.44)$$

Material parameters from the measured wave parameters R^* and T^* can be obtained in a simple manner. Multiplying the numerator and denominator of the right sides of Equations 3.43 and 3.44 by $\mu_0 \gamma_0 \mu^* \gamma$:

$$R^* = \frac{\left(\dfrac{\gamma_0 \mu^*}{\gamma \mu_0} - \dfrac{\gamma \mu_0}{\gamma_0 \mu}\right)(e^{i\gamma d} - e^{-i\gamma d})}{e^{-i\gamma d}\left[2(e^{i\gamma d} + e^{-i\gamma d}) + \left(\dfrac{\gamma \mu_0}{\gamma_0 \mu^*} + \dfrac{\mu^* \gamma_0}{\mu_0 \gamma}\right)(e^{i\gamma d} - e^{-i\gamma d})\right]}, \quad (3.45)$$

$$T^* = \frac{4}{e^{-i\gamma d}\left[2(e^{i\gamma d} + e^{-i\gamma d}) + \left(\dfrac{\gamma \mu_0}{\gamma_0 \mu^*} + \dfrac{\mu^* \gamma_0}{\mu_0 \gamma}\right)(e^{i\gamma d} - e^{-i\gamma d})\right]}. \quad (3.46)$$

Dividing Equation 3.45 by Equation 3.46:

$$e^{i\gamma d} - e^{-i\gamma d} = \frac{4R^*}{T^*} \times \frac{1}{\left(\dfrac{\gamma_0 \mu^*}{\gamma \mu_0} - \dfrac{\gamma \mu_0}{\gamma_0 \mu^*}\right)}. \quad (3.47)$$

Squaring Equation 3.47 and adding 4 to both sides yields

$$e^{i\gamma d} - e^{-i\gamma d} = \left(\left[\frac{4R^*}{T^*}\left(\frac{\gamma_0 \mu^*}{\gamma \mu_0} - \frac{\gamma \mu_0}{\gamma_0 \mu^*}\right)^{-1}\right]^2 + 4\right)^{1/2}. \quad (3.48)$$

Now, defining:

$$Z_0 = \frac{\omega\mu_0}{\gamma_0}, \quad Z_1 = \frac{\omega\mu}{\gamma}, \quad Z = \frac{Z_1}{Z_0}. \tag{3.49}$$

Substituting Equation 3.48 for Equation 3.46 and taking into account Equation 3.49:

$$T^{*2} - R^{*2} = e^{i2\gamma_0 d} - 2R^* e^{i\gamma_0 d} \times \frac{Z + 1/Z}{Z - 1/Z},$$

or

$$\frac{Z + 1/Z}{Z - 1/Z} = \frac{e^{i2\gamma_0 d} + R^{*2} - T^{*2}}{2R^* E^{i\gamma_0 d}} = C \tag{3.50}$$

and

$$Z = \left(\frac{C + 1}{C - 1}\right)^{1/2}. \tag{3.51}$$

From Equation 3.47 and Equation 3.49:

$$e^{i\gamma d} - e^{-i\gamma d} = \frac{4R^*}{T^*} \frac{1}{Z - 1/Z}. \tag{3.52}$$

Now, Equation 3.50 generates $1/(Z - 1/Z)$. Substituting this into Equation 3.52 yields

$$e^{i\gamma d} - e^{-i\gamma d} = \frac{2}{T^*} \frac{e^{i2\gamma_0 d} + R^{*2} - T^{*2}}{e^{i\gamma_0 d}(Z + 1/Z)}. \tag{3.53}$$

Substituting Equation 3.53 for Equation 3.46 gives

$$e^{i\gamma d} + e^{-i\gamma d} = \left[\frac{2}{T^*} e^{i\gamma_0 d} - \frac{1}{T^*} \frac{e^{i2\gamma_0 D} + R^{*2} - T^{*2}}{e^{i\gamma_0 d}}\right]. \tag{3.54}$$

Addition of Equations 3.54 and 3.53 gives an expression which yields the propagation constant γ in a material:

$$e^{i\gamma d} = \frac{1}{2T^*} \left(2e^{i\gamma_0 d} + \left[\frac{1}{Z + 1/Z} - 1\right] \frac{e^{i2\gamma_0 d} + R^{*2} - T^{*2}}{e^{i\gamma_0 d}}\right) = D,$$

or

$$\tag{3.55}$$

$$\gamma = \frac{1}{id} \ln D.$$

The impedance Z in the plane $x = -d/2$ (Figure 3.12) and the propagation constant γ in the dielectric medium 2 can be found using Equations 3.51 and 3.55 after the measurement of complex reflection and transmission coefficients, R^* and T^*. Then the value of complex magnetic permeability of the material under investigation can be found and is given by

$$\mu^* = \frac{\mu_0 \gamma Z}{\gamma_0}. \tag{3.56}$$

The value of complex dielectric permittivity of the material, from Equation 3.38, is given by

$$\varepsilon^* = \frac{\gamma^2 + \left(\dfrac{\pi}{a}\right)^2}{\omega^2 \varepsilon_0 \mu_0 \mu^*}. \tag{3.57}$$

There are several particular cases of this method of determining the complex dielectric permittivity.

Reactive load. One can find R^* and T^* in the short-circuited waveguide with the sample (Figure 3.13), which can be considered as a four-port. Parameters of the scattering matrix are a_1, a_2, b_1 and b_2. Measuring the complex reflection coefficient, R_{exp} and φ_{exp}, at $l = 0$, $\lambda/4$, $\lambda/6$ and $\lambda/12$ one can find the reflection R^* and transmission T^* coefficients. Here, λ is the wavelength in a waveguide.

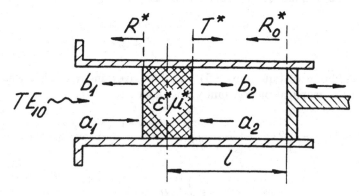

Figure 3.13. Finding of R^* and T^*.

For the four-port, shown in Figure 3.13, one can write:

$$b_1 = S_{11}a_1 + S_{12}a_2, \tag{3.58}$$

$$b_2 = S_{21}a_1 + S_{22}a_2. \tag{3.59}$$

The reflection and transmission coefficients are given by:

$$R^* = Re^{i\varphi_R} = \frac{b_1}{a_1}. \tag{3.60}$$

and

$$T^* = Te^{i\varphi_T} = \frac{b_2}{a_1}. \tag{3.61}$$

The reflection coefficient from the short is

$$R_0^* = \frac{a_2}{b_2}, \tag{3.62}$$

where $R_0 = -1$ at $l = 0$, and $R_0 = 1$ at $l = \lambda/4$.
From Equations 3.59 and 3.62:

$$b_2 = \frac{S_{21}a_1}{1 - S_{22}R_0^*}, \tag{3.63}$$

and

$$a_2 = \frac{R_0^* a_1 S_{21}}{1 - S_{22}R_0^*}. \tag{3.64}$$

From Equations 3.58 and 3.64 one finds

$$b_1 = S_{11}a_1 + \frac{a_1 R_0^* S_{12} S_{21}}{1 - S_{22}R_0^*}$$

and, from Equation 3.60:

$$R^* = S_{11} + \frac{R_0^* S_{21} S_{12}}{1 - S_{22}R_0^*}. \tag{3.65}$$

Loading the symmetrical four-port by a matched load, one obtains

$$R_0^* = 0, \quad R^* = S_{11}, \quad T^* = S_{21}$$

and

$$S_{12} = S_{21}, \quad S_{22} = S_{11}.$$

Then, Equation 3.65 is given by

$$R_{exp}^* = R^* + \frac{R_0^* T^{*2}}{1 - R^* R_0^*}. \tag{3.66}$$

At the four positions of the short:

$$\begin{aligned}
1_1 &= 0, & R_0^* &= -1 \\
1_2 &= \lambda/12, & R_0^* &= -e^{-i\pi/3}, \\
1_3 &= \lambda/6, & R_0^* &= -e^{-i2\pi/3}, \\
1_4 &= \lambda/4, & R_0^* &= 1.
\end{aligned}$$

Instead of Equation 3.66, the following set of equations is obtained:

$$\begin{aligned}
R_{exp\,1}^* &= R^* - \frac{T^{*2}}{1 + R^*}, \\[2mm]
R_{exp\,2}^* &= R^* - \frac{T^{*2} e^{-i\pi/3}}{1 + R^* e^{-i\pi/3}}, \\[2mm]
R_{exp\,3} &= R^* - \frac{T^{*2} e^{-i2\pi/3}}{1 + R^* e^{-i2\pi/3}}, \\[2mm]
R_{exp\,4} &= R^* + \frac{T^{*2}}{1 - R^*}.
\end{aligned} \tag{3.67}$$

From this set of equations (3.67) the coefficients R^* and T^* are obtained using a computer. Then from Equations 3.56 and 3.57, both μ^* and ε^* of the sample are obtained.

The complex dielectric permittivity for nonmagnetic materials can be determined using network analyzers only from R^* or T^* measurements, or from moduli of reflection and transmission measurements.

Equations 3.43 and 3.44 can be rewritten as

$$R^* = \frac{i(\gamma_0^2 \mu^2 - \gamma^2 \mu_0^2) \sin \gamma d}{e^{-i\gamma_0 d}[2\mu_0 \mu \gamma_0 \gamma \cos \gamma d + i(\mu_0^2 \gamma^2 + \mu^2 \gamma_0^2) \sin \gamma d]}, \tag{3.68}$$

$$T^* = \frac{2\mu_0 \mu \gamma_0 \gamma}{e^{-i\gamma_0 d}[2\mu_0 \mu \gamma_0 \gamma \cos \gamma d + i(\mu_0^2 \gamma^2 + \mu^2 \gamma_0^2) \sin \gamma d]}. \tag{3.69}$$

Infinite sample. Equations 3.68 and 3.69 can be simplified when the sample possesses high loss. In that case one can neglect the reflection from the back plane $x = d/2$ of the sample (Figure 3.12) and the sample may be considered as infinite, i.e., $\alpha d \gg 1$, where α is the damping term of the propagation constant, $\gamma = \beta - i\alpha$, and the wave is written in the form $E \sim e^{i(\omega t - \gamma d)}$. In this case it follows

$$R^* = \frac{a[c + b(1 - 2e^{-i\beta d} \times e^{-2\alpha d})]}{q(b + c)^2}$$

and

$$T^* = \frac{2be^{-\alpha d} \times e^{-i\beta d}}{q(b + c)^2},$$

where

$$a = \gamma_0^2 \mu^{*2} - \gamma^2 \mu_0^2, \quad c = \mu_0^2 \gamma^{*2} + \gamma_0^2 \mu^{*2},$$

$$b = 2\mu_0 \mu^* \gamma_0 \gamma^*, \quad q = e^{-i\gamma_0 d}.$$

Having measured R^* and T^*, from Equations 3.56 and 3.57 both ε^* and μ^* can be found. When

$$e^{-2\alpha d} \ll 1,$$

the R^* is:

$$R^* = \frac{a}{q(b + c)}.$$

There are various simplifications of this method. For instance, measuring the scattering parameter S_{21} over a band of frequencies, the permittivity and loss can be expressed (Lanagan et al. 1988) in terms of the slopes of S_{21} phase and magnitude, respectively:

$$\varepsilon' = \left[\left(\frac{\Delta\varphi}{\Delta\nu} \right) \frac{c}{2\pi d} \right]^2$$

and

$$\tan\delta = \left[\frac{\Delta S_{21}}{\Delta\nu} \cdot \frac{8.686c}{\pi d(\varepsilon)^{1/2}} \right],$$

where c is the speed of light in free space, $\Delta S_{21}/\Delta\nu$ and $\Delta\varphi/\Delta\nu$ are the slopes of magnitude and phase with respect to the frequency. The phase difference is in radians, and the magnitude difference is in dB.

The method can be used for low and medium permittivity and high loss materials. When permittivity is high the impedance of the sample

$$Z = \left[\frac{[1 - (\lambda/\lambda_c)]}{\varepsilon_0 \varepsilon^* - (\lambda/\lambda_c)} \right]^{1/2} \tag{3.70}$$

becomes small, and moduli of reflection and transmission coefficients are $R \simeq 1$ and $T \simeq 0$. In Equation 3.70, λ_c is the cut-off wavelength. In that case, the sample transmits an insufficient microwave signal for generating accurate and reproducible data. But in such cases, reflection can be decreased and transmission enhanced by the use of matching sections with $\varepsilon_m < (\varepsilon)^{1/2}$ on either side of the specimen in the waveguide. The network analyzer is first calibrated for reflection and transmission using a standard calibration kit. Dielectric permittivity and loss tangent can be calculated using the following expressions (Lanagan et al. 1988):

$$\varepsilon' = \lambda_0^2 \left(\frac{1}{\lambda_0^2} + \frac{\beta^2 - \alpha^2}{4\pi^2} \right)$$

and

$$\tan \delta = \frac{2\alpha\beta}{\beta^2 - \alpha^2 + 4\pi^2/\lambda_c^2},$$

where λ_0 is the free space wavelength,

$$\beta = \frac{\text{Measured phase (in degrees)} + 2N \times 180°}{d},$$

$$\alpha = \frac{\text{Measured loss (in dB)}}{d}.$$

Dielectric permittivity should be estimated first to choose the appropriate value of the integer N in the equation given above. When $\varepsilon > 1000$ the results of measurements strongly depend on the parameters of matching sections.

Resonant transmission. When in materials of high permittivity and low loss ($\tan \delta < 0.3$) the wavelength in the sample is of the same order as the thickness of the sample d, that is,

$$d = N \frac{\lambda_\varepsilon}{2} \quad or \quad \beta_\varepsilon = N \frac{\pi}{d}, \, N = 1, 2, 3\ldots,$$

where λ_e is the wavelength in a waveguide filled up with the sample, one can obtain resonant transmission of electromagnetic waves by the sample which follows from Equation 3.69 and is given by

$$T^*_{res} = -\frac{b \times e^{i\pi N}}{q[b\cosh\alpha d + c\sinh\alpha d]}.$$

The reflection coefficient then becomes

$$R^* = -\frac{a\sinh\alpha d}{q[b\cosh\alpha d + c\sinh\alpha d]}.$$

While measuring the transmission, the sample thus acts as a Fabry–Perot etalon, yielding a fringe pattern when the phase factor $\beta_e d$ is varied. The phase factor can be varied by sweeping the frequency and by changing the temperature. Using backward-wave oscillators and sweeping a frequency, one finds the permittivity from the peak spacing. The absorption coefficient is determined from the peak value of the resonant transmission coefficient. In that case, phase measurements are unnecessary. Dielectric permittivity and loss can be obtained also according to

$$\varepsilon' = \left(\frac{\lambda}{2d\Delta\lambda}\right)^2 + \left(\frac{\lambda}{\lambda_c}\right)^2$$

and

$$\tan\delta = \lambda\frac{\alpha}{2\pi}(\varepsilon)^{1/2},$$

where $\Delta\lambda$ is the distance between the neighboring maxima of the transmission.

The sample with plane parallel faces may also be mounted between the flanges of two rectangular waveguides. The use of the free-standing samples eliminates the mechanical problems associated with mounting crystals inside the waveguide, and facilitates the interchange of samples. With this technique, high values of the permittivity can be accurately measured (Rytz et al. 1985) by using samples with sufficiently small thickness.

Resonant reflection. When the sample with plane parallel faces is placed in the short-circuited waveguide (Figure 3.14) and the thickness of the sample, d, is a multiple of $\lambda_e/4$, resonant absorption in the sample

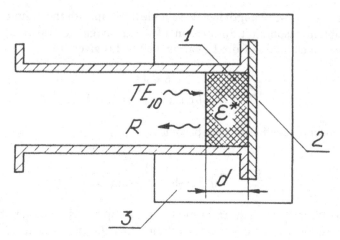

Figure 3.14. Sample (1) in rectangular waveguide terminated by a metallic short (2), and (3) is the thermostat.

occurs and a reflection coefficient is smallest. The resonance condition is given by

$$d = (2N + 1)\frac{\lambda_\varepsilon}{4}, \quad N = 0, 1, 2, 3\ldots$$

The resonant reflection coefficient follows from Equation 3.68 and is given by

$$R_{\text{res}} = \frac{a \cosh \alpha d}{q[b \sinh \alpha d + c \cosh \alpha d]}.$$

The transmission coefficient then becomes

$$T = \frac{b \times i e^{i\pi N}}{q[b \sinh \alpha d + c \cosh \alpha d]}.$$

Sweeping a frequency, from the distance between the neighboring minima one can find the value of λ_ε and, consequently, dielectric permittivity (Petrov 1972):

$$\varepsilon' = \frac{2\lambda^2}{\lambda_\varepsilon^2[1 + (1 + A^2)]^{1/2}} + \left(\frac{\lambda}{\lambda_c}\right)^2,$$

where $A = \tan \delta/[1 - 1/\varepsilon(\lambda/\lambda_c)^2] \approx \tan \delta$. When $\tan \delta < 0.1$ and the permittivity is high, the above expression can be simplified and is given by

$$\varepsilon' = \left(\frac{\lambda}{\lambda_\varepsilon}\right)^2 = \left(\frac{N\lambda}{4d}\right)^2.$$

The value of loss can be found from the resonant reflection coefficient, or approximately from

$$\tan \delta = \frac{4\lambda_\varepsilon K}{N\pi\lambda},$$

where K is the value of the standing wave coefficient at the resonance. In that case, phase measurements are also unnecessary. The advantage of this method is that it is easy to put the section of the waveguide with the dielectric specimen to a heater or cooler.

Ferroelectric films. The method of filling the cross section of the rectangular waveguide can be used for determining the complex dielectric permittivity of thin ferroelectric films on a dielectric substrate (Figure 3.15). The reflection and transmission coefficients in that case (Lapinskas et al. 1990) are given by

$$R^* = \frac{e^{-i2\gamma_1 d_1}m - n}{1 - e^{-i2\gamma_1 d_1}mn}$$

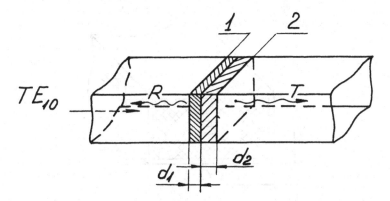

Figure 3.15. Thin film (1) on the dielectric substrate (2) in rectangular waveguide: d_1 is the thickness of the film, d_2 is the thickness of the substrate.

and

$$T^* = \frac{8\gamma_1\gamma_2 e^{-i(\gamma_1 d_1 - \gamma_2 d_2)}}{\left\{ \left(\frac{\gamma_1}{\gamma_0} + 1 \right)(1 - e^{-i2\gamma_1 d_1} mn)[e^{i2\gamma_2 d_2}(\gamma_2 + \gamma_0)(\gamma_1 + \gamma_2) + (\gamma_2 - \gamma_0)(\gamma_1 - \gamma_2)] \right\}},$$

where

$$m = \frac{e^{i2\gamma_2 d_2}\dfrac{\gamma_1 - \gamma_2}{\gamma_1 + \gamma_2} + \dfrac{\gamma_2 - \gamma_0}{\gamma_2 + \gamma_0}}{e^{i2\gamma_2 d_2} + \dfrac{\gamma_2 - \gamma_0}{\gamma_2 + \gamma_0}\dfrac{\gamma_1 - \gamma_2}{\gamma_1 + \gamma_2}}, \quad n = \frac{\gamma_1 - \gamma_0}{\gamma_1 + \gamma_0},$$

$$\gamma_1 = \left[\omega^2 \varepsilon_1^* \varepsilon_0 \mu_0 - \left(\frac{\pi}{a} \right)^2 \right]^{1/2}, \quad \gamma_2 = \left[\omega^2 \varepsilon_2^* \varepsilon_0 \mu_0 - \left(\frac{\pi}{a} \right)^2 \right]^{1/2},$$

and γ_0 is given by Equation 3.37; d_1 and d_2 are the thicknesses of the ferroelectric film and dielectric substrate, respectively; ε_1^* and ε_2^* are the complex permittivity of the ferroelectric film and dielectric substrate, respectively; γ_1, γ_2 and γ_0 are the propagation constants in the ferroelectric film, dielectric substrate and empty wavequide, respectively.

In a wide frequency range the values of permittivity and tan δ can be calculated by iteration method, using a computer, from the results of measurements of moduli of the reflection and transmission coefficients.

The above methods of determination of complex permittivity of solids and liquids from the measurements of the transmission and/or reflection coefficient of the dominant mode of a uniformly-filled wave-

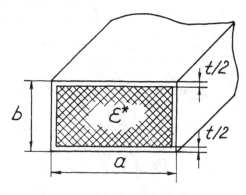

Figure 3.16. Rectangular waveguide showing gap between sample and waveguide walls.

guide section are often used while dealing with materials that are produced as thin sheets or disks. Samples are machined to completely fill the cross section of a rectangular waveguide. However, this is difficult to achieve in practice, because of an unavoidable gap between the sample and the waveguide walls (Figure 3.16). The importance of the gap increases when the magnitude of the complex permittivity is high. It may cause large errors in measurements of high-permittivity or high-loss materials.

An empirical formula to correct for the air gap, considering the effect of this gap on the dominant mode only, and thereby neglecting all higher-order modes generating at the planes of discontinuity between the empty and filled waveguides, can be written (Champlin and Glover 1966)

$$\varepsilon_m^* = \frac{\varepsilon^*}{1 + (\varepsilon^* - 1)\left(\dfrac{t}{b}\right)},$$

where ε_m^* is measured complex permittivity. One can see that ε_m^* saturates, approaching the limiting value b/t, regardless of the permittivity of the material. For ferroelectrics in a dielectric dispersion region, even a gap of a few microns in the millimeter wave range causes large errors in measurements of dielectric parameters. Therefore, the specimens should be electroded with sputtered gold or coated with silver around the faces of flange contact.

Microwave sounding. This nondestructive method is based on the measurements of the modulus and the phase of the complex reflection coefficient from the open end of a rectangular waveguide which is in contact with the polished surface of the infinite crystal (Decreton and Gardiol 1974). This method offers the advantage of being able to measure dielectric parameters of large isotropic bodies without cutting them. This method can also be used for anisotropic materials. Taking into account the anisotropy of the crystal, the relation between the electric and magnetic fields in the aperture of a waveguide is given by (Grigas et al. 1988a):

$$H_x(x,y) = \frac{i}{\omega\mu_0}\left\{\left(k_y^2 + \frac{\partial^2}{\partial x^2} + \frac{\partial^2}{\partial y^2}\right)\frac{1}{2\pi}\int_0^a\int_0^b E_y(x',y')\frac{e^{-ik_y r}}{r}\,dx\,dy'\right.$$
$$\left. - \frac{\partial^2}{\partial y^2}\frac{1}{2\pi}\int_0^a\int_0^b E_y(x',y')\frac{e^{-ik_z r}}{r}\,dx'\,dy'\right\},$$

where

$$r = [(x - x')^2 + (y - y')^2]^{1/2},$$
$$k_y = \omega(\varepsilon_0 \mu_0 \varepsilon_{yy}^*)^{1/2},$$
$$k_z = \omega(\varepsilon_0 \mu_0 \varepsilon_{zz}^*)^{1/2}.$$

The numerical calculations show that for ferroelectrics with the permittivity $\varepsilon_{yy}' > 100$ and any value of ε_{zz}', in the dispersion region, when $\tan \delta > 0.3$, diffraction at the end of the waveguide can be neglected and the Fresnel formula for the infinite dielectric body in rectangular waveguide can be used:

$$R^* = \frac{\gamma_0 - k_y}{\gamma_0 + k_y},$$

where γ_0 is given by Equation 3.37.

Since the impedance of the dielectric body (Equation 3.70) for ferroelectrics with a large value of permittivity is small and the reflection coefficient is $R \simeq 1$, then the matching section (see *Infinite sample*) or transformer of the reflection coefficient in the shape of an iris with a symmetric circular aperture can be used. The iris is placed at a distance of a half-wave from the sample. The S-matrix parameters of the iris are found by using three standard loads. The reflection coefficient from the sample is then given by

$$R^* = \frac{R_i^* - S_{11}}{R_i^* S_{22} + S_{12}^2 - S_{11} S_{22}},$$

where R_i^* is the reflection coefficient of an iris-terminated waveguide structure loaded with a sample.

In principle, combining the above methods allows for the determination of the values of permittivity up to 10^4 and $\tan \delta$ from 10^{-3} to several units.

However, a great number of anisotropic single crystals of ferroelectrics and related materials cannot be grown of sufficient dimensions to totally fill a cross section of a waveguide—especially in the centimeter wave range. Nor are the above methods applicable for a number of one-dimensional needle-shaped crystals, which can be grown of almost perfect structure. Further the possibilities of microwave dielectric spectroscopy of small samples shall be discussed in relation to anisotropic single crystals of different materials.

3.3.2. Thin Cylindrical Rod in Rectangular Waveguide

The method of using a thin cylindrical rod instead of filling the cross section of the waveguide offers the advantage of being able to measure any complex dielectric permittivity of a small amount of material. For many single crystals it is better to prepare a sample in the form of a perfect cylinder and to place it in the center of a rectangular waveguide, than to completely fill the cross section of a waveguide. This method allows the investigation of virgin needle-shaped single crystals of a large family of compounds which can be grown of a nearly cylindrical shape with a diameter ranging from several microns to several millimeters.

For low-permittivity materials, the cylindrical rod in a quasi-static approximation can be considered as a four port with lumped parameters and, by measuring a complex reflection coefficient, one can calculate dielectric parameters (Grigas and Meškauskas 1971). Due to the non-homogeneous distribution of the electric field in the rod, the resonant absorption takes place with increasing frequency, the frequency of which is given by

$$v_0 = \frac{c}{2\pi r_0 (\varepsilon')^{1/2}},$$
(3.71)

where r_0 is the radius of the rod, and c is the speed of light in free space. For crystals with $\varepsilon \geq 100$, the resonant frequency lies in the centimeter or millimeter wave range. When ε' in ferroelectrics depends on temperature according to the Curie–Weiss law, the resonant frequency behaves as a soft mode frequency (see Chapter 5).

Next, the complex reflection coefficient of the cylindrical dielectric rod placed in the rectangular waveguide, shall be calculated taking into account the nonhomogeneous distribution of the microwave field in the rod. Various exact solutions have been proposed for the electromagnetic wave scattering by a dielectric rod in the waveguide (Grigas and Shugurov 1969; Nielsen 1969; Bhartia 1977; Sahalos 1985). This problem is important also in connection with the wide use of dielectric rods in waveguides as different functional devices of microwave technique.

First, consider a dielectric cylinder of complex permittivity ε^* and radius r_0, placed centrally ($l = 0$) in a guide so that the cylinder axis is parallel to the electric field vector of the dominant TE_{10}-mode (Figure 3.17). Consider the most general case (Grigas and Shugurov 1969) when a wavelength in the cylinder may be of any value, including the value of the same order as its dimensions (i.e., $\beta_0 = k(\varepsilon)^{1/2} r_0 \leq 1$, where k is the wave number of the TE_{10}-mode). This complicated case requires rigor-

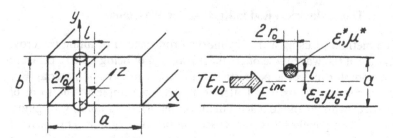

Figure 3.17. Cylindrical dielectric rod in rectangular waveguide.

ous solution of Maxwell's equations. From the Lorentz lemma, following from the rigorous solution of Maxwell's equations, the amplitude reflection coefficient of the TE_{10}-mode is given by

$$R^* = \frac{i(\varepsilon^* - 1)}{2\gamma} \int_v (E^c, E^{inc}) dv, \qquad (3.72)$$

where γ is the propagation constant of the TE_{10}-wave (Equation 3.38), E^c is the field inside the cylinder, and E^{inc} is the field of the incoming TE_{10}-mode in the waveguide outside the cylinder. Vectors of E^c and E^{inc} are parallel to the y-axis, which coincides with the cylinder axis. Integration is over the volume of the cylinder. The electric field in the cylinder has only the y component (i.e., $E_y^c(x, z) = E_y^c$), $E_x = E_z = 0$.

For a cylindrical coordinate system the fields E_{yc} and E_{yinc} in the cylindrical modes are

$$E_y^c = ik^2(\varepsilon^*)^{1/2} \sum_{m=-\infty}^{\infty} B_m^c J_m(\beta) e^{im\varphi},$$

$$E_y^{inc} = ik^2 \sum_{m=-\infty}^{\infty} B_m^{inc} J_m(\alpha) e^{im\varphi}, \qquad (3.73)$$

$$B_m^c = \tfrac{1}{2} A \cos\varphi_1 [i^m e^{im\varphi_1} + (-1)^m e^{-im\varphi_1}],$$

where J_m are Bessel functions, φ and φ_1 are the polar coordinates, $\varphi_1 < \pi/2$ is the angle between the wave number $k(\pi/2, \gamma, 0)$ and the x-axis, A is the normalization constant, $\alpha = kr$, $\beta = k(\varepsilon)^{1/2}r$. In the particular case where $m = 0$

$$B_0^c = \frac{1}{k}\left(\frac{2}{ab}\right)^{1/2}.$$

Substituting Equation 3.73 into Equation 3.72 yields

$$R^* = \frac{i(\varepsilon^* - 1)}{2\gamma} I,$$

where

$$I = -2\pi k^4 (\varepsilon^*)^{1/2} b \sum_{m=-\infty}^{\infty} B_m^c B_m^{inc} \int_0^{r_0} J(\beta) J_m(\alpha) r \, dr. \tag{3.74}$$

First, consider the case $\alpha_0 = k r_0 \ll 1$, which is important for the needle-shaped single crystals of high-dielectric permittivity (e.g., SbSI etc.). In the expansion of one of the Bessel functions in Equation 3.74, one can consider the term $(m = 0)$ solely. Substitution of its received integral:

$$\int_0^{r_0} J_m(\beta) r^{m+1} \, dr = \frac{r_0^{m+1}}{k(\varepsilon^*)^{1/2}} J_{m+1}[k(\varepsilon^*)^{1/2} r_0]$$

into Equation 3.74 yields

$$R^* = i\pi(\varepsilon^* - 1) \frac{k}{\gamma} b^2 r_0 [B_0^c B_0^{inc} J_1(\beta_0) - B_1^c B_1^{inc}(\alpha_0) J_2(\beta_0)$$
$$+ B_2^c B_2^{inc}(\alpha_0)^2 J_3(\beta_0) + \cdots]. \tag{3.75}$$

The coefficients B_m^c are determined from the boundary conditions for the tangential components of the electric E_y and magnetic H_φ fields inside and outside the cylinder:

$$H_\varphi^c = k^2 \varepsilon^* \sum_{m=-\infty}^{\infty} B_m^c J_m'(\beta_0) e^{im\varphi}. \tag{3.76}$$

The reflected field has form similar to Equation 3.76. However, instead of cylindrical Bessel functions, one utilizes the Hankel functions of the second order, $H_m^{(2)}$.

Applying the boundary conditions at the surface of the cylinder, and ignoring the influence of the side walls of the waveguide, yields a set of equations for the determination of the unknown expansion coefficients. These are

$$(\varepsilon^*)^{1/2} B_m^c J_m(\beta_0) = B_m^{inc} J_m(\alpha_0) + B_m^0 H_m^{(2)}(\alpha_0),$$

$$\varepsilon^* B_m^c J_m'(\beta_0) = B_m^{inc} J_m'(\alpha_0) + B_m^0 H_m^{(2)}(\alpha_0), \tag{3.77}$$

$$B_m^{sc} = B_m^{inc} \frac{J_m(\beta_0) J_m'(\alpha_0) - (\varepsilon^*)^{1/2} J_m'(\beta_0) J_m(\alpha_0)}{(\varepsilon^*)^{1/2} H_m^{(2)}(\alpha_0) J_m'(\beta_0) - H_m^{(2)}(\alpha_0) J_m(\beta_0)},$$

where J_m' and H_m' are derivatives of the Bessel and Hankel functions.

Solving the set of Equations 3.77:

$$B^c_m = B^{inc}_m \frac{2i}{\pi\alpha_0} \times \frac{1}{(\varepsilon^*)^{1/2}H_m(\alpha_0)J'_m(\beta_0) - H^{(2)'}_m(\alpha_0)J_m(\beta_0)}. \qquad (3.78)$$

When $\varepsilon = 1$, $B^c_m = B^{inc}_m$ and $B^{sc}_m = 0$ (i.e., the incident wave has not become deformed). Equation 3.78 is obtained considering that

$$J'_m(y)H^{(2)}_m(y) - J_m(y)H^{(2)'}_m(y) = \frac{2i}{\pi y}.$$

This formula is valid only for the short interval of time while the scattered cylindrical wave, reflected from the side walls of the waveguide, comes back to the axis of the cylinder.

For the rigorous solution of the problem, the dielectric cylinder is replaced by multipole sources, located at the center of the cylinder. The scattered field is calculated using the images of these with respect to the guide walls. Figure 3.18 shows two cylindrical waves. The relation between their amplitudes B_m and B_{-m} is found from the condition of disappearance of the tangential component E_y on the reflective walls of the waveguide. Since $\varphi + \varphi' = \pi$:

$$B_m = -B'_{-m}. \qquad (3.79)$$

Equation 3.79 is valid when the magnetic field has only a tangential component. Thus, the boundary conditions at the surface of the cylinder can be written with the same amplitudes of the electric and magnetic field as in the set of Equations 3.77. While calculating E_y on the surface of the cylinder, the radius r_0 shall be neglected in comparison to the distance to

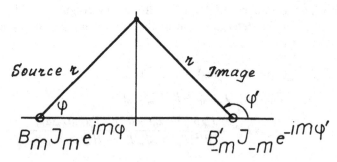

Figure 3.18. Cylindrical waves of the source and the reflector.

the reflections. Hence, it is enough to calculate the fields induced by the reflections on the cylinder axis. For all the waves coming from the right reflections, the angles are $\varphi' = \pi$, while, for the waves coming from the left reflections $\varphi' = 0$. Mathematically, this means that in expansion of a cylindrical wave in cylindrical waves with another axis, one should take into account the first term only. The following terms will contain α_0 as a factor. In that approximation, the second and successive terms in the square brackets of Equation 3.75 may be neglected. In other words, in the set of equations (3.77) it is enough to specify the equations corresponding to the value $m = 0$. Note also that

$$\frac{B_m^0}{B_m^{inc}} \sim \alpha_0^{2m+1},$$

so that all terms in the reflected wave may be neglected, except the first, having the B_0^c amplitude.

The reflection coefficient corresponding to the value $m = 0$ shall be calculated. Then, the boundary conditions for the electric field can be written as

$$(\varepsilon^*)^{1/2} B_0^c J_0(\beta_0) = B_0^{inc} J_0(\alpha_0) + B_0^0 H_0(\alpha_0) + 2 B_0^0 \sum_{m=1}^{\infty} (-1)^m H_0^{(2)}(mka).$$

For the magnetic field these conditions are the same as given by the second equation of the set (3.77). Therefore, omitting mathematical calculations, one can write

$$B_0^c = -i \frac{2 B_0^{inc}}{\pi \alpha_0 \Delta_1},$$

where the determinant Δ_1 is given by

$$\Delta_1 = \varepsilon^* J_1(\beta_0) \left[H_0^{(2)}(\alpha_0) + 2 \sum_{m=1}^{\infty} (-1)^m H_0^{(2)}(mka) \right] - (\varepsilon^*)^{1/2} J_0(\beta_0) H_1(\alpha_0).$$

Thus, the complex reflection coefficient with regard to the non-homogeneous distribution of the microwave field in the cylinder is given by the simple formula

$$R_{m=0}^* = -\frac{4(\varepsilon^* - 1) J_1(\beta_0)}{\pi \left[\left(\frac{2a}{\lambda} \right)^2 - 1 \right]^{1/2} \Delta_1}. \tag{3.80}$$

The negative sign means that the reflection coefficient for the magnetic component is a complex conjugate with the reflection coefficient of the electric component of the dominant TE_{10} wave. Moduli are the same, obviously.

Equation 3.80 can be rewritten as

$$A^c = \frac{H_0(\alpha_0) + 2\sum_{m=1}^{\infty}(-1)^m H_0^{(2)}(mka)}{\alpha_0 H_1(\alpha_0)} - \frac{2k^2(B_0^{inc})^2}{R^*\gamma\alpha_0 H_1^{(2)}(\alpha_0)},$$

$$A^c = \frac{J_0(\beta_0)}{\beta_0 J_1(\beta_0)}. \tag{3.81}$$

For the given frequency and radius of the cylinder the value A^c is constant. From the plot A^c vs. $k(\varepsilon^*)^{1/2}r_0$, and the measurements of the reflection coefficient one can find the value of a dielectric permittivity in a very similar way. When $A^c \to 0$, the impedance of the cylinder ap-

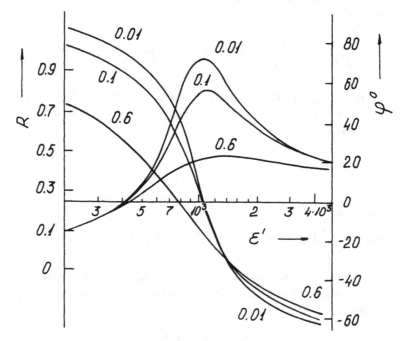

Figure 3.19. Dependence of modulus and phase of the reflection coefficient on permittivity at a frequency of 36 GHz. Radius of cylinder $r = 0.32$ mm. Numbers at the curves mean values of $\tan \delta$.

proaches zero and a resonant absorption in the rod (i.e., dielectric resonance) occurs. The frequency of the resonance can be obtained from Equation 3.81 and it coincides with the frequency given by Equation 3.71. Equation 3.80 is valid and allows calculation of dielectric parameters even at $\beta_0 = k(\varepsilon^*)^{1/2}r_0 \geq 1$ (i.e., at the frequencies above the dielectric resonance). Figure 3.19 shows the dependence of the complex reflection coefficient on the radius of the cylinder. The radius of the cylinder at the higher frequencies should be taken thinner because in a vicinity of the dielectric resonance the reflection coefficient and phase are extremely sensitive to the changes in all parameters of the dielectric rod. In thin needle-shaped crystals, dielectric resonance occurs at very high values of permittivity (Figure 3.20). The reflection coefficient and phase also depend strongly on the loss of the rod. Thus, properly selecting a radius of a rod, one can determine various values of dielectric permittivity and loss from the measurements of modulus and phase of a reflection coefficient.

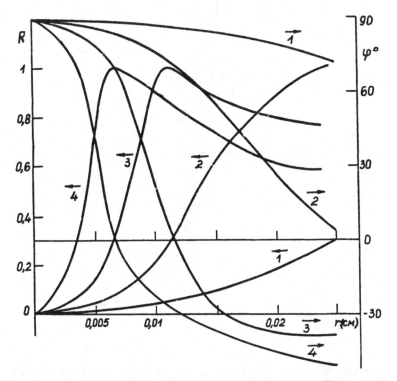

Figure 3.20. Dependence of modulus and phase of reflection coefficient on the radius of cylinder ($\varepsilon = 100$, $\tan \delta \ll 1$) at frequencies (GHz): 1) 9; 2) 20; 3) 40; 4) 70.

From the resonant value of the reflection coefficient one can find dielectric permittivity and loss of the lossy rod even without the phase measurements.

The value of a reflection coefficient can be changed by moving the rod from the center of waveguide along the wide wall. In this case, conditions of interaction of the incident field with the rod and current sources images with respect to the guide walls change. Moving the rod can excite the dielectric resonance (Figure 3.21) even if $\beta_0 < 1$ and increase the sensitivity of the reflection coefficient and phase on dielectric parameters of the rod.

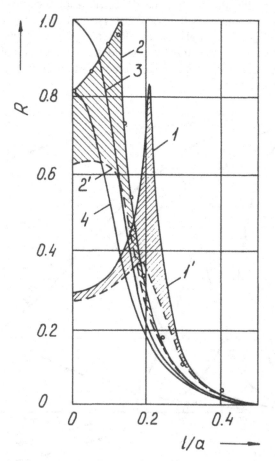

Figure 3.21. Reflection coefficient vs. l/a: curves 1 and 1′ are for $\varepsilon = 50$, $\tan \delta \approx 0$ and 0.2; curves 2 and 2′ are for $\varepsilon = 100$, $\tan \delta \approx 0$ and 0.2; curves 3 and 4 are for $\varepsilon = 130$ and 230, $\tan \delta \approx 0$; $l/a = 0$ corresponds to the center of the guide.

The formula of the complex reflection coefficient, when the dielectric cylinder is at the arbitrary distance l (Figure 3.17) from the center of the guide, is given by

$$R^* = \frac{4(\varepsilon^*)^{1/2}\cos^2\dfrac{\pi l}{a}}{\pi\left[\left(\dfrac{2a}{\lambda}\right)^2 - 1\right]^{1/2}} \times \frac{J_1(\beta_0)}{\Delta_2}, \tag{3.82}$$

where

$$\Delta_2 = (\varepsilon^*)^{1/2}J_0'(\beta_0)\left[H_0^{(2)}(\alpha_0) + 2\sum_{m=1}^{\infty}H_0^{(2)}(mka)\right] - \sum_{m=0}^{\infty}H_0[(2m+1)a+2l]k$$

$$- \sum_{m=0}^{\infty}H_0^{(2)}[(2m+1)a-2l]k - J_0(\beta_0)H_0^{(2)'}(\alpha_0).$$

Equation 3.82 can be used to calculate dielectric parameters of the dielectric cylinder placed at any position l in the rectangular waveguide, the axis of the cylinder being parallel to the electric field vector. This equation can be used also for calculation of microwave devices using dielectric, ferroelectric or semiconductive rods in the waveguide.

In the case when the condition $\alpha_0 \ll 1$ is insufficiently fulfilled, an additive correction of the reflection coefficient (Equation 3.80) will be found corresponding to $m = \pm 1$. The corresponding boundary conditions are given by

$$(\varepsilon^*)^{1/2}B_1^c J_1(\beta_0) - B_1^{sc}H_1^{(2)}(\alpha_0) = B_1^{inc}J_1(\alpha_0),$$

$$(\varepsilon^*)^{1/2}B_{-1}^c J_1(\beta_0) - B_1^{sc}H_1^{(2)}(\alpha_0) = B_1^{inc}J_1(\alpha_0),$$

$$\varepsilon^* B_1^c J_1'(\beta_0) - B_1^{sc}\left\{H_1^{(2)}(\alpha_0) + \sum_{m=1}^{\infty}(-1)^m\left[H_1^{(2)}(mka) + \frac{1}{mka}H_1^{(2)}(mka)\right]\right\}$$

$$+ B_1^{sc}\sum_{m=1}^{\infty}(-1)^m\left[H_1^{(2)'}(mka) - \frac{1}{mka}H_1^{(2)}(mka)\right] = B_1^{inc}J_1'(\alpha_0),$$

$$\varepsilon^* B_{-1}J_1'(\beta_0) + B_1^{sc}\sum_{m=1}^{\infty}(-1)^m\left[H_1^{(2)}(mka) - \frac{1}{mka}H_1^{(2)}(mka)\right]$$

$$- B_1^{sc}\left\{H_1^{(2)}(\alpha_0) + \sum_{m=1}^{\infty}(-1)^m\left[H_1^{(2)}(mka) + \frac{1}{mka}H_1^{(2)}(mka)\right]\right\}$$

$$= B^{inc_1}J_1'(\alpha_0),$$

where

$$B_1^c = B_{-1}^c$$

and

$$B_1^{sc} = B_{-1}^{sc}.$$

Then the determinator is given by

$$\Delta_3 = -(\varepsilon^*)^{1/2} J_1(\beta_0) \left[H_1^{(2)\prime}(\alpha_0) + \sum_{m=1}^{\infty} (-1)^m \frac{1}{mka} H_1^{(2)}(mka) \right]$$
$$+ \varepsilon^* J_1'(\beta_0) H_1(\alpha_0).$$

In that approximation

$$B_1^c = i \frac{2 B_1^{inc}}{\pi \alpha_0 \Delta_3}.$$

The additive correction of the reflection coefficient, corresponding to $m = 1$, is given by

$$R_{m=1}^* = \frac{4(\varepsilon^* - 1)}{\pi \left[\left(\frac{2a}{\lambda} \right)^2 - 1 \right]^{1/2}} \times \frac{\alpha J_2(\beta_0)}{\Delta_3}, \qquad (3.83)$$

where

$$J_2(\beta_0) = \frac{2}{\beta_0} J_1(\beta_0) - J_0(\beta_0).$$

The total reflection coefficient may be recast in the form

$$R^* = \frac{4(\varepsilon^* - 1)}{\pi \left[\left(\frac{2a}{\lambda} \right)^2 - 1 \right]^{1/2}} \times \left[\frac{J_1(\beta_0)}{\Delta_1} + \frac{\alpha_0 J_2(\beta_0)}{\Delta_3} \right]. \qquad (3.84)$$

Numerical calculations show that the additive correction given by Equation 3.83 changes the modulus of the reflection coefficient about 1%, when $\varepsilon \le 200$. When permittivity is higher, this correction becomes considerable.

Equation 3.84 enables us to determine any values of dielectric parameters of thin rods or needle-shaped crystals from the measurements of the complex reflection coefficient in the centimeter and millimeter wave range.

Determination of parameters of semiconductors and negative values of permittivity. A number of semiconductors (see Chapter 4) can be grown in a form of thin needle-shaped crystals. Such crystals can be used as microwave power gauges and other devices. Figure 3.22 shows the

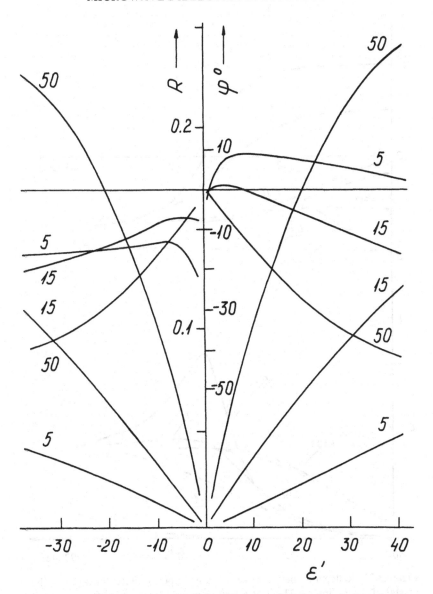

Figure 3.22. Dependence of reflection coefficient and phase on ε and $\tan \delta$ of the needle-shaped crystal of radius 0.024 mm at a frequency of 38 GHz. Numbers at the curves mean values of $\tan \delta$.

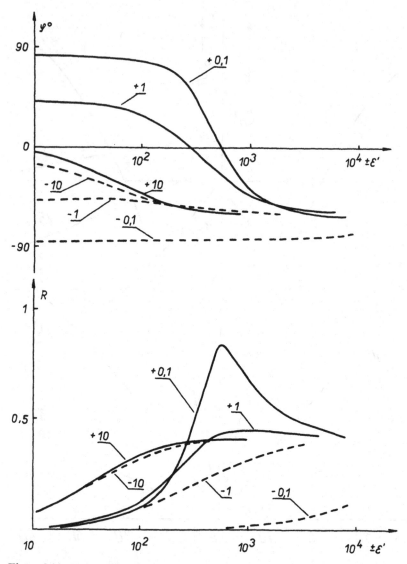

Figure 3.23. Reflection coefficient and phase vs. permittivity and tan δ of the crystal of the radius $r = 23\,\mu$m at a frequency of 70 GHz. Numbers at curves mean values of tan δ. Dotted lines correspond to negative values of permittivity.

possibilities to determine both values of dielectric permittivity and tan δ from the measurements of the complex reflectivity in the millimeter wave range. These values of permittivity and tan δ are characteristic for most semiconductors used in microwave techniques such as Ge, Si, InSb etc. The reflection coefficient is sensitive to the changes of permittivity and tan δ, which can be found with high accuracy. From these parameters one can determine conductivity and other semiconductive parameters.

The reflection coefficient (Equations 3.80 or 3.84) is asymmetrical with respect to the sign of the dielectric permittivity. Changing the ε to $-\varepsilon$, the Bessel function changes as

$$J(i\beta_0) \to I(\beta_0).$$

As a result, one can find low (Figure 3.22) and high, positive and negative values (Figure 3.23) of dielectric permittivity of the crystals possessing different loss. A negative dielectric permittivity may be only in a region of anomalous dielectric dispersion, where losses are high. In ferroelectrics, the value of permittivity becomes negative when microwave frequency approaches a soft mode frequency. Equations 3.80, 3.82 or 3.83 can be used for diagnostics of semiconductors by means of plasma or magnetoplasma resonance (see Section 3.3.7).

3.3.3 Dielectric Cylinder of any Size in Rectangular Waveguide

In many cases, it is difficult to prepare a cylindrical sample of the material under study that fulfills the condition $kr_0 \ll 1$, especially in the millimeter wave range. Usually, samples of $kr_0 < 1$ are used. Grigas et al. 1990a; Brilingas et al. 1990 describes the rigorous solution for the reflection and transmission coefficients of the TE_{10}-wave for the dielectric cylinder of any radius placed in a rectangular waveguide (Figure 3.17).

The components of the electric and magnetic field in the cylinder can be expanded similarly to Equation 3.73 as a sum of the partial waves:

$$E_y^c \sim \sum_{m=-\infty}^{\infty} A_{1,m} J_m(\beta) e^{im\varphi},$$

$$H_\varphi^c \sim -i\left(\varepsilon^* \frac{\varepsilon_0}{\mu_0}\right)^{1/2} \sum_{m=-\infty}^{\infty} A_{1,m} J_m'(\beta) e^{im\varphi}. \tag{3.73a}$$

Similarly, the scattered field outside the cylinder is given by

$$E_y \sim \sum_{m=-\infty}^{\infty} A_{2,m} H_m^{(2)}(\alpha) e^{im\varphi},$$

where $A_{1,m}$ and $A_{2,m}$ are the complex expansion coefficients.

A set of complex linear equations for the determination of the unknown complex coefficients is found by applying the boundary conditions at the surface of the cylinder $r = r_0$. They may be recast in the form

$$\sum_{m=-\infty}^{\infty} S_{m,n} A_{2,n} - X_m A_{2,m} = A_{0,m},\qquad (3.85)$$

where

$$X_m = \frac{(\varepsilon^*)^{1/2} H_m^{(2)}(\alpha_0) J'_m(\beta_0) - H_m^{(2)'}(\alpha_0) J_m(\beta_0)}{(\varepsilon^*)^{1/2} J_m(\alpha_0) J'_m(\beta_0) - J'_m(\alpha_0) J_m(\beta_0)}$$

$$S_{m,n} = (-1)^{m+n}(S_{1,n+m} - S_{3,n-m}) + S_{2,n+m} - S_{3,n-m},$$

$$A_{0,m} = -\tfrac{1}{2}\left[i^m e^{i(m\varphi_1 - k_x l)} + (-i)^m e^{-i(m\varphi_1 - k_x l)}\right],$$

$$k_x = \frac{\pi}{a}, \quad \cos\phi_1 = \frac{\pi}{k_0 a},$$

and the coefficients S_1, S_2 and S_3 involve the images of the cylinder with respect to the narrow waveguide walls and are given by

$$S_{1,n+m} = \sum_{p=0}^{\infty} H_{n+m}^{(2)}\{k_0[(2p+1)a + 2d]\},$$

$$S_{2,n+m} = \sum_{p=0}^{\infty} H_{n+m}^{(2)}\{k_0[(2p+1)a - 2d]\},$$

$$S_{3,n-m} = \sum_{p=1}^{\infty} H_{n-m}^{(2)}(2k_0 a p).$$

The coefficients $A_{1,m}$ are expressed as

$$A_{1,m} = \frac{J_m(\alpha_0)\left[A_{0,m} - \sum_{n=-\infty}^{\infty} S_{n,m} A_{2,n}\right] + H_m(\alpha_0) A_{2,m}}{(\varepsilon^*)^{1/2} J_m(\beta_0)}.$$

The amplitude reflection and transmission coefficients are obtained by determining the electric and magnetic field inside the cylinder as previously described. They can be written as

$$R^* = C^*(-TE_{10}), \quad T^* = 1 + C^*(+TE_{10}),\qquad (3.86)$$

where

$$C^*(\pm TE_{10}) = \frac{i(\varepsilon^* - 1)}{2\gamma} I_\pm,$$

$$I_- = -\frac{4\pi k_0}{a(\varepsilon^* - 1)} \beta_0 \sum_{m=-\infty}^{\infty} (-1)^m A_{1,m} A_{0,m} [(\varepsilon^*)^{1/2} J_{m+1}(\beta_0) J_m(\alpha_0)$$

$$- J_m(\beta_0) J_{m+1}(\alpha_0)],$$

$$I_+ = I_- | A_{0,-m} \rightarrow A_{0,m}^-, \quad A_{0,m}^- = A_{0,-m} | \varphi_1 \rightarrow -\varphi_1.$$

For a finite number of terms in the summation, the system (3.85) can be solved using a computer. It should be noted that, when the radius of the cylinder decreases, the contribution of the higher waves becomes negligible and the zero term ensures sufficient accuracy. This case is similar to the one obtained in Section 3.3.2 and is the most interesting for practical purposes. It has been used for a long time in our laboratory in dielectric spectroscopy of ferroelectrics and related materials.

The obtained equations (3.86) have no limitations in regards to the radius and the permittivity of the cylinder. Bhartia's (1977) equations fail to give good results when the cylinder is thick. In comparison with Nielsen's (1969) case, they require less computations to find the dielectric parameters, as the summation of series is performed only once.

In a general case, the number of waveguide modes, which should be taken into account by solving Equations 3.85 and 3.86, depends on a radius of the cylinder. A computer program, taking into account only even terms $m = 0, \pm 2, \pm 4, \pm 6, \pm 8, \pm 10$, showed good agreement between computational and experimental results right up to the complete filling of the cross section of the waveguide ($kr_0 \simeq 2.1$) with the cylindrical samples (Figure 3.24). In the value of dielectric permittivity of p-germanium, negative contribution of free charges ($n = 1.7 \times 10^{20} \, m^{-3}$) is taken into account. At a definite kr_0, a resonant transmission occurs and a lossless cylinder acts as a matched load in a guide. The results show that the electric field in the cylinder, in general, is highly non-homogeneous. The spatial distribution of the electric field in the cylinder can be calculated using Equation 3.73a.

Figure 3.25 shows the spatial distribution of the electric field of the TE_{10}-mode in the cylindrical germanium samples of various radii along the z- and x-directions (see Figure 3.17) at the frequency 18.2 GHz. The electric field is nonhomogeneous along both the z- and x-directions. The thinner the cylinder, the more the electric field is concentrated in it. Also, one should note that the higher the conductivity, the weaker the electric

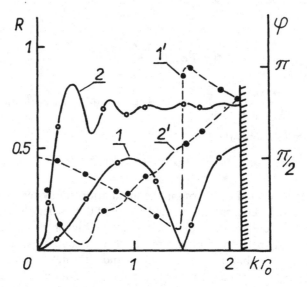

Figure 3.24. Dependence of reflection coefficient (1 and 2) and phase (1' and 2') on kr_0 at the frequency 18.2 GHz for teflon with $\varepsilon = 2.02$ and $\tan \delta = 0.005$ (curves 1 and 1') and p-Ge with $\varepsilon = 15.8$ and $\tan \delta = 0.31$ (curves 2 and 2'). Curves are theoretical, points are experimental.

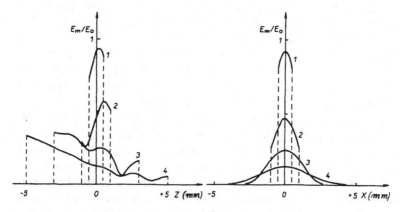

Figure 3.25. Spatial distribution of the microwave electric field along the z- and x-directions in the Ge-cylinders of various radii r: 1) 0.5; 2) 1; 3) 3; 4) 5 mm. E_0 is the electric field strength in the center of the empty guide.

field inside the cylinder. However, in thin needle-shaped ferroelectric single crystals of SbSI-type, the electric field is nearly homogeneous even in the millimeter wave range (Figure 3.26). Besides, the electric field in the lossless sample can be stronger than in a waveguide because of concen-

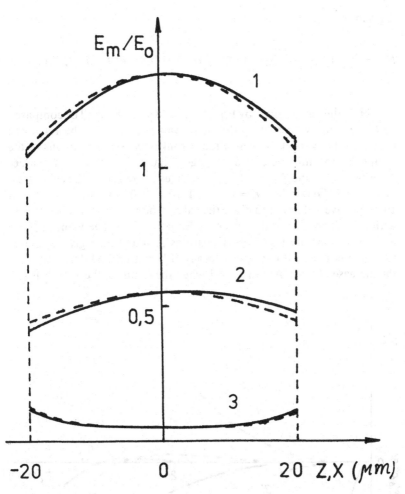

Figure 3.26. Spatial distribution of the microwave electric field in a cylinder with $\varepsilon = 100$ of radius $20\,\mu$m at 70 GHz: 1) $\tan\delta = 0.1$; 2) $\tan\delta = 1$, and 3) $\tan\delta = 10$. Solid lines along the z-direction, and dotted lines along the x-direction.

tration of the electric field, while in a lossy sample the electric field is weak due to the wave damping.

Certain situations require knowledge of the strength of the electric field inside the sample: 1) investigations of hot current carriers; 2) their transport properties and energy loss processes at high microwave electric fields in semiconductors; and 3) their use as the microwave electric field or power sensing elements. The value of the mean field in the sample is given by

$$\overline{E^*E} = \frac{2}{r}\sum_m A_{1,m}\frac{A^*_{1,m}}{[(k^2-k^{*2})]}[kJ_{m+1}(kr_0)J_m(k^*r_0) - k^*J_m(kr_0)J_{m+1}(k^*r_0)].$$

$$(3.87)$$

Now the results given by Equations 3.84 and 3.86 shall be compared with the same values of modulus, R, and phase, φ, of the reflection coefficient for various values of the permittivity, loss and radius of the sample in the millimeter wave range. Figure 3.27 shows the values $\Delta\varepsilon' = (\varepsilon' - \varepsilon'_c)/\varepsilon'$ and $\Delta\varepsilon'' = (\varepsilon'' - \varepsilon''_c)/\varepsilon''$ versus the radius of the cylindrical sample at 70 GHz, where $\varepsilon' = 10$, $\varepsilon'' = 1$; 10 and 100, or $\tan\delta = 0.1$; 1, and 10, respectively. The ε' and ε'' are the values obtained from Equation 3.84 with the values of R and φ given by Equation 3.86. The values $\Delta\varepsilon' = 0$ and $\Delta\varepsilon'' = 0$ signify that both Equations 3.84 and 3.86 give the same values. It is apparent that the accuracy of Equation 3.84 decreases with the increase of $\tan\delta$. At $\tan\delta = 0.1$ when the radius of the sample is less

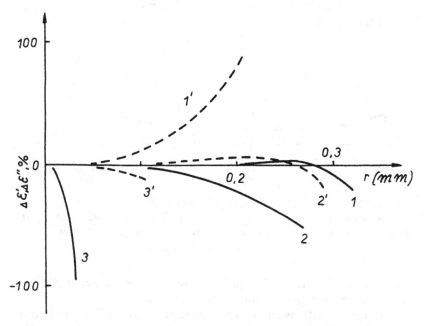

Figure 3.27. Divergence of real, $\Delta\varepsilon'$, and imaginary, $\Delta\varepsilon''$, parts of permittivity calculated from Equation 3.84 on the radius of cylindrical sample with $\varepsilon' = 10$ at 70 GHz. Solid lines are for $\Delta\varepsilon'$, dotted lines for $\Delta\varepsilon''$. (1, 1') $\tan\delta = 0.1$; (2, 2') $\tan\delta = 1$; (3, 3') $\tan\delta = 10$.

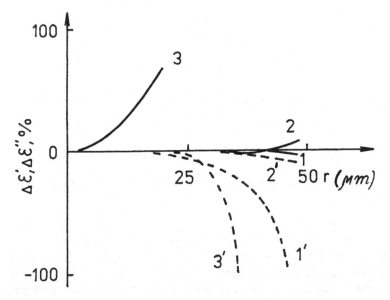

Figure 3.28. Designations are the same as in Figure 3.27, only permittivity of the sample is $\varepsilon = 1000$.

than $310 \, \mu m$, one can obtain ε' with an accuracy better than 10%. When $\tan \delta = 1$, the radius should be $r < 160 \, \mu m$, and when $\tan \delta = 10$, the required radius is $r < 10 \, \mu m$. Figure 3.28 shows similar results for the sample with $\varepsilon' = 1000$. When the radius of high dielectric permittivity cylindrical samples in the millimeter wave region is very small, one may use the simple Equations 3.80, 3.82 or 3.84. In other cases, Equations 3.86 are more suitable for the calculation of dielectric parameters of the cylindrical samples.

In toto, the derived Equations 3.80, 3.82 and 3.84 can be used for microwave dielectric spectroscopy of needle-shaped single crystals with a diameter ranging from several microns to several millimeters.

3.3.4. Layered Cylinder in Rectangular Waveguide

The complex reflection and transmission coefficients for a layered cylinder in a rectangular waveguide shall be found for microwave dielectric investigations of: 1) hollow needle-shaped single crystals, which can be grown from vapor phase; 2) some semiconductors and ferroelectrics, including SbSI and Sb_2S_3-type compounds (Chapter 4); and 3) liquids or biological systems which show interesting excitations at microwaves.

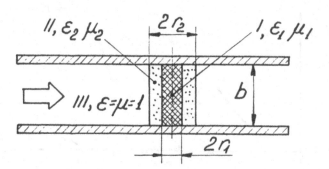

Figure 3.29. Layered cylinder between two infinite metal planes.

First, consider a homogeneous plane wave incident to a two-layered circular cylinder placed between two infinite parallel metal planes, which are separated from each other by a distance b (Figure 3.29). For the cylindrical coordinate system with the cylinder center as origin, the components E_y and H_y can be written as an expansion in partial waves (Kalesinskas and Shugurov 1991).

In the inner cylinder I:

$$E_y = \sum_{m=-\infty}^{\infty} A_m^1 J_m(k_1 r) e^{im\varphi} \cos hy,$$

$$H_y = \sum_{m=-\infty}^{\infty} B_m^1 J_m(k_1 r) e^{im\varphi} \sin hy. \tag{3.88}$$

In the outer cylinder II:

$$E_y = \sum_{m=-\infty}^{\infty} [A_m^{II} J_m(k_2 r) + C_m^{II} N_m(k_2 r)] e^{im\varphi} \cos hy,$$

$$H_y = \sum_{m=-\infty}^{\infty} [B_m^{II} J_m(k_2 r) + D_m^{II} N_m(k_2 r)] e^{im\varphi} \sin hy,$$

where $k_1^2 = k^2 \varepsilon_1 \mu_1 - h^2$; $k_2 = k^2 \varepsilon_2 \mu_2 - h^2$; $h = n\pi/b$; $n = 0, 1, 2 \ldots$; A_m, B_m, C_m and D_m are the expansion coefficients, and k here is the free space wave number.

Equations similar to Equation 3.88 describe incident fields where Bessel functions depend on kr and have the known amplitudes A_m^{inc} and

B_m^{inc}. They also describe scattered fields in the waveguide (III) with the Bessel functions replaced by the corresponding Hankel functions $H^{(2)}$ (kr) and the amplitudes A_m^{sc} and B_m^{sc}.

All the other field components from the Maxwell equations can be expressed through the E_y and H_y. For example, in the inner cylinder (I):

$$E_r = \frac{1}{k_1^2} \sum_{m=-\infty}^{\infty} \left[k\mu_1 \frac{m}{r} B_m J_m(k_1 r) - hk_1 A_m J'_m(k_1 r) \right] \times e^{im\varphi} \sin hy,$$

$$H_\varphi = \frac{1}{k_1^2} \sum_{m=-\infty}^{\infty} \left[\frac{imh}{r} B_m J_m(k_1 r) - ik\varepsilon_1 k_1 A_m J'_m(k_1 r) \right] e^{im\varphi} \cos hy,$$

$$E_\varphi = \frac{1}{k_1^2} \sum_{m=-\infty}^{\infty} \left[ik\mu_1 k_1 B_m J'_m(k_1 r) - \frac{imh}{r} A_m J_m(k_1 r) \right] e^{im\varphi} \sin hy,$$

$$H_r = \frac{1}{k_1^2} \sum_{m=-\infty}^{\infty} \left[hk_1 B_m J'_m(k_1 r) - k\varepsilon_1 \frac{m}{r} A_m J_m(k_1 r) \right] e^{im\varphi} \cos hy.$$

Similar equations can be written for the outer cylinder. By equating tangential components of E_y, E_φ, H_x and H_φ on the surface $r = r_1$, coefficients of the outer cylinder are obtained and expressed through those of the inner cylinder:

$$A_m^{II} = a_{11} A_m^I + a_{12} B_m^I, \quad B_m^{II} = b_{11} B_m^I + b_{12} A_m^I,$$

$$C_m^{II} = a_{21} A_m^I + a_{22} B_m^I, \quad D_m^{II} = b_{21} B_m^I + b_{22} A_m^I,$$

where

$$a_{11} = \frac{\pi r_1 k_2}{2} J_m(k_1 r_1) N'_m(k_2 r_1) - \frac{\pi r_1 \varepsilon_1 k_2^2}{2\varepsilon_2 k_1} J'_m(k_1 r_1) N_m(k_2 r_1),$$

$$a_{21} = -\frac{\pi r_1 k_2}{2} J_m(k_1 r_1) J'_m(k_2 r_1) + \frac{\pi r_1 \varepsilon_1 k_2^2}{2\varepsilon_2 k_1} J'_m(k_1 r_1) J_m(k_2 r_1),$$

$$a_{12} = \frac{\pi h m}{2k\varepsilon_2} \left(\frac{k_2^2}{k_1^2} - 1 \right) J_m(k_1 r_1) N_m(k_2 r_1),$$

$$a_{22} = -\frac{\pi h m}{2k\varepsilon_2} \left(\frac{k_2^2}{k_1^2} - 1 \right) J_m(k_1 r_1) J_m(k_2 r_1),$$

The coefficients b_{ik} can be obtained from the coefficients a_{ik} by substituting $\varepsilon \to \mu$. Obviously, when $\varepsilon_1 = \varepsilon_2$ and $\mu_1 = \mu_2$, the coefficients $a_{11} = b_{11} = 1$, and all the others vanish.

Similarly, equating the tangential field components on the surface $r = r_2$ results in the set of equations

$$A_m^I[a_{11}J_m(k_2r_2) + a_{21}N_m(k_2r_2)] + B_M^I[a_{12}J_m(k_2r_2) + a_{22}N_m(k_2r_2)]$$
$$- A_m^{sc}H_m^{(2)}(k_3r_2) = A_m^{inc}J_m(k_3r_2),$$

$$A_m^I[b_{12}J_m(k_2r_2) + b_{22}N_m(k_2r_2)] + B_M^I[b_{11}J_m(k_2r_2) + b_{21}N_m(k_2r_2)]$$
$$- B_m^{sc}H_m^{(2)}(k_3r_2) = B_m^{inc}J_m(k_3r_2),$$

$$A_m^I\left\{-\frac{hm}{r_2}\left(\frac{1}{k_2^2} - \frac{1}{k_3^2}\right)[a_{11}J_m(k_2r_2) + a_{21}N_m(k_2r_2)] + \frac{k\mu_2}{k_2}[b_{12}J_m'(k_2r_2)\right.$$
$$\left. + b_{22}N_m'(k_2r_2)]\right\} + B_m^I\left\{-\frac{hm}{r_2}\left(\frac{1}{k_2^2} - \frac{1}{k_3^2}\right)[a_{12}J_m(k_2r_2) + a_{22}N_m(k_2r_2)]\right.$$
$$\left. + \frac{k\mu_2}{k_2}[b_{11}J_m'(k_2r_2) + b_{21}N_m'(k_2r_2)]\right\} - \frac{k}{k_3}B_m^{sc}H_m^{(2)'}(k_3r_2)$$
$$= \frac{k}{k_3}B_m^{inc}J_m'(k_3r_2), \tag{3.89}$$

$$A_m^I\left\{-\frac{k\varepsilon_2}{r_2}[a_{11}J_m'(k_2r_2) + a_{21}N_m'(k_2r_2)] + \left(\frac{1}{k_2^2} - \frac{1}{k_3^2}\right)\frac{hm}{r_2}[b_{1m}J_m(k_2r_2)\right.$$
$$\left. + b_{22}N_m(k_2r_2)]\right\} + B_m^I\left\{-\frac{k\varepsilon_2}{k_2}[a_{12}J_m'(k_2r_2) + a_{22}N_m'(k_2r_2)]\right.$$
$$\left. + \left(\frac{1}{k_2^2} - \frac{1}{k_3^2}\right)\frac{hm}{r_2}[b_{11}J_m(k_2r_2) + b_{21}N_m(k_2r_2)]\right\} + \frac{k}{k_3}A_m^sH_m^{(2)'}(k_3r_2)$$
$$= -\frac{k}{k_3}A_m^{inc}J_m'(k_3r_2),$$

where $k_3^2 = k^2 - h^2$.

The solution of the set of Equations 3.89 yields the amplitudes A_m^I and B_m^I of the field in the inner cylinder, and the amplitudes of the scattering field A_m^{sc} and B_m^{sc} through the amplitudes A_m^{inc} and B_m^{inc} of the incident wave.

In a similar way one can find the scattering by a multilayered cylinder.

Now, using the set of Equations 3.89 one can find the scattering by the layered cylinder placed at the arbitrary distance l in a rectangular waveguide (Figure 3.17). The presence of sidewalls of the waveguide makes the solution of this problem more complicated, as compared to the previous case. Similarly, as described in the previous Sections, the layered cylinder should be replaced by multipole sources located at the center of the cylinder, and, using the images of these with respect to the waveguide walls, the scattered field can be found. In fact, the reflections from the walls are taken into account by substituting into Equations 3.89 as follows:

$$A_m^{inc} \to A_m^{inc} + \sum_{p=-\infty}^{\infty} B_p^{sc}(-1)^{m+p} S_{m+p}^- - \sum_{p=-\infty}^{\infty} A_p^{sc} S_{m+p}^+$$

$$+ \sum_{p=-\infty}^{\infty} A_p^{sc}[1 + (-1)^{p+m}] S_{p-m},$$

$$B_m^{inc} \to B_m^{inc} + \sum_{p=-\infty}^{\infty} B_p^{sc}(-1)^{m+p} S_{m+p}^- + \sum_{p=-\infty}^{\infty} B_p^{sc} S_{m+p}^+$$

$$+ \sum_{p=-\infty}^{\infty} B_p^{sc}[1 + (-1)^{m+p}] S_{p-m},$$

where

$$S_{p-m} = \sum_{n=1}^{\infty} H_{p-m}^{(2)}(2nk_3 a),$$

$$S_{p+m}^- = \sum_{n=0}^{\infty} H_{p+m}^{(2)}\{k_3[(2n+1)a - 21]\}, \qquad (3.90)$$

$$S_{p+m}^+ = \sum_{n=0}^{\infty} H_{p+m}^{(2)}\{k_3[(2n+1)a + 21]\}.$$

Having determined the E and H fields in the cylinder, the reflection coefficient of any waveguide mode of the magnetic type as with Equation 3.72 can be written:

$$C_{\pm n}^* = \frac{i}{2\gamma_n} \int_v \{(\varepsilon^* - 1)(EE_{\pm n}) - (\mu^* - 1)(HH_{\mp n})\} dv, \qquad (3.91)$$

if eigenwaves are normalized according to Kacenelenbaum (1966). For the incident dominant TE_{10}-mode propagating along the positive z-direction, the coefficients are

$$B_m^{inc} = \quad , \quad h = 0$$

and

$$A_m^{inc} = -ik\left(\frac{2}{ab}\right)^{1/2} \cos\left(\frac{\pi l}{a} - m\varphi - \frac{m\pi}{2}\right),$$

$$\tan \varphi = \frac{\gamma a}{\pi}.$$

For the cylinder placed along the x-direction perpendicular to the electric field vector, the coefficients are

$$A_m^{inc} = 0, \quad h = \pi/a$$

and

$$B_m^{inc} = (-1)^m i\gamma\left(\frac{2}{ab}\right)^{1/2}$$

In both cases, reflection and transmission coefficients are expressed by Equations 3.86 with the C^* given by Equation 3.91.

For a numerical study, summation of the series in Equations 3.90 can be detached into a separate subroutine, which is used only once for every frequency. When the cylinder is placed symmetrically $(l = 0)$, values of the series can be obtained from an approximation in Chebyshev polynomials of the previously computed series in a given interval. That considerably shortens computation time. Cutting off the infinite series in Equations 3.90 with the general number diminishing as $n^{-1/2}$, at $n = 10^5$ gives accuracy in the order of 3×10^{-3} with respect to the first number of the series and allows an accuracy of 10^{-5} to be obtained at the same n (Kalesinskas and Shugurov 1991).

The number of partial waves to be taken depends on the radii, magnetodielectric properties of the cylinder, and frequency. For the $k_1 r_1$, $k_2 r_2 < 10$ and $r_2 < a/4$, 11 waves ($-5 \leq m \leq 5$) ensure that the computation accuracy of the reflection and transmission coefficients will be better than 10^{-4}.

By comparison the computed $R^*(v_i)$, $T^*(v_i)$ and the measured reflection and transmission coefficients, $R_m^*(v_i)$, $T_m^*(v_i)$ at various frequencies v_i, and by calculation the root-mean-square deviations:

$$\delta_R = \frac{1}{n}\left\{\sum_{i=1}^{n} [R^*(v_i) - R_m^*(v_i)]^2\right\}^{1/2},$$

$$\delta_T = \frac{1}{n}\left\{\sum_{i=1}^{n} [T^*(v_i) - T_m^*(v_i)]^2\right\}^{1/2},$$

and using the function $f(\varepsilon^*, \mu^*) = \delta_R + \delta_T$ minimum finding procedure, one can determine dielectric and/or magnetic parameters of layered materials at microwaves (e.g., magnetodielectrics), which are widely used as radiation-absorbing materials of a wide frequency band.

3.3.5. Square Post in Rectangular Waveguide

In certain cases, it is difficult to process a single ferroelectric, semiconductor or superionic crystal into the cylindrical shape, especially if it is one- or two-dimensional. Some crystals, for instance, of antimonite- or hollandite-type structures grow well along one direction with a square or nearly square cross section and have various cross-sectional sizes. Furthermore, their complex dielectric permittivity and/or conductivity along this axis is, in most cases, the highest and shows various excitations at microwaves.

However, it is difficult to obtain rigorous solutions for the reflection and transmission coefficient for a post with a cross section of arbitrary shape (Auda and Harrington 1984; Ise and Koshiba 1986), as was done for the cylindrical rod. This is because in the former case the Helmholtz partial differential equation for the microwave fields should be solved under nonhomogeneous boundary conditions at the surface of the post. Therefore, various replacement schemes of the post or approximations for determining electric parameters of semiconductors (such as silicon, germanium etc.) or superionic conductors at microwaves have been used. Often it was considered that the reflection or transmission coefficient for the square post was equal to those for a cylindrical rod with equivalent area. A number of authors have dealt with a post by approximate analytical or numerical methods.

Consider a square post placed centrally in a rectangular waveguide, similar to the case shown in Figure 3.17. The complex reflection and transmission coefficients can be found using an approximate analytical solution for the Helmholtz equation by introducing the discrete boundary conditions (Yoshikado and Taniguchi 1989). The wave equation satisfied by the electric field in the post, E^c, is given by

$$\left(\frac{\partial^2}{\partial x^2} + \frac{\partial^2}{\partial z^2} - i\omega\mu_0\sigma^* \right) E^c = 0, \tag{3.92}$$

where $\sigma^* = i\omega\varepsilon_0\varepsilon^*$ is the complex conductivity along the y-axis, which is also equal to $\sigma^* = \sigma + i\omega\varepsilon_0\varepsilon'$, σ being the real part of the conductivity (see Section 8.1).

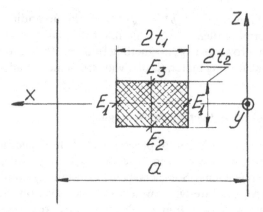

Figure 3.30. A square post in a rectangular waveguide.

Using the boundary conditions for the E^c at the surface of the post (Figure 3.30):

$$E^c\left(\frac{a}{2} \pm t_1, 0\right) = E_1,$$

$$E^c\left(\frac{a}{2}, -t_2\right) = E_3,$$

$$E^c\left(\frac{a}{2}, t_2\right) = E_2,$$

the approximate solution of Equation 3.92 is obtained:

$$E^c = \left[\frac{E_1}{\cosh c_1 t_1}\cosh c_2 z + \frac{E_2 - E_3}{2 \sinh c_2 t_2}\sinh c_2 z\right]\cosh c_1\left(x - \frac{a}{2}\right),$$

where $c_1^2 + c_2^2 = i\omega\mu_0\sigma^*$. In this case, the complete form of the exact Green's function for the nonhomogeneous boundary conditions in Equation 3.92 is not obtained. The wave equation for the scattered electric field, E^{sc}, is

$$\left(\frac{\partial^2}{\partial x^2} + \frac{\partial^2}{\partial z^2} + k_0^2\right)E^{sc} = i\omega\mu_0\sigma_n^* E^c\left[H\left(x - \frac{a}{2} + t_1\right) - H\left(x - \frac{a}{2}t_1\right)\right]$$

$$\times [H(z + t_2) - H(z - t_2)], \qquad (3.93)$$

where $\sigma_n^* = \sigma^* - i\omega\varepsilon_0$, and H is the Heaviside function (Kraus and Carver 1973). Due to a symmetry of the post, only the odd $TE_{(2n-1)0}$-

modes are excited by the post. The solutions of Equation 3.93 are given by

$$E^{sc} = \frac{2}{a} \sum_{n=1}^{\infty} K_{2n-1}^{-} e^{\Gamma_{2n-1}z} \sin \frac{(2n-1)\pi x}{a}, \quad z \le -t_2$$

$$E^{sc} = \frac{2}{a} \sum_{n=1}^{\infty} (I_{2n-1} \cosh c_2 z + J_{2n-1} \sinh c_2 z + L_{2n-1}^{-} e^{\Gamma_{2n_1}z} + L_{2n-1}^{+} e^{-\Gamma_{2n-1}z})$$

$$\times \sin \frac{(2n-1)\pi x}{a},$$

$$0 \le x \le \frac{a}{2} - t_1, \quad \frac{a}{2} + t_1 \le xa, \quad -t_2 \le z \le t_2,$$

$$E^{sc} = \frac{2}{a} \sum_{n=1}^{\infty} K_{2n-1}^{+} e^{-\Gamma_{2n-1}z} \sin \frac{(2n-1)\pi x}{a}, \quad z \le t_2$$

where $\Gamma_n^2 = n^2 (\pi/a)^2 - k_0^2$. The boundary conditions, whereby E^{sc} and $\partial E^{sc}/\partial t$ must be continuous at $z = \pm t_2$, give

$$K_{2n-1}^{\pm} = \frac{1}{\Gamma_{2n-1}} [(\Gamma_{2n-1} I_{2n-1} \mp c_2 J_{2n-1}) \cosh c_2 t_2 \sinh \Gamma_{2n-1} t_2$$

$$- (c_2 I_{2n-1} \mp \Gamma_{2n-1}) \sinh c_2 t_2 \cosh \Gamma_{2n-1} t_2],$$

$$L_{2n-1}^{\pm} = -\frac{e^{\Gamma_{2n-1}t_2}}{2\Gamma_{2n-1}} [(\Gamma_{2n-1} I_{2n-1} \mp c_2 J_{2n-1}) \cosh c_2 t_2 \qquad (3.95)$$

$$+ (c_2 I_{2n-1} \mp \Gamma_{2n-1} J_{2n-1}) \sinh c_2 t_2],$$

where

$$I_{2n-1} = i\omega\mu_0 \sigma_n^* F_{2n-1} \frac{1}{c_2^2 - \Gamma_{2n-1}^2} \times \frac{E_1}{\cosh c_1 t_1},$$

$$J_{2n-1} = i\omega\mu_0 \sigma_n^* F_{2n-1} \frac{1}{c_2^2 - \Gamma_{2n-1}^2} \times \frac{E_2 - E_3}{2 \sinh c_2 t_2},$$

$$F_{2n-1} = (-1)^{n-1} \left\{ \frac{1}{c_1 + i(2n-1)\dfrac{\pi}{a}} \times \sinh[c_1 a + i(2n-1)\pi] \right.$$

$$\left. + \frac{1}{c_1 - i(2n-1)\dfrac{\pi}{a}} \times \sinh[c_1 a - i(2n-1)\pi] \right\}.$$

Thus, the scattered electric field depends on both E_1 and $E_2 - E_3$. The boundary conditions at the surface of the post are

$$E^{inc} + E^{sc} = \begin{cases} E_1, & x = \dfrac{a}{2} \pm t_1, & z = 0, \\[2mm] E_2, & x = \dfrac{a}{2}, & z = -t_2, \\[2mm] E_3, & x = \dfrac{a}{2}, & z = t_2, \end{cases} \tag{3.96}$$

where the incident electric field is $E^{inc} = E_0 \sin\left(\dfrac{\pi x}{a}\right) e^{-\Gamma_1 z}$. From Equations 3.94 through 3.96, E_1 and $E_2 - E_3$ are given by

$$E_1 = E_0 A,$$

$$E_2 - E_3 = -E_0 B,$$

where

$$A = \cosh \Gamma_1 t_2 \frac{\cosh c_1 t_1}{\cosh c_2 t_2} \left[1 - i\omega\mu_0 \sigma_n^* \frac{2}{a \cosh c_1 t_1} \sum_{n=1}^{\infty} \frac{(-1)^n F_{2n-1} e^{-\Gamma_{2n-1} t_2}}{\Gamma_{2n-1}(c_2^2 - \Gamma_{2n-1}^2)} \right.$$

$$\left. \times (\Gamma_{2n-1} \cosh c_1 t_1 \sinh \Gamma_{2n-1} t_2 - c_2 \sinh c_2 t_2 \cosh \Gamma_{2n-1} t_2) \right]^{-1}$$

and

$$B = 2\sinh \Gamma_1 t_2 \left[1 + i\omega\mu_0 \sigma_n^* \frac{1}{a \sinh c_2 t_2} \sum_{n=1}^{\infty} \frac{(-1)^n F_{2n-1} e^{-\Gamma_{2n-1} t_2}}{\Gamma_{2n-1}(c_2^2 - \Gamma_{2n-1}^2)} \right.$$

$$\left. \times (c_2 \cosh c_2 t_2 \sinh \Gamma_{2n-1} t_2 - \Gamma_{2n-1} \sinh c_1 t_1 \cosh \Gamma_{2n-1} t_2) \right]^{-1}.$$

For $z \ll -t_2$ and $z \gg t_2$ a scattered field has only the TE_{10}-mode. Therefore, the reflection and transmission coefficients are given by

$$R^* = \sigma_n^* P^- e^{2\Gamma_1 z},$$

and

$$T^* = 1 + \sigma_n^* P^+, \tag{3.97}$$

where

$$P^{\pm} = \frac{2i\omega\mu_0 F_1}{a\Gamma_1(c_2^2 - \Gamma_1^2)} \left[\left(\Gamma_1 \frac{A}{\cosh c_2 t_2} \pm c_2 \frac{B}{2\sinh c_2 t_2} \right) \cosh c_2 t_2 \sinh \Gamma_1 t_2 \right.$$

$$\left. - \left(c_2 \frac{A}{\cosh c_2 t_2} \pm \Gamma_1 \frac{B}{2\sinh c_2 t_2} \right) \sinh c_2 t_2 \cosh \Gamma_1 t_2 \right].$$

Thus, if R^* and/or T^* are measured, the complex conductivity and/or dielectric permittivity of the lossy dielectric post can be calculated by solving Equations 3.97. It should be noted that the solution of these equation is not valid for large posts having a high conductivity or dielectric permittivity. Though the validity of Equations 3.97 cannot be verified analytically, it can be verified by checking the discrete boundary conditions given by Equations 3.96 and the continuous boundary conditions at the surface of the post.

Comparison of measured values at 9.3 GHz (Yoshikado and Taniguchi 1989) with the values of conductivity and dielectric permittivity for known materials has shown that Equations 3.97 can be used for determination of electric and dielectric parameters of the post from various materials at microwaves if the boundary conditions are satisfied.

3.3.6. Determination of Dielectric Parameters: Dielectric Spectrometers

In the centimeter and millimeter wave ranges, the complex reflection and transmission coefficients of the TE_{10}-wave produced by a nonmagnetic sample in a rectangular waveguide, the axis of this sample being parallel to the electric field vector, are defined by: 1) a complex dielectric permittivity (or complex conductivity); 2) dimensions of the sample; and, 3) frequency and width of a wide wall of a waveguide, i.e., $R = f\,(\varepsilon^*, r, v, a)$. The dielectric parameters $\varepsilon^*(v, T)$ can be calculated with a computer by solving the nonlinear complex equation $\varepsilon^* = f(R, T)$ or $\varepsilon^* = f(R^*)$. The latter case gives a better convergence, but it leads to an indetermination, which arises when determining ε^* from the moduli R and T (for $n = 0$ only). In this case, the proper value of permittivity is chosen analytically.

The complex reflection coefficient $R^* = Re^{i\varphi}$ can be expressed as follows:

$$R = f_1(\varepsilon', \varepsilon'')$$
$$\varphi = \varphi_2(\varepsilon', \varepsilon''). \tag{3.98}$$

Having experimentally measured the values R and φ, one can obtain a set of two nonlinear equations with respect to ε' and ε'', which can be solved easily by a computer.

Measurements of moduli of the reflection and transmission coefficients by network analyzers are sometimes more convenient experimentally, especially when sweeping a frequency. In that case, we get the set of two nonlinear equations with respect to the moduli of R^* and T^*:

$$R = f_1(\varepsilon', \varepsilon''),$$
$$T = f_2(\varepsilon', \varepsilon''). \tag{3.99}$$

Equations 3.98 or 3.99 can be solved by a modified Newton's method. Then, the set of nonlinear equations is transformed to the set of linear equations:

$$\varepsilon'_{n+1} = \frac{ED_2 - FD_1}{C_1 D_2 - D_1 C_2},$$

$$\varepsilon''_{n+1} = \frac{FC_1 - EC_2}{C_1 D_2 - D_1 C_2}, \tag{3.100}$$

where

$$C_1 = \frac{f_1(\varepsilon'_n, \varepsilon''_n) - f_1(\varepsilon'_{n-1}, \varepsilon''_n)}{\varepsilon'_n - \varepsilon'_{n-1}},$$

$$D_1 = \frac{f_1(\varepsilon'_n, \varepsilon''_n) - f_1(\varepsilon'_n, \varepsilon''_{n-1})}{\varepsilon''_n - \varepsilon''_{n-1}},$$

$$C_2 = \frac{f_2(\varepsilon'_n, \varepsilon''_n) - f_2(\varepsilon'_{n-1}, \varepsilon''_n)}{\varepsilon'_n - \varepsilon'_{n-1}},$$

$$D_2 = \frac{f_2(\varepsilon'_n, \varepsilon''_n) - f_2(\varepsilon'_n, \varepsilon''_{n-1})}{\varepsilon''_n - \varepsilon''_{n-1}},$$

$$E = R - f_1(\varepsilon'_n, \varepsilon''_n) + C_1 \varepsilon'_n + D_1 \varepsilon''_n,$$

$$F = T(\text{or } \phi) - f_2(\varepsilon'_n, \varepsilon''_n) + C_2 \varepsilon'_n + D_2 \varepsilon''_n.$$

Preliminary values of ε'_{min}, ε'_{max} and ε''_{min}, ε''_{max} are selected approximately. Calculations stop when

$$R - f_1(\varepsilon', \varepsilon'') < \delta R,$$

and

$$T(\text{or } \phi) - f_2(\varepsilon', \varepsilon'') < \delta T(\text{or } \delta \phi)$$

One can select, for instance, $\delta R \leq 0.001$ and $\delta T \leq 0.001$, or $\delta \varphi \leq 0.1^0$ and calculate with the acceptable accuracy the dielectric parameters of the measured samples.

Figure 3.31 shows the dielectric spectrometer setup for the measurements of the moduli of the reflection and transmission coefficients

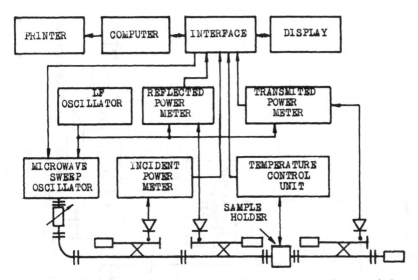

Figure 3.31. Dielectric spectrometer setup for reflection and transmission measurements. From Grigas et al. (1990a).

(Grigas et al. 1990a). Using the backward-wave oscillators as variable frequency sources or the klystrons as fixed frequency sources, and changing only the waveguides terminating with a matched load, the frequency range 8 to 140 GHz can be covered. The accuracy of measurements depends on the magnitude of the coefficients.

The cosine data window can be used to smooth out the frequency dependences of the reflection and transmission coefficients. This data window enables measurements not only at frequencies v (at which the coefficients R and T are calculated), but also at frequencies in the intervals Δv on either side of the main ones. The measured results $P(v)_i$, (Δv_i) are multiplied by the weight function and then averaged. Thus, the measured quantity $P(v_j)$ proportional to the R and T coefficients is calculated according to

$$P(v_j) = \frac{1}{2S} \sum_{n=1}^{l+1} \left(1 - \cos \frac{2\pi n}{l+1} \right) P\left[v_i - \left(\frac{l+1}{2} - n \right) \Delta v \right],$$

where

$$S = \frac{1}{2} \sum_{n=1}^{l+1} \left(1 - \cos \frac{2\pi n}{l+1} \right)$$

is the rate setting constant, and l is the number of additional measurements. The method of smoothing makes it possible to decrease the

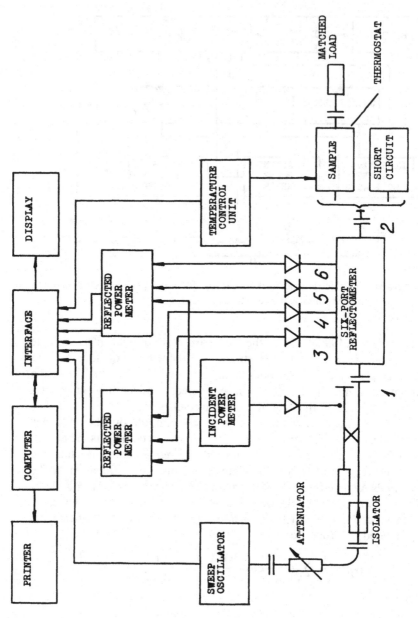

Figure 3.32. Dielectric spectrometer setup for complex reflection measurements.

influence of both the parasite reflections in the waveguide and the noises of the registering devices.

The complex reflection coefficient can be determined with high resolution using the six-port (Oldfield et al. 1985) or seven-port waveguide reflectometers.

Figure 3.32 shows an example of the automatic spectrometer based upon a six-port reflectometer for the short millimeter wave range 80 to 120 GHz (Banys et al. 1992a). The reflectometer can be calibrated at every frequency and temperature by measuring the reflection coefficient for $i = 1, 2, 3 \ldots 10$ positions of a movable short circuit. The calibration algorithm is derived from the equation

$$nP_i = \frac{C_{1i} + C_{2i}x + C_{3iy}}{1 + D_1 x + D_2 y},$$

where P_i is the signal of the i-th power meter, x and y are the real and imaginary parts of the complex reflection coefficient of the short circuit, $C_{1i}, C_{2i}, C_{3i}, D_1$ and D_2 are real constants. The constants are determined by the method of least squares. Knowing these constants, the modulus and phase of the complex reflection coefficient of unknown sample can be calculated from the equation by the least-squares method.

The accuracy of the measurements is 0.001 in a modulus of reflection coefficient and 0.5^0 in phase across the majority of the band.

3.3.7. Applications of Microwave Helicons

A number of conductive materials such as semimetallic Bi or $Bi_{1-x}Sb_x$ compounds, and narrow-gap (Pb, Sn, Ge)Te semiconductors etc. possess high lattice dielectric permittivity. Some of them exhibit ferroelectric phase transitions (Jantsch 1983) with zone-center soft-mode behavior (Section 4.4.3) and dielectric anomalies. The problems resulting from the high conductivity of these materials can be avoided by measuring the permittivity in the space charge regime of p-n junctions, Schottky barriers or MIS structures. These materials are opaque to microwaves, but it is possible to make them transparent by the application of an external magnetic field. The magnetic-field-induced transparency leads to a variety of electromagnetic phenomena, which provide a convenient nondestructive tool for the study of free charge carriers and lattice dielectric parameters.

Consider the interaction of the plane electromagnetic wave with a magnetized free carrier plasma that moves in the background of the lattice dielectric constant ε_L. Microwave response of this system can be obtained by solving Maxwell's equations, which, assuming a plane wave solution in the form $\exp i(\omega t - (kx))$, leads to the wave equation

$$k \times k \times E - \omega^2 \mu_0 \varepsilon^* E = 0, \tag{3.101}$$

where k is the complex wave number

$$\varepsilon^* = \varepsilon_0 \left(\varepsilon_L \hat{I} - i \frac{\sigma^*}{\omega \varepsilon_0} \right)$$

$\hat{\varepsilon}^*$ is the tensor of complex dielectric permittivity, $\hat{\sigma}^*$ is the tensor of complex conductivity of the plasma in the magnetic field B, and $\hat{I}$ is the unit matrix. Solution of the wave equation (3.101) gives the dielectric permittivity of the plasma (Palik and Furdyna 1970) in the form

$$\varepsilon^* = \varepsilon_L \left\{ 1 - \frac{2\omega_p^2 [\omega(\omega + i\upsilon) - \omega_p^2]}{\omega\{2(\omega - i\upsilon)[\omega(\omega + i\upsilon) - \omega_p^2] - \omega\omega_c^2 \sin^2 \varphi \atop \pm \omega_c[\omega^2\omega_c^2 \sin^4 \varphi + 4[\omega(\omega + i\upsilon) - \omega_p^2]^2 \cos^2 \varphi]^{1/2}\}} \right\},$$

$$\tag{3.102}$$

where

$$\omega_p = \left(\frac{e^2 n}{\varepsilon_0 \varepsilon_L m^*} \right)^{1/2}$$

and

$$\omega_c = \frac{eB}{m^*}$$

are the plasma and cyclotron frequency, respectively, φ is the angle between the wave vector and the external magnetic field, υ is the collision frequency, m^* and n are the effective mass and concentration of charge carriers, respectively, and B is the magnetic induction.

Equation 3.102 can be simplified for a single-component plasma of the charge carriers having an isotropic effective mass for a circularly polarized wave in Faraday geometry ($k \parallel B \perp E$, $\varphi = 0$):

$$\varepsilon^* = \varepsilon_L \left\{ 1 - \frac{\omega_p^2}{\omega[(\omega \pm \omega_c) + i\upsilon]} \right\}, \tag{3.103}$$

where the signs ($+$) and ($-$) refer to the two waves with opposite circular polarization. The directions of rotation of the field vectors of the ordinary wave ($+$) and the free carriers in the magnetic field are opposite, while these directions of the extraordinary wave ($-$) coincide. The latter extraordinary wave is known as a *helicon*.

Separation of the real and imaginary part of Equations 3.103, under the condition $\omega \pm \omega_c \gg \upsilon$, yields

$$\varepsilon' = \varepsilon_L \left(1 - \frac{\omega_p^2}{\omega(\omega \pm \omega_c)} \right)$$

and

$$\varepsilon'' = \frac{\varepsilon_L \omega_p^2 \upsilon}{\omega(\omega \pm \omega_c)}. \tag{3.104}$$

It follows from Equations 3.104 that the ordinary wave corresponds to the "evanescent" mode: a wave so polarized is almost totally reflected ($\varepsilon' < 0$) by the medium at $\omega < \omega_p$. The negligible amount of power admitted into the medium is strongly damped. Helicons, on the other hand, correspond to the dielectric conditions when, excited by the helicons, the medium behaves like a dielectric with a positive dielectric permittivity and small loss in a strong external magnetic field ($\omega_c > \omega$). Transparency of the conducting medium with respect to the helicons provides an opportunity to probe the medium with electromagnetic waves. In the majority of narrow-gap high mobility semiconductors, $\omega_c \approx 10^{13}$ Hz at $B = 1\ T$. Because the helicons have no low frequency limit, they can be excited by high frequencies and microwaves.

The physical reason for the small damping of the helicons is that the current component, which is in phase with the electric field of the helicon and determines dissipative loss, is only a small fraction of the total current.

The condition of propagation of the ordinary waves ($\varepsilon' > 0$) is given by

$$\frac{\omega_p^2}{\omega(\omega + \omega_c)} = 1 \tag{3.105}$$

and it is called magnetoplasma resonance. At microwaves, even in a weak magnetic field, $\omega_c \gg \omega$. Equation 3.105 can be rewritten as

$$\frac{\omega_p^2}{\omega \omega_c} = 1 \tag{3.106}$$

which means that at this frequency, displacement and conduction currents are equal. When $\omega_p^2/\omega\omega < 0$, the plasma behaves as a dielectric. Equality (3.106) in conductive materials can be fulfilled only at very high frequencies (e.g., the propagation of the ordinary waves is possible only at short microwaves or even higher frequencies).

In Voigt geometry ($k \perp B \perp E$, $\varphi = 90°$), it follows from Equation 3.102 that the dispersion of the ordinary waves does not depend on the magnetic field. But for the helicons

$$\varepsilon' = \varepsilon_L \left[1 - \frac{\omega_p^2(\omega_p^2 - \omega^2)}{\omega^2(\omega_p^2 - \omega^2 + \omega_c^2)} \right] \qquad (3.107)$$

When a magnetic field is strong (i.e., when $\omega_c^2 \gg \omega_p^2 \gg \omega^2$, Equation 3.107 may be recast in the form

$$\varepsilon' = \varepsilon_L \left[1 - \frac{\omega_p^4}{\omega^2 \omega_c^2} \right] \qquad (3.108)$$

which shows that the condition of excitation of the helicons in Voigt geometry is

$$\frac{\omega_p^4}{\omega^2 \omega_c^2} = 1$$

and it coincides with the beginning of excitation of the ordinary waves (Equation 3.106) in Faraday geometry. Thus, the propagation of the ordinary waves in Faraday and the helicons in Voigt geometry is possible only beyond the magnetoplasma edge ($\omega_p^2 = \omega\omega_c$ or $\varepsilon' = 0$) (Figure 3.33). In a magnetic field lower than the magnetoplasma edge, these waves are strongly reflected.

Propagation of microwave helicons can be studied directly using a number of techniques (see the survey by Brazis et al. 1979). The simplest technique involves direct transmission (or reflection) of microwaves normally incident on a slab of thickness d (see Section 3.3.2). When d is less than the skin depth ($\alpha d < 1$), both the transmitted and the reflected signals show well-resolved Fabry–Perot oscillations as a function of magnetic field B. This is a consequence of the strong dependence of ε' (and thus of the wavelength within the slab) on B.

From Equations 3.104 we get

$$\varepsilon' = \left(\frac{\lambda}{\lambda_h} \right)^2 = \varepsilon_L \left(1 + \frac{en}{\omega\varepsilon_0\varepsilon_L B} \right), \qquad (3.109)$$

where λ_h is the wavelength of the helicon, which can be found from Fabry–Perot transmission patterns (see Section 3.3.2, *Resonant transmission*), and

$$d = N \frac{\lambda_h}{2} = N \frac{\lambda_0}{2(\varepsilon')^{1/2}}. \qquad (3.110)$$

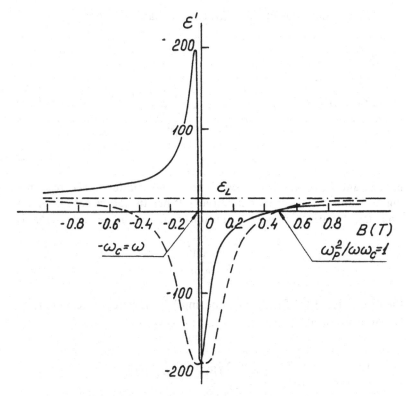

Figure 3.33. Dielectric permittivity of semiconductive plasma vs. magnetic induction at a frequency of 35 GHz. Parameters of plasma: concentration $n = 1 \times 10^{20}\,\mathrm{m}^{-3}$, mobility $\mu = 50\,\mathrm{m}^2/\mathrm{v \cdot s}$, $m^*/m_0 = 0.014$ and $\varepsilon_L = 16$. The solid line corresponds to Faraday, dashed to Voigt geometry. Arrows show the cyclotron and magnetoplasma resonances.

It follows directly that the magnetic field B_N corresponding to the N-th helicon transmission maximum, satisfies the relation

$$B_N^{-1/2} = \frac{N\lambda_0}{2d}\left(\frac{\omega\varepsilon_0}{\mathrm{ne}}\right)^{1/2} = \frac{N\pi}{d(\omega\mu_0\mathrm{ne})^{1/2}}.$$

When the values of $B_N^{-1/2}$, corresponding to consecutive extrema, are plotted against consecutive arbitrary integers, one obtains a straight line whose slope gives n directly.

The solution of Equation 3.109 for two resonances (e.g., for N and

$N + M$), give the value of concentration of charge carriers:

$$n = \frac{\varepsilon_0 \pi^2 c^2 M^2}{e\omega d^2} \times$$

$$\times \frac{B_N B_{N+M}\left\{(B_N + B_{N+M}) + 2\left[B_N B_{N+M} + \left(\frac{\omega^2 d^2 \varepsilon_L}{M^2 \pi^2 c^2}\right)(B_N - B_{N+M})^2\right]^{1/2}\right\}}{(B_N - B_{N+M})^2}$$

$$(3.111)$$

here B_N and B_{N+M} are the values of the magnetic induction corresponding N-th and $(N + M)$-th order extrema. When the lattice permittivity is small in comparison to the contribution of the free charge carriers, that is

$$\varepsilon_L \ll \frac{en^2}{\varepsilon_0 \omega B} \qquad (3.112)$$

then Equation 3.111 for the neighboring Fabry–Perot extrema may be recast in the form

$$n = \frac{\varepsilon_0 \pi^2 c^2}{e\omega d^2} \times \frac{1}{(B_{N+1}^{-1/2} - B_N^{-1/2})^2}.$$

The imaginary part of permittivity (Equation 3.104), which can be found by measuring the imaginary part of a propagation constant, contains information on mobility of free charge carriers which is given by (Laurinavičius et al. 1987)

$$\mu = \frac{\pi(B_{N+1}^{+3/2} - B_N^{-3/2})}{2\ln\left[\frac{A_N}{A_{N+1}}\left(\frac{B_{N+1}}{B_N}\right)^{1/2}\right](B_{N+1}^{-1/2} - B_N^{-1/2})},$$

where A_N and A_{N+1} are the corresponding amplitudes of Fabry–Perot extrema. When the condition (3.112) is not fulfilled, the mobility is given by

$$\mu = \frac{end}{2\varepsilon_0 c(\varepsilon_L)^{1/2}} \times \frac{\dfrac{1}{B_{N+M}(1 + K/B_{N+M})^{1/2}} - \dfrac{1}{B_N^2(1 + K/B_N)^{1/2}}}{\ln\left(\dfrac{A_N}{A_{N+M}} \times \dfrac{(1 + K/B_N)^{1/2}}{(1 + K/B_{N+M})^{1/2}}\right)},$$

where

$$K = \frac{en}{\varepsilon_0 \varepsilon_L \omega}.$$

In general, the Fabry–Perot oscillations in magnetotransmission are more damped as compared to those in magnetoreflection because, in the first case, helicon should make an additional pass through the slab resulting in an additional loss (Brazis et al. 1979).

When $\alpha d > 1$, Fabry–Perot patterns vanish. To obtain phase information in the absence of resonances, we can beat the signals transmitted through the slab with a reference signal which bypasses the slab. This interference pattern is sometimes referred to as a Rayleigh interferogram (Figure 3.34). The interferometer (Brazis et al. 1979) consists of the reference signal arm and the sample arm. When the semiconductor sample with plane parallel faces is placed between the special dielectric waveguide probes, it provides a local excitation and reception of a narrow helicon beam. The sample can be moved by a scanning mechanism. Such a helicon beam allows determination of local parameters of conductive materials (Laurinavičius et al. 1987) and their spatial distribution. Changes in the magnetic field cause changes in the amplitude and phase of the helicon, which interferes with the reference signal. The resulting interferogram is also shown in Figure 3.34. The condition of the existence of the interference extrema is

$$d = \left(\frac{N}{2} + \delta \right) \lambda_h,$$

where δ is the phase change determined by the phase shifter. The interferogram period and amplitudes contain information on concentration and mobility, respectively.

Dielectric permittivity can also be determined from Fabry–Perot patterns of helicons. Equations 3.109 and 3.110 lead to

$$2d = N\lambda_h = N \frac{\lambda}{\left(\varepsilon_L + \dfrac{en}{\omega \varepsilon_0 B} \right)}. \tag{3.113}$$

When

$$\varepsilon_L \gg \frac{en}{\omega \varepsilon_0 B}$$

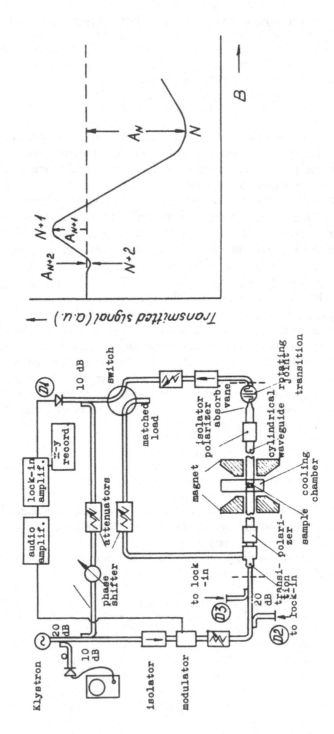

Figure 3.34. Diagram of the helicon spectrometer setup and a Rayleight interferogram. From Brazis et al. (1979).

the lattice dielectric permittivity from Equation 3.113 is given by

$$\varepsilon_L = \left(\frac{\pi c N}{d\omega}\right)^2. \tag{3.114}$$

Equation 3.114 is valid when dimensions of the sample are large in comparison to the wavelength of the helicon. Otherwise

$$\varepsilon_L = \left(\frac{\pi c}{\omega}\right)^2 \times \left(\frac{N^2}{d^2} + \frac{4}{\lambda_0^2}\right)$$

By this method the dielectric permittivity of Ge, Si and GaAs has been determined (Kliefoth 1972).

When the contribution of the free charge carriers in Equation 3.109 cannot be neglected, the permittivity can be calculated when the concentration n previously is measured. In such a way, the dielectric permittivity of PbTe (Baryshnikov et al. 1977) and $Bi_{i-x}Sb_x$ (Oelgart et al. 1977) has been studied. In the latter compound, the dielectric permittivity at 70 GHz for $x = 7.2\%$ was found to be $\varepsilon_L = 360$.

Magnetoplasma resonance also can be used for the determination of the dielectric permittivity. From Equation 3.106:

$$\varepsilon_L = \frac{en}{\varepsilon_0 \omega B}. \tag{3.115}$$

A value of the dielectric permittivity can be obtained when the concentration of free charge carriers is determined by measuring the frequency and the value of magnetic induction at which the magneto-plasma resonance occurs. Both the magnetotransmission and mag-netoreflection methods can be used for magnetoplasma resonance experiments. However, the latter method is experimentally simpler and more convenient. The sample can be pressed to the open end of a waveguide. This method has been used by Nishi et al. (1980) to study lattice instability in PbTe-SnTe compounds. Ichiguchi et al. (1980) have shown high sensitivity of magnetotransmission in a stripline for deter-mination of dielectric and electronic properties of $Pb_{1-x}Sn_xTe$.

3.4. SUBMILLIMETER WAVE RANGE

For a long time the submillimeter wave range has been inaccessible for a direct dielectric spectroscopy from both microwave and infrared regions. This phenomenally rich range was successfully brought under spectroscopical investigations of materials only after development of Fourier transform spectroscopy (FTS) (see e.g., Petzelt and Grigas 1973). However, conventional FTS was unable to solve all the problems involved in the determination of the optical constants of materials or parameters of phonons in the submillimeter wave range. FTS, as well as inelastic neutron and light scattering techniques, enable direct determination of only one part of the response function (i.e., only one optical constant).

By now, a large number of distinct techniques have been developed for the determination of optical constants in the submillimeter wave range. Two of these, dispersive Fourier transform spectroscopy (DFTS) and monochromatic backward-wave spectroscopy (BWS), are the most fruitful.

3.4.1. Dispersive Fourier Transform Spectroscopy

DFTS, developed as a far-infrared technique, is now used from the near millimeter wavelength region to the ultraviolet. The specimen may be transparent, translucent or opaque. The light source is usually a high-pressure, mercury-vapor lamp for which intensity, unfortunately, falls as $v4$ in the long wavelength region and, thus, usually require a cryogenically operated sensitive InSb or other bolometer. The only significant difference between DFTS and FTS is the position of the specimen during the measurements. In DFTS, the specimen, just as in microwave spectrometers (see, e.g., Figure 3.2), is placed within one of the two active arms of the interferometer (Figure 3.35) so that the radiation in that arm either passes through or is reflected from the specimen, as appropriate for the required measurement. This ensures that both the specimen attenuation and phase shift information are present in the recorded interferogram and can be calculated from its complex Fourier transform. Consequently, both optical constants can be directly determined.

Consider the general approach of the DFTS measurement of a translucent solid (Birch 1987). The configuration leads to the reference $I_0(x)$ and specimen $I_s(x)$ interferograms, as a function of the path difference, x, between the two arms of the interferometer. If the insensity spectrum giving rise to this interferogram is $S_0(k)$ as a function of wave number, $k = 2\pi/\lambda$, the spectrum is related to the measured reference interferogram by the Fourier integral

$$S_0(k) = \int_l^{-l} I_0(x) \cos 2xkx \, dx$$

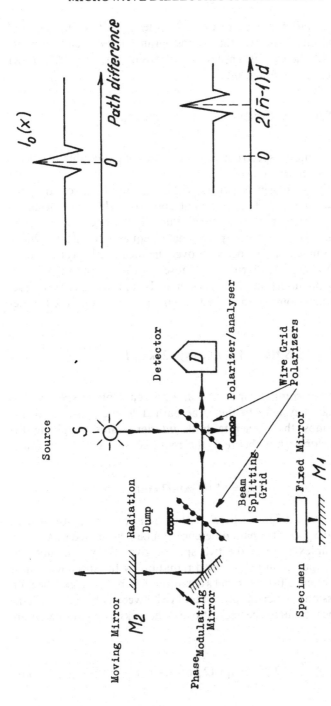

Figure 3.35. Schematic representation of the DFTS interferometer. The polarizer/analyzer grid both polarizes the incident beam and analyzes the emergent beam before it reaches the detector. The beam-splitter grid splits the polarized beam and later combines the two partial beams (From Afsar 1984). The reference, $I_0(x)$, and specimen, $I_s(x)$, interferograms are also shown.

for which the interferogram is recorded between path difference limits of $\pm l$. In a real interferometer the interferogram may not be symmetric about $x = 0$. In that case, it is necessary to consider the complex Fourier transform of the interferogram

$$S_0^*(k) = S_0(k)\exp(i\varphi_0) = \int_l^{-l} I_0(x)\exp(i2\pi kx)\,dx,$$

in which the complex spectrum, $S_0^*(k)$, consists of a modulus spectrum, S_0, and a phase spectrum, φ_0.

When the specimen is placed within the fixed mirror arm, the interferogram changes. The dominant zero path difference fringe is displaced to positive path difference values. The fringe is centered on a value of $2(n - 1)d$, where n is approximately equal to the mean value of the refractive index of the specimen over the measured spectral range, and d is its thickness. The term -1 is caused by the thickness of vacuum displaced by the insertion of the specimen. To find the specimen spectrum, $S_s^*(k)$, first compute a shifted spectrum as the complex Fourier transform:

$$S_s'^*(k) = \int_l^{-l} I_s(x)\exp(i2\pi kx')\,dx'.$$

In order to compare specimen and reference phase spectra, it is necessary that they refer to the same path difference position in the sampling combs that generate both interferograms. Applying the Fourier transform shift theorem, the required specimen spectrum is obtained:

$$S^*(k) = S_s^{*'}(k)\exp(i2\pi k\delta),$$

where δ is the path difference between the origins of computations, used for $S_s^*(k)$ and $S_0^*(k)$. This path difference can be determined exactly.

The complex ratio of the two spectra gives the phase shift and attenuation imposed on the detected radiation by the specimen. These can also be obtained from Fresnel's equations for the complex reflection and transmission coefficients of an interface between two media with the result that the measured spectral ratio and the unknown parameters are related by

$$\frac{S_s^{*'}(k)}{S_0^*(k)} = T^{*2}(k)\exp[i(2\varphi_t(k) - 4\pi kd - 2\pi k\delta)], \qquad (3.116)$$

in which $T^*(k)$ is the complex amplitude transmission coefficient

$$T^*(k) = T^*(k)\exp[i\varphi_t(k)] = (1 - R^{*2}(k)\exp(-\alpha d/2)(i2\pi knd), \quad (3.117)$$

where n is the refractive index, α is the power absorption coefficient, and $R^*(k)$ is the complex amplitude reflection coefficient of the vacuum-specimen interface

$$R^* = \frac{1 - n^*}{1 + n^*}. \quad (3.118)$$

Substituting Equations 3.117 and 3.118 into Equation 3.116 and equating modulus and phase terms on both sides of the result, yields

$$n = 1 + \frac{1}{4\pi kd}[ph\{S_s^{*\prime}(k)\} - ph\{S_0^*(k)\}] + \frac{\delta}{2d} \quad (3.119)$$

and

$$\alpha = \frac{1}{d}\ln\left[\frac{16n^2}{(1 + n)^4\left\{\frac{|S_s^{*\prime}(k)|}{|S_0^*(k)|}\right\}}\right], \quad (3.120)$$

where, in Equation 3.119, the terms $ph\{\ \}$ refer to the phase of the complex spectrum in the brackets.

Equations 3.119 and 3.120 are not exact because it was assumed that the phase of the reflection coefficient of Equation 3.118 was π radians. However, by taking the n and α values given by Equations 3.119 and 3.120 and calculating the corresponding R^* values by iteration, improved n and α values can be calculated.

DFTS is widely used for measurements of optical constants of gases, liquids and solids in the temperature region between 4.2 and 300 K. Measurements on fairly low-loss and low-permittivity solid materials mainly dominate in the optical region 100 to 450 GHz, close to the atmospheric windows at 96, 140 and 220 GHz. These materials might be used in various telecommunications related applications at those frequencies (Afsar and Button 1985; Afsar 1985) or on commercially available microwave materials between 90 and 1200 GHz (Birch 1983).

A problem with temperature measurements is that the thickness of the plane parallel-sided specimen should be known at each temperature. However, it is possible to use the DFTS itself to obtain the thickness information in addition to the refraction and absorption spectra (Birch 1987).

The DFTS reflection method is a sensitive probe also used to study the lattice dynamics of highly absorbing solids (Parker 1987), revealing detailed structure in the optical constants. A backward-wave dielectric spectroscopy is also very efficient for the lattice dynamic studies in the submillimeter wave region.

3.4.2. Backward-Wave Dielectric Spectroscopy

Backward-wave dielectric spectroscopy is the extension of the millimeter wave dielectric spectroscopy to the quasi-optical frequency range. It uses the same tunable electronic oscillators of monochromatic radiation as in the centimeter and millimeter wave ranges. It uses the same quasi-optical transmission lines, mirrors, lenses, polarizers, etc., as in DFTS.

As in the cases considered previously, one can deduce quantitatively the values of the dielectric parameters from the measured parameters only if the radiation-specimen interaction occurs in a well-defined geometry that can be analytically modelled. For BWS, just as with DFTS, this implies a plane, parallel-sided infinite specimen normally illuminated by a plane wave. BWS allows the determination of complex transmission or reflection coefficients.

Recall the macroscopic expressions which relate dielectric parameters to experimentally measured quantities. Waves propagating in a medium satisfy a scalar wave equation $\gamma = k_0(\varepsilon^*)^{1/2}$ (see Equation 3.38), where k_0 is the free space wave number, $\gamma = \beta - i\alpha$ and

$$\alpha = k_0 \left(\frac{|\varepsilon| - \varepsilon'}{2} \right)^{1/2},$$

$$\beta = k_0 \left(\frac{|\varepsilon| + \varepsilon'}{2} \right)^{1/2},$$

$$|\varepsilon| = (\varepsilon'^2 - \varepsilon''^2)^{1/2}.$$

For a plane wave $E_i^\sim \exp i(\omega t - \gamma d)$, normally incident on an infinite half-space, a complex amplitude reflection coefficient is given by

$$R_i^* = \frac{1 - (\varepsilon^*)^{1/2}}{1 + (\varepsilon^*)^{1/2}} = \frac{k_0 - \gamma}{k_0 + \gamma} = R_i e^{i\varphi_i}.$$

The magnitude and phase of the reflected wave can be expressed explicitly by

$$R_i = \left[\frac{(k_0 - \beta)^2 + \alpha^2}{(k_0 + \beta)^2 + \alpha^2} \right]^{1/2},$$

$$\varphi_i = \arctan \left[\frac{2k_0\alpha}{k_0^2 - \beta^2 - \alpha^2} \right].$$

Power reflection coefficient is

$$R_p = \frac{k_0^2 + \beta^2 + \alpha^2 - 2k_0\beta}{k_0^2 + \beta^2 + \alpha^2 + 2k_0\beta} = R_i^2. \tag{3.121}$$

The standard expressions for the amplitude transmission and reflection coefficients for a plane wave normally incident on an infinite plane of thickness d are given by

$$T^* = \frac{(1 - R_i^{*2})\exp(-i\gamma d)}{1 - R_i^{*2}\exp(-2i\gamma d)} = T\exp(i\varphi_t),$$

and

$$R^* = -\frac{R_i^*[1 - \exp(-2i\gamma d)]}{1 - R_i^{*2}\exp(-2i\gamma d)} = R\exp(i\varphi_r).$$

The measurable quantities R, T, φ_r, and φ_t can be obtained in terms of α and β from these equations. For example, the power transmittance is given by

$$T^2 = \frac{\exp(-2\alpha d)x[1 - 2R_p\cos(\beta d - 2\varphi_t)] + R_p^2}{1 - 2R_p\exp(-2\alpha d)\cos 2(\beta d - \varphi_t) + R_p^2\exp(-4\alpha d)}.$$

The BWS setup for the complex transmission measurements shown in Figure 3.36 allows the determination of the dielectric parameters of solids with fairly good accuracy (Volkov et al. 1990). To obtain the transmission spectrum, $T(v)$, the part of the spectrometer taken in the dashed area is not used. A frequency of the backward-wave oscillator (BWO) is varied by changing step-by-step its voltage. The opto-acoustic detector records frequency dependent reference and specimen signals one after another. The spectrum $T(v)$ is obtained by dividing the specimen spectrum by the reference at every value of the voltage which determines the frequency of BWO. The accuracy of the obtained spectrum depends on a level of coincidence of these two spectra at every frequency.

The phase spectrum is obtained using a Mach–Zehnder-type interferometer. By sweeping a frequency, a follow-up system moves one of the mirrors so that the interferometer remains in a balanced state. The quantity now measured is a translation of the mirror. The frequency dependence of the reference phase, $\varphi_r(v)$, and the specimen phase, $\varphi_s(v)$, is

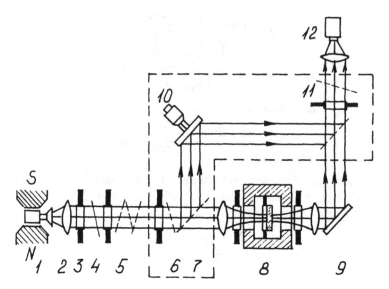

Figure 3.36. Backward wave dielectric spectrometer: 1) backward wave lamp, 2) teflon lenses, 3) limiting diaphragm, 4) modulator, 5) wire grid attenuator, 6) polarizer, 7) beam-splitting grid, 8) thermostat with the specimen, 9) mirror, 10) compensator of phase shift, 11) analyzer, 12) detector. From Volkov et al. (1990).

related by changing a voltage of BWO point by point. The phase spectrum is obtained by subtracting these spectra (i.e., $\varphi(v) = \varphi_s - \varphi_r(v)$). Dielectric parameters of the specimen are calculated using transmission amplitude and phase spectra. Note that the plane waves approximation for specimens of finite dimensions in a short millimeter range is very rough however. Dipole scattering at least should be taken into account in the transmission or reflection formulas.

Still, BWS enables measurements of the transmittance of lossy samples of an order of 10^{-5} when thicknesses of the samples are of an order of $10\,\mu m$. With four or five BWOs one can cover the 150 to $900\,\text{GHz}$ (5 to $30\,\text{cm}^{-1}$) frequency range.

When loss is small ($\alpha d \ll 1$) and βd is large, the transmission is oscillatory as a function of β and d, displaying maxima when $\beta d = N\pi$ (see Section 3.3.1, *Resonant transmission*).

These Fabri–Perot oscillations are clearly resolved as long as $\alpha d < 1$ and considerably increase the accuracy of determination of ε' and ε''. Wire grid polarizers allow 99.99% beam polarization to be obtained. The reader is referred to the comprehensive review by Volkov et al. (1990) for more details on this technique and its development.

4 LOW-DIMENSIONAL SEMICONDUCTORS

4.1. INTRODUCTION

The number of materials undergoing ferrodistortive phase transitions (PT) continually increases. Such phase transitions and soft lattices are found, for example, in dielectric, magnetic, semiconductive, organic, and liquid crystals, usually with a considerable degree of ionic or weaker chemical bonding. This chapter considers *a new family* of crystals—low-dimensional antimony and bismuth chalcogenides—which have been treated as semiconductors of tight covalent chemical bonding.

Microwave investigations of these crystals have revealed unusual dielectric properties for such semiconductors. Since these crystals are less familiar to the reader, their properties will be discussed to a greater extent beyond the microwave spectroscopic investigations. The peculiarities of their structures and highly anisotropic chalcogenide ion polarizabilities provide for various types of instabilities in lattice and electronic subsystems, and are of interest to materials scientists.

The earth's crust contains only 5×10^{-5} percent of antimony, but there are more than 100 crystalline antimony compounds in nature. About 80 of them are sulfide compounds. There are also more than 50 bismuth sulfides. So far, only a small part of these minerals have been obtained in laboratories. In these compounds, both antimony and bismuth may be of $+3$, -3 (and $+5$) valency. A great number of these antimony and bismuth chalcogenides are low-symmetry, low-dimensional crystals. The distinctive feature of these layered or chained crystals is their easy mechanical cleavage along the planes parallel to the layers. This feature implies that the chemical bonding within each layer is much stronger than the bonding between layers. An individual layer may be composed of either a single sheet of atoms or a set of atomic sheets or chains.

The unique feature of these crystals is the nonequivalent positions of atoms in the weakly-bonded sheets of atoms. In these crystals, interatomic distances vary in a wide range, and the coordination number is not equal to the normal valency of atoms. As a result, these crystals have the "friable" crystal lattice of a complicated chemical bonding. The most outstanding structure of this family of crystals is the *stibnite-(antimonite)*-type structure. A stibnite-type layer is the basic feature of many antimony and bismuth chalcogenides and halogenides.

The structural anisotropy directly affects electronic and lattice properties (Wieting and Schlüter 1979). Therefore, these chalcogenides are interesting semiconductors. The semiconductive properties of some of these crystals are well studied.

Lattice properties of these crystals shall be discussed from the standpoint of structural phase transitions, which have been revealed, first of all, by applying microwaves.

4.2. STRUCTURES AND CHEMICAL BONDING

4.2.1. $A_2^V B_3^{VI}$ Compounds

The sesquichalcogenides of As, Sb and Bi have different crystal structures and can be divided into three groups. The first group is layered monoclinic isomorphous arsenic semiconductors of C_{2h}^5 symmetry (Zallen 1975): As_2S_3 (orpiment), As_2Se_3, and As_3Te_3. In the primitive unit cell, they have two layers parallel to the (010) plane, which are coupled by nearly van der Waals forces. A single layer is composed of two molecules, where each As atom has three nearest-neighbor S atoms bonded by covalent forces.

The crystals of the second group included in the orthorhombic stibnite family are isomorphous Sb_2S_3, Sb_2Se_3 and Bi_2S_3. The structure of stibnite (Ščavničar 1960; Bayliss and Nowacki 1972) is formed of infinite $(Sb_4S_6)_\infty$ ribbons (two per unit cell) along the c-axis (needle axis). These ribbons, related by the [001] two-fold screw axis, can be regarded as forming zigzag layers which are roughly perpendicular to the b-axis (Figure 4.1). The layers are held together by the nearly van der Waals forces.

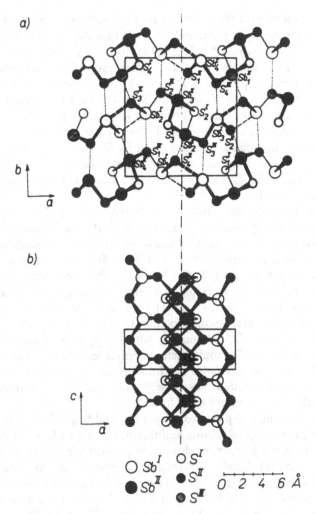

Figure 4.1. Arrangement of atoms of Sb_2S_3 in the (001) plane (a) and projection of a double chain in the (010) plane (b). From Petzelt and Grigas (1973).

The ribbon consists of two types of chains along the c-axis. The bonds within the chains are of different length, owing to the two different kinds of coordination exhibited by both Sb and S atoms. One half of the antimony atoms, Sb_1, is bonded to three S atoms located at the corners of a trigonal pyramid and exhibit the usual trivalency as the Sb_1—S bond length approaches the sum of covalent radii.

The other half of the antimony atoms, quintvalent Sb_{II}, has fivefold coordination in a square pyramid. Sb_{II} is slightly displaced (0.17 Å) out of the base center. Two of the three values for the Sb_{II} — S bond lengths considerably exceed the sum of covalent radii. Therefore, chemical bonding in $Sb_{II}S_5$ polyhedra is ionic-covalent. The Sb_{II} atoms have positive charge ($e_{ef} = 0.3e$), while the S_1 atoms have the same negative charge. Ionicity of the chemical bonding is about 10 percent.

Thus, the side chains formed by Sb_1S_3 pyramids are covalently bound, while the middle chains formed by $Sb_{II}S_5$ pyramids are ionic-covalently bound. Formally, the formula of stibnite can be written as $Sb_2S_3 = Sb_1^{III}Sb_{II}^VS_3$. This anomaly in such a simple compound is known as *stibnite's paradox*.

The spatial configuration of the electronic cloud of the antimony atoms gives an explanation of its stereochemical behavior. Sb_1 atoms in a trigonal pyramidal arrangement have three covalent σ-bonds (formed by p-orbitals) of multiplicity equal to 1. Each Sb_{II} atom has: 1) a σ-bond with the S_1 atom of multiplicity 1 formed by the linear combination of p_x- and p_y-orbitals; 2) two resonant σ-bonds of multiplicity 1/2 and C_{2v} symmetry with the S_{III} atoms formed by the linear combination of p_y- and p_z-orbitals; and 3) three resonant bonds of multiplicity 1/4, 1/4 (σ-bonds) and 1/2 (π-bonds) of C_{2v} symmetry with the S_I atoms formed by the sp_zd_z-hybridization. The multiplicity of bonds less than 1 is the characteristic feature of *metallic bonding*. Moreover, each S atom has a *lone pair* of p-electrons, while Sb atoms have the lone pairs of s-electrons. This leads to the *donor-acceptor* bond formation.

By obtaining two p-electrons from Sb_1 and Sb_{II}, sulfur atoms have two lone pairs of s^2p^2-electrons. Sulfur atoms may transfer these pairs to Sb_{II} atoms, thereby becoming donors. Sb_{II} atoms use s^2p^2-electrons via a donor-acceptor bond with vacant $5d$-orbitals, thereby becoming acceptors. Two s^2 electrons form the lone pair of the Sb^{3+} electrons.

Thus, the chemical bonding of the Sb_2S_3 crystal is complex. The same applies to Sb_2Se_3 and Bi_2S_3 crystals (see Ščavničar 1960).

The layered Bi_2Se_3, Sb_2Te_3 and Bi_2Te_3 semiconductors with the rhombohedral primitive unit cell (Wieting and Schlüter 1979) belong to the third group. Each layer is composed of five planes of atoms. The chemical bonding of atoms in the layers is also ionic-covalent and partially metallic.

In conclusion, the crystals of the first group of the $A_2^VB_3^{VI}$ family have pure covalent bonding, while those of the second and third groups have ionic-covalent chemical bonding. From a crystallochemical point of view, these crystals are similar to the narrow-gap semiconductive ferroelectrics mentioned in Section 3.3.7, in which the covalent interaction is supposed to be a decisive factor for ferroelectricity (Bussman-Holder et al. 1983).

4.2.2. Complex Sulfides

Stibnite-type chains, which have two types of coordination of antimony atoms with similar Sb—S distances, also exist in numerous complex antimony (and bismuth) sulfides: for example in *berthierite*, $FeSb_2S_4$, *chalcostibite*, $CuSbS_2$, *livingstonite*, $HgSb_4S_8$, *jamesonite*, $FePb_4Sb_6S_{14}$, and others.

Figure 4.2 shows a (001) and (010) projection of the $FeSb_2S_4$ structure. The structure consists of infinite double chains $(Sb_2S_4)^{2-}_\infty$ and ordinary chains $(SbS_2)^-_\infty$ running along the shortest c-axis and linked together by Fe atoms (Buerger and Hann 1955). In the ordinary chains, as in stibnite, each Sb atom is bonded to three sulfur atoms by covalent bonds and forms a trigonal pyramid. In the double chains, the antimony atoms are surrounded by five sulfur atoms and form distorted square pyramids. The chemical bonding is similar to the bonding of stibnite.

In the $CuSbS_2$ structure, the chains $(SbS_2)^-_\infty$ are linked together by Cu atoms in the double chains so that the coordination of antimony atoms also becomes a distorted square pyramid of different bond length. Cu atoms form CuS_4 chains.

Livingstonite, $HgSb_4S_8$, has a more complicated structure, which consists of four chains in a primitive unit cell (Srikrishanan and Nowacki 1975). Figure 4.3 shows one eighth of the unit cell, which contains four nonequivalent positions of antimony atoms and two positions of Hg atoms. The coordination of Hg atoms is a distorted octahedron, while the coordination of Sb is different: Sb_I atoms form trigonal pyramids with an additional S atom; the coordination of Sb_{II} atoms are distorted square pyramids similar to the ones in stibnite; Sb_{III} atoms form trigonal pyramids; and the coordination of Sb_{IV} atoms is similar to Sb_I atoms.

The polyhedra are bonded again to the infinite chains running along the short axis. The trigonal pyramids form three symmetrically different $(SbS_2)^-_\infty$ chains. These are linked together into the double $(Sb_2S_4)^{2-}_\infty$ chains (similar to berthierite). These double chains are connected in different ways: one chain through the common sulfur atoms, and another one through Hg atoms, so that the structural formula can be written as $(HgSb_2S_4)_\infty(Sb_2S_4)^{2-}_\infty$.

In jamesonite, $FePb_4Sb_6S_{14}$, antimony atoms have threefold coordination in three nonequivalent positions, and they form weakly bonded Sb_4S_{14} layers. In these layers, each Sb atom is bonded to five sulfur

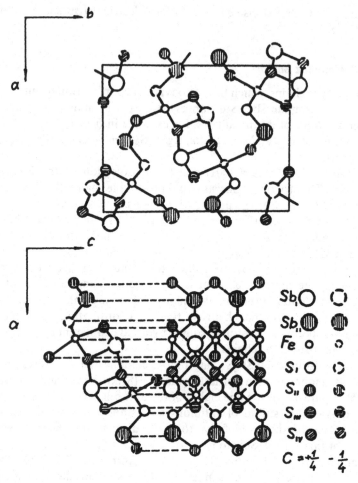

Figure 4.2. Projection of the structure of $FeSb_2S_4$ on (001) and (010) planes. From Buerger and Hann (1955).

atoms as in stibnite, berthierite or livingstonite. The primitive unit cell contains two Sb_6S_{14} groups, which make space for Fe and two kinds of lead atoms, Pb_I and Pb_{II}. The coordination of Fe is the same as in berthierite and Fe—S bonding is mainly ionic. The coordination of Pb_I and Pb_{II} is a distorted octahedra (Niizeki and Buerger 1957).

Such a coordination of Pb atoms is also found in *bournonite*-($CuPbSbS_3$) and *seligmanite* ($CuPbAsS_3$) crystals (Edenharter et al. 1970). However, antimony has threefold coordination in an isolated

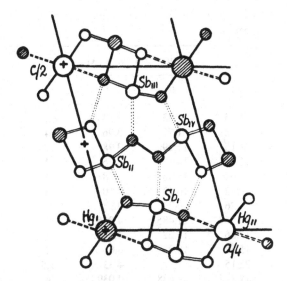

Figure 4.3. Fragment of the structure of $HgSb_4S_8$ on (010) plane. Dashed lines indicate weak Hg-S bonds. Dotted lines indicate weak Sb-S bonds. From Srikrishanan and Nowacki (1975).

trigonal pyramid and forms noncentrosymmetric crystal lattices. The lattice constants and the space group of some of these compounds are given in Table 4.1. The short axis coincides with the direction of the chains and is identified as the c-axis.

In *stephanite* (Ag_5SbS_4), antimony also has threefold coordination in a trigonal pyramid, but the weak bonds are present in other polyhedra (Ribar and Nowacki 1970).

Bismuth, as well as antimony, forms a number of complex sulfides. The structure of some known complex bismuth sulfides is a superstructure of *aikinite* ($CuPbBiS_3$) (Kohatsu and Wuensch 1971), where Bi and Pb atoms have fivefold coordination in a square pyramid. The infinite chains running along the short axis are formed by Bi and Cu polyhedra, while Pb atoms are between these two chains (Figure 4.4). The structure of aikinite is similar to the structure of Sb_2S_3 and Bi_2S_3. Despite the presence of Pb and Cu atoms, the primitive unit cell of aikinite is enlarged by only 2.9 percent along the a-axis and by 1.5 percent along the b- and c-axes as compared to the Bi_2S_3 unit cell. This confirms that stibnite-like lattices are "friable".

Bi_2S_3 and $CuPbBiS_3$ form superstructures of stoichiometric composition with the same symmetry. However, the ordering of Cu atoms in $CuPbBi_3S_6$ lowers the symmetry to C_{2v}^9.

Table 4.1. The lattice constants and space group at room temperature of stibnite-like chalcogenides

Crystal	Lattice Constants (Å)				Space Group
	a	b	c	z	
Sb_2S_3	11.25	11.33	3.84	4	D_{2h}^{16}
Sb_2Se_3	11.62	11.77	3.96	4	"
Bi_2S_3	11.15	11.29	3.98	4	"
$FeSb_2S_4$	11.44	14.12	3.76	4	"
$CuSbS_2$	6.01	14.46	3.78	4	"
$SbSJ$	8.52	10.13	4.10	4	"
$SbSBr$	8.26	9.79	3.97	–	"
$HgSb_4S_8$	30.567	21.485	4.015	8	C_{2h}^{6}
$AgSbS_2$	12.86	13.22	4.41	8	C_{2h}^{3}
$FePb_4Sb_6S_{14}$	15.57	18.98	4.03	2	C_{2h}^{5}
$Pb_6Sb_{14}S_{27}$	22.15	–	4.33	1,5	C_6^6
$CuPbBiS_3$	11.319	11.638	4.039	4	D_{2h}^{16}
$CuPbSbS_3$	8.153	8.692	7.793	4	C_{2v}^{7}
$CuPbAsS_3$	8.076	8.737	7.634	4	C_{2v}^{7}
Ag_5SbS_4	7.837	12.467	8.538	4	C_{2v}^{12}

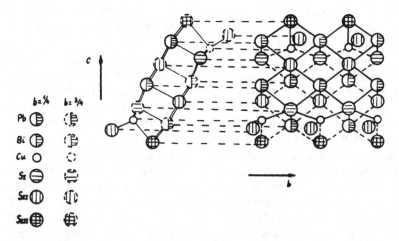

Figure 4.4. Projection of a chain of $CuPbBiS_3$ on (100) plane. From Kohatsu and Wuensch (1971).

Table 4.2. Bond lengths of fivefold coordination in square pyramids

Crystal	Bond	Bond Length (Å)	Another Coordination Number
Sb_2S_3	Sb_{II}-S	$\underline{2.49(1)}$; 2.68(2); 2.83(2)	3
Sb_2Se_3	Sb_{II}-Se	$\underline{2.58(1)}$; 2.78(2); 2.98(2)	3
SbSJ	Sb-S, J	$\underline{2.57(1)}$; 2.86(2); 3.08(2)	–
SbSBr	Sb-S, Br	$\underline{2.49(1)}$; 2.67(2); 2.94(2)	–
$HgSb_4S_8$	Sb_I-S	$\underline{2.49}$; 2.54; 2.70; 2.94; 2.95	3
	Sb_{III}-S	$\underline{2.47}$; 2.55; 2.62; 2.95; 2.96	–
	Sb_{IV}-S	$\underline{2.54}$; 2.59; 2.66; 2.88; 2.98	–
$FeSb_2S_4$	Sb_I-S	$\underline{2.48(1)}$; 2.58(2); 2.93(2)	3
$CuSbS_2$	Sb-S	$\underline{2.44(1)}$; 2.57(2); 3.11; 3.66	3
$FePb_4Sb_6S_{14}$	Sb_1-S	$\underline{2.41}$; 2.52; 2.67; 2.85; 2.90	3
	Sb_{II}-S	$\underline{2.43}$; 2.59(2); 3.04; 3.07	–
	Sb_{III}-S	$\underline{2.44}$; 2.57; 2.81; 2.94; 3.18	–
$Pb_4Sb_4S_{11}$	Sb_I-S	$\underline{2.437}$; 2.724(2); 2.909(2)	3
$CuPbBiS_3$	Bi-S	$\underline{2.46}$; 2.50; 2.70; 2.94(2)	–
$NaSbF_4$	Sb-F	$\underline{1.93}$; 2.03; 2.08; 2.19; 2.51	–
	Sb_I-F	$\underline{1.98}$; 2.06(2); 2.18(2)	–
	Sb_{II}-F	$\underline{1.99}$; 1.99(2); 2.29(2)	–
$(NH_4)_2SbCl_5$	Sb-Cl	$\underline{2.36}$; 2.62(4)	–

In addition to the main threefold coordination of antimony or bismuth atoms with the covalent chemical bonds in the complex sulfides, another part of these atoms has fivefold coordination in a distorted square pyramid with different bond lengths (Table 4.2). The shortest bond to the vertex of the pyramid is underlined. Thus, the *stibnite paradox* is quite a common feature among antimony (and bismuth) chalcogenides.

Groups of atoms, such as (Sb_2S_4), (Sb_4S_6), and others, can also be considered as radicals, and the complex sulfides can be considered as sulfosalts. The "building particles" of sulfides are sulfur atoms with their individual properties, and in sulfosalts these particles are radicals with their collective properties. Similar radicals cause similar physical properties of these crystals, which are similar to molecular crystals with a considerable amount of covalent bonding. In general, the complex sulfides contain the stibnite-like structure as long as the amount of other heavy atoms is not large.

4.2.3. $A^V B^{VI} C^{VII}$ Compounds

Tables 4.1 and 4.2 also include antimony sulfoiodide, SbSI, and antimony sulfobromide, SbSBr, which belong to the $A^V B^{VI} C^{VII}$ family, where A is Sb or Bi, B is S or Se, and C is Cl, Br or I. Of the twelve possible compounds, all but SbSCl, SbSeCl, and BiSeCl crystallize with the space group symmetry D_{2h}^{16} and are shown to be isomorphous. However, only the crystal structure of SbSBr has been studied in detail (Christofferson and McCullough 1959).

The structure consists of weakly bonded double chains running along the c-axis. In fact, the chains are quite similar to the middle part of the stibnite chains, the side $Sb_I S_3$ trigonal pyramids being replaced by S and I (or Br) atoms (Figure 4.5). The coordination of antimony is analogous to the coordination of Sb_{II} atoms in $Sb_2 S_3$ (i.e., Sb atoms have fivefold coordination in a square pyramid, and the Sb atom is also displaced by 0.28 Å).

The chemical bonding of SbSI is similar to the chemical bonding within the chains of stibnite formed by $Sb_{II} S_5$ pyramids. Thus, the crystal

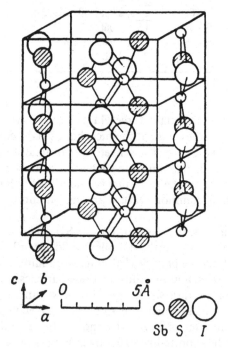

Figure 4.5. Structure of SbSI.

structure of SbSI-type crystals is similar to the structure of stibnite. The crystal structure of the later is only complicated by adding the side chains formed of trigonal pyramids characteristic of the trivalent antimony threefold coordination. The middle part of the chains, formed of $Sb_{II}S_5$ pyramids, is identical to SbSI-type chains. Known quasi-one-dimensional ferroelectrics from this family are SbSI and SbSBr. One can expect that similar crystallochemical properties of these crystals should determine their similar physical properties.

4.2.4. Complex Halogenides

Trivalent antimony and bismuth also form a large number of halogenides. However, the structure of only a few of them has been studied. The crystal lattices of fluorides, which like those of sulfides are anisotropic, low-symmetrical and friable, have been studied to a greater extent. Table 4.2 includes only three compounds of halogenides and shows that the antimony atom also has fivefold coordination in a square pyramid with an Sb—F bond length similar to the Sb—S bond in sulfides. These square pyramids form complex anions $(Sb_4F_{16})^{4-}$ in $KSbF_4$, or $(SbF_5)^{2-}$ in M_2SbF_5, where M = K, Na, Cs, NH_4 and Rb, as well as infinite chains $(SbF_4)^-_\infty$ in $KSbF_4$. The reader should refer to the comprehensive review by Polynova and Porai-Koshic (1966) for more details on the stereochemistry of trivalent antimony in various compounds, including fluorides, chlorides, oxides, and others.

In all compounds where a 5-coordinated complex of square pyramidal shape is formed, the lone pair of electrons on the remaining corner of the octahedron is present (Figure 4.6). Due to the stereochemical role of the lone pair of electrons, the geometry of antimony and bismuth compounds is distorted. It is very likely that the stereochemical activity of the lone pairs of electrons in the antimony and bismuth compounds causes the weakly bonded structural units, channels or holes in the crystals along the direction of the lone pairs. These holes may be empty or occupied by weakly bonded ions. In other words, the lone pairs of electrons favor the formation of the "friable", soft and labile structures of antimony and bismuth compounds. The role of the lone pair of electrons in the antimony, bismuth and arsenic compounds was reviewed in depth by Volkova and Udovenko (1988). However, halogenides, unlike the sulfides, are ionic dielectrics.

In these low-dimensional crystals, the weakly bound structural units, ionic-covalent chemical bonding, channels and holes, and stereochemically active lone pairs of electrons (which, in contrast to the valent electrons, are bonded to only one atomic frame enabling them to

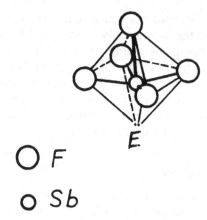

$$\bigcirc\ F \qquad E$$

$$\circ\ Sb$$

Figure 4.6. Configuration of $(SbF_5)^{2-}$ anion showing presence of the lone pair of electrons on one corner of the octahedron.

change their positions) make these compounds good candidates for various phase transitions and related anomalous physical properties.

4.3. SEMICONDUCTIVE PROPERTIES

Antimonite (Sb_2S_3) and bismuthinite (Bi_2S_3) are frequently found minerals in nature. Therefore, their semiconductive properties were discovered as far back as 1907 by Jaeger. Most early investigations dealt with the electric, photoelectric and optic properties of natural crystals (Elliot 1915; Tyndall 1923; Voigt 1929; Ibuki and Yoshimatsu 1955). Later, these crystals were grown from the vapor phase (Karpus and Mikalkevičius 1962) and the melt (Mikalkevičius and Grigas 1966).

Due to the complicated chemical bonding, the electrical properties of these crystals are very susceptible to nonstoichiometric variations and impurities leading to a very large variation of electrical parameters. The electrical d.c. conductivity of Sb_2S_3 changes from 10^{-3} to 10^{-7} S/m in the melt-grown crystals (Rinkevičius et al. 1967) and from 10^{-5} to 10^{-8} S/m in the vapor-grown crystals (Karpus and Mikalkevičius 1962). The conductivity of Sb_2Se_3 changes from 10^{-1} to 10^{-6} S/m (Mikalkevičius and Grigas 1966), and the conductivity of Bi_2S_3 changes from 10^{-4} to 10^{-5} S/m (Black et al. 1957). The conductivity can be of either n- or p-type, but, in fact, the conductivity is bipolar (Kavaliauskiene et al. 1973). The drift mobility of holes in Sb_2S_3 at 300 K is 10^{-4} m^2/V·s. The Hall mobility is 10^{-3} m^2/V·s (Lipskis et al. 1971). The drift mobility of holes in Sb_2Se_3 is 4.5×10^{-3} m^2/V·s, and that of electrons is 1.5×10^{-3} m^2/V·s. The drift mobility of electrons in Bi_2S_3 is

$2 \times 10^{-2} \, m^2/V \cdot s$. The third group, Bi_2Te_3-type crystals, are narrow-gap semiconductors with a considerably larger charge carrier mobility (Black et al. 1957).

The energy gap for Sb_2Se_3 crystals is $E_g = 1.2 \, eV$, the gap for Bi_2S_3 crystals is $E_g = 1.3 \, eV$, and that of Sb_2S_3 crystals is $E_g = 1.7 \, eV$ (Black et al. 1957). The absorption edge of Sb_2S_3 crystal is complicated (Audzijonis and Karpus 1978). An interband absorption for these crystals starts from excitation of electrons in narrow bands formed by the lone pairs of p-electrons of sulfur. Thus, for indirect transitions, $E_g = 1.54 \, eV$. For direct transitions, $E_g = 1.72 \, eV$. The energy gap, determined from the photoconductivity measurements, corresponds to $E_g = 1.55 \, eV$ in Sb_2S_3 (Ibuki and Yoshimatsu 1955).

It was found long ago that the electric conductivity of Sb_2S_3 crystals depends on the frequency of the electric field. This dependency indirectly indicates strong polarization effects. However, the value of static dielec-

Table 4.3. Semiconductive parameters of the sulfides at 295 K

Crystal	Conductivity (S/m)	Type	Energy Gap (Ev)	Mobility (10^{-4} m²/V·s)	References
$FeSb_2S_4$	7×10^{-2}	p	–	3.6	Grigas et al. (1975a)
$CuSbS_2$ (natural)	2×10^{-2}	p	–	1.2	Grigas et al. (1975b)
$CuSbS_2$ (melt grown)	2.3	p	1.73	–	"
$HgSb_4S_8$	7×10^{-6}	p	1.6	–	Brilingas et al. (1977a)
$CuPbBiS_3$	1×10^{-3}	p	1.03	5	Grigas et al. (1975c)
$CuPbSbS_3$	2×10^{-3}	n	2	50 $\mu_H = 45$	Brilingas et al. (1976)
$CuPbAsS_3$	6.5×10^{-3}	–	1.33	$\mu_H = 0.9$	Orliukas et al. (1979)
$FePb_4Sb_6S_{14}$	5×10^{-2}	p	–	–	Brilingas et al. (1977)
$CuPbBi_5S_9$-Bi_3S_3	1×10^{-2}	–	0.86	$\mu_H = 1$	Brilingas et al. (1977b)
$5PbS \times 4Sb_2S_3$	2×10^{-5}	n	1	$\mu_H = 1$	Brilingas et al. (1977c)
$6PbS \times 7Sb_2S_3$	2×10^{-8}	p	1.38	–	Brilingas et al. (1976)

tric permittivity after the measurements by Ibuki and Yoshimatsu (1955) was generally accepted to be equal to the electronic one (i.e., $\varepsilon(0) = \varepsilon_e$). Only microwave investigations by Grigas and Karpus (1967) revealed high and anisotropic dielectric permittivity of Sb_2S_3-type semiconductors.

Some of the semiconductive parameters of complex sulfides are enumerated in Table 4.3.

Semiconductive properties of $A^V B^{VI} C^{VII}$-type crystals have been reviewed by Fridkin (1980). The halogenides are dielectrics.

4.4. LATTICE DYNAMICS AND DIELECTRIC PROPERTIES OF Sb_2S_3-TYPE CRYSTALS

In low-dimensional crystal structures, phonons are affected by the magnitude and anisotropy of large directional variation in crystal bonding forces. If a primitive unit cell contains s layers, in addition to the internal modes, interlayer force constants control the frequencies of $3(s-1)$ rigid-layer optical modes (Wieting and Schlüter 1979). The crystal lattice is stable with respect to small deformations if the frequencies of all the phonon modes are real.

4.4.1. Lattice Stability

The microscopic approach to lattice dynamics has been a subject of considerable activity for many years (see e.g., Horton and Maradudin 1975). Several first-principle and parameter-free approaches have been developed to demonstrate the physical origin of crystal lattice or soft mode instability (see e.g., Bussman-Holder et al. 1983). A crystal stability condition should be recalled briefly in order to define notations used later.

A crystal is regarded as a periodic array of N unit cells, each of which contains n atoms. If the position of the j-th atom in a unit cell is designated as $r(lj) = r(l) + r(j)$, and the displacement from the equilibrium positions of ions and electrons is designated as $u(lj)$ and $w(lj)$, respectively, then an increment of a potential energy in the harmonic approximation due to these displacements is

$$V = -\frac{1}{2}\sum_{lj\alpha}\left\{\sum_{l'j'\beta}\left[V_{\alpha\beta}^R(lj,l'j')u_\alpha(lj) + u_\beta(l'j') + Y_j^{-1}V_{\alpha\beta}^T(lj,l'j')u_\alpha(lj)p_\beta(l'j')\right.\right.$$
$$\left. + Y_j^{-1}V_{\alpha\beta}^T(lj,l'j')u_\beta(l'j')p_\alpha(lj) + Y_j^{-1}Y_j^{-1}V_{\alpha\beta}^S(lj,l'j')p_\alpha(lj)p_\beta(l'j')\right]$$
$$\left. + \alpha_j^{-1}p_\alpha^2(lj) + [p_\alpha(lj) + Z_ju_j(lj)]E_j(lj)\right\}, \tag{4.1}$$

where α and β are different components of vectors, Z_j and Y_j are numbers defining effective charges of the j-th ion and electron, $V_{\alpha\beta}^R$, $V_{\alpha\beta}^T$ and $V_{\alpha\beta}^S$ are the short-range force constants between the ions, ions and electrons, and electrons, respectively. The term $\alpha_j^{-1}p_\alpha^2(lj)$ takes into account the electrostatic interaction of electrons with their ions. The last term is the coulombic interaction of all electrons and ions of the crystal. The electronic polarizability of the j-th ion is α_j. The dipole moment that arises when electrons are displaced with respect to their frame is $p(lj) = Y_j w(lj)$. $E_j(lj)$ is the effective electric field acting on the j-th ion.

Assuming that electrons are inertialess at phonon frequencies, the equations of motion of ions and electrons can be written as

$$M_j \ddot{u}(lj) = -\frac{\partial V}{\partial u_\alpha(lj)}, \quad 0 = -\frac{\partial V}{\partial p_\alpha(lj)}, \tag{4.2}$$

where M_j is the mass of the j-th ion. The displacement of ions and electrons can be considered in terms of plane waves:

$$u(lj) = U(j)\exp i[qr(lj) - \omega t], \tag{4.3}$$

where q is the wave vector. Substituting Equation 4.3 into Equation 4.2 yields

$$\omega^2 M_j U_\alpha(j) = \sum_{j'\beta} [R_{\alpha\beta}(jj') + Z_j C_{\alpha\beta}(jj')Z_j]U_\beta(j')$$
$$+ \sum_{j'\beta} [T_{\alpha\beta}(jj') + Z_j C_{\alpha\beta}(jj')Y_{j'}]W_\beta(j'),$$
$$0 = \sum_{j'\beta} [T_{\alpha\beta}(jj') + Z_j C_{\alpha\beta}(jj')Y_{j'}]U_\beta(j') \tag{4.4}$$
$$+ \sum_{j'\beta} [V_{\alpha\beta}(jj') + Y_j C_{\alpha\beta}(jj')Y_{j'}]W_\beta(j'),$$

where designations for $R_{\alpha\beta}(jj')$, $T_{\alpha\beta}(jj')$, and $S_{\alpha\beta}(jj')$ are of the form

$$R_{\alpha\beta}(jj') = -\sum V_{\alpha\beta}^R(lj, l'j')\exp iq[r(l'j') - r(lj)],$$
$$V_{\alpha\beta}(jj') = S_{\alpha\beta}(jj') + \delta_{\alpha\beta}\delta_{jj'}\alpha_j^{-1}Y_j^2.$$

$C_{\alpha\beta}(kk')$ are the structural coefficients of the internal electric field, and $\delta_{\alpha\beta}$ and δ_{ij} are the Kronecker deltas.

Equations 4.4 in a matrix form are

$$\omega^2 M_d U_c = (R + Z_d C Z_d)U_c + (T + Z_d C Y_d)W_c,$$
$$0 = (T + Y_d C Z_d)U_c + (V + Y_d C Y_d)W_c. \tag{4.5}$$

The equation for the ions is derived by eliminating W_c from Equations 4.5:

$$\omega^2 M_d U_c = \Phi U_c, \tag{4.6}$$

where

$$\Phi = R + Z_d C Z_d - (T + Z_d C Y_d)(V + Y_d C Y_d)^{-1}(T + Y_d C Z_d)$$

is the dynamical matrix of the force constants describing the interaction of the sublattices.

The set of Equations 4.6 has nonzero solutions only when the determinant $[\Phi - \omega^2 M_d] = 0$. That is

$$[\Phi_{\alpha\beta}(jj') - \omega^2 M_j \delta_{\alpha\beta} \delta_{ij}] = 0. \tag{4.7}$$

For each value of q, the secular equation (4.7) has $3n$ real solutions of ω (i.e., $3n$ phonon modes when the unit cell contains n atoms). The necessary condition for stable vibrational spectra of a crystal is that real and nonzero values of all the optical frequencies be given by Equation 4.7. A frequency of zero signifies a change of structure of the crystal. The condition for the existence of real solutions is $\text{Det}^{3(n-1)}|\Phi| > 0$. When one of the main minors $|\Phi|$ of the determinant becomes equal to zero, the instability of the crystal lattice, or a phase transition, occurs. Of course, the harmonic approximation is too rough, but introduction of anharmonicity strongly complicates the problem (see e.g, Silverman 1964).

Introducing into Equations 4.5 an external electric field, and using dielectric relations, one obtains (Cochran 1960) the equation for the static dielectric permittivity at $q = 0$:

$$\varepsilon(0) - \varepsilon_\infty = \frac{e^2}{\varepsilon_0 V} [Z_r][\text{Det } \Phi]^{-1}[Z_c], \tag{4.8}$$

where Z_r and Z_c are the effective dynamical charges of ions taking into account an overlap of electron orbitals and interaction of corresponding sublattices of ions; V is the volume (Equation 6.15).

It follows from Equation 4.8 that as $\text{Det } \Phi$ approaches zero, $\varepsilon(0) \to \infty$ (i.e., ferroelectric phase transition results due to instability of the crystal lattice with respect to one of the normal vibrations). The reason for this is a compensation of long- and short-range forces in an ionic crystal. Such a vibration in a layered crystal can be either an internal or rigid-layer vibration.

Chalcogenides and halogenides are electronically highly polarizable. According to the polarizability model (Bussman-Holder et al. 1983), the central role in ferroelectric phase transitions of chalcogenides should exhibit highly anisotropic chalcogenide ion polarizability. The local, nonlinear, electron-ion interaction of the chalcogenide ions can provide the driving mechanism for ferroelectricity. So far, this model has not been applied to the stibnite type crystals. Nevertheless, the influence of an electronic subsystem on the lattice stability can be found even from the following equation:

$$\varepsilon(0) = \sum_{i=1}^{n} \frac{\Delta_i}{\Delta} x_i, \tag{4.9}$$

where $\Delta = \mathrm{Det}(E_{ij}^l x_i - \delta_{ij})$, Δ_i is obtained by replacing the i-th column in Δ by -1; E_{ij}^l are the coefficients of the local Lorentz field, and x_i is the polarizability. Calculation of the coefficients E_{ij}^l and the surfaces of x_i (Gavelis 1977) shows that an Sb_2S_3-type structure is indeed favorable for ferroelectric phase transitions. Also Equation 4.9 enables the calculation of incremental values of x_i for a given ion to undergo the phase transition ($\varepsilon(0) \to \infty$). This increase of x_i is designated as $\Delta x_i = x_i - x_i^e$ (x_i^e is the electronic polarizability) and is proportional to the charge and displacement of an ion. The calculations show that a small displacement of the Sb_{II} atoms can induce the phase transition in the Sb_2S_3 crystal. Larger displacements of Sb_{II} atoms in Sb_2Se_3 are necessary to induce the phase transition because of different values of x_{Sb}^e in these crystals.

Rao et al. (1982) by computer simulations have shown that a small displacement of Sb_{II} atoms induces soft modes in phonon spectra of Sb_2S_3 along Σ_2, Λ_2 and Δ_2 directions, and some of these become imaginary.

Spontaneous polarization of the layered crystals is

$$P_s = P_e + P_i + P_{rl},$$

where P_e, P_i and P_{rl} are the electronic, ionic and rigid-layer vibration contributions, respectively. Equation 4.9 and experimental values of $\varepsilon(0)$ and the polarizability of atoms coordinated with the effective charges of ions enable one to obtain the displacement of Sb_{II} atoms along the c-axis. The displacement $\Delta z = 0.05\,\text{Å}$ yields $P_e = 1\,\mu\text{C/cm}^2$ and $P_i = 0.3\mu\text{C/cm}^2$. The displacement $\Delta z = 0.1\,\text{Å}$ gives $P_e = 2.1\,\mu\text{C/cm}^2$ and $P_i = 0.6\,\mu\text{C/cm}^{2\cdot}$ However, the strongest Lorentz field manifest in Sb_{II} atoms is much weaker than in SbSI crystals.

4.4.2. Lattice Vibration Analysis

Consider Sb_2S_3-type lattice vibration group theoretical analysis in the Γ point of the Brillouin zone (Petzelt and Grigas 1973). The irreducible representations of the Sb_2S_3 crystal lattice vibrations are

$$5A_u + 5B_{1u} + 5B_{2g} + 5B_{3g} + 10B_{2u} + 10B_{3u} + 10A_g + 10B_{1g}.$$

The long-wavelength phonon modes for both possible point groups, D_{2h} (mmm) and C_2v (mm2) together with the correlations between them and with the selection rules for infrared (and submillimeter) absorption and Raman scattering are summarized in Table 4.4. There are 4 and 14 dielectrically active modes in the $E \parallel c$ spectrum, and 9 and 14 modes in both $E \parallel a$ and $E \parallel b$ spectra for D_{2h} and C_{2v} symmetry, respectively. Therefore, the number of one-phonon absorption peaks enables one to distinguish between both symmetries.

By taking into account the fact that the bond between $(Sb_4S_6)_n$ ribbons is much weaker than that between the atoms in the ribbon, one can divide the normal vibrations into rigid-layer modes and interlayer modes. These may be compared with the external (molecular-like) and internal modes of molecular crystals. The rigid-layer modes are determined by weak interlayer forces and are of lower frequencies, whereas internal modes in both ribbons should differ only slightly. Thus, the vibrations in the lattice consisting of only one type of a ribbon can be first analyzed. Such a lattice contains only 10 atoms in the unit cell and the symmetry reduces to C_{2h}^2 or C_2^2 (polar) space groups. The symmetry vectors in this case can be obtained simply from those in Table 4.4 taking only the atoms in one ribbon (i.e., 1 and 4 or 2 and 3). One can see that the symmetry-type pairs A_g and B_{1g}, A_u and B_{1u}, B_{2g} and B_{3g}, and B_{2u} and B_{3u} are nonequivalent. The frequencies of modes belonging to these symmetry-type pairs are roughly the same. Infrared absorption shows nearly the same transversal internal mode frequencies for $E \parallel b$ and $E \parallel a$ polarizations. However, the oscillator strength and longitudinal mode frequencies can be different in both directions.

The irreducible representations and properties of the translational and librational molecular-like vibrations for both the D_{2h} and C_{2v} symmetries have been discussed by Grigas and Batarunas (1980).

Group theoretical analysis of lattice modes for wave vectors within the Brillouin zone, together with neutron and x-ray scattering studies, were performed by Rao et al. (1982).

Properties of the long-wavelength lattice vibrations of As_2S_3 and Bi_2Te_3-type crystals have been reviewed by Wieting and Schlüter (1979).

Table 4.4. Classification of normal vibration modes in Sb_2S_3-type structure at the center of Brillouin zone

Symmetry Type		Symmetry Vectors (M_i stands for Sb_i^I, Sb_i^{II}, S_i^I, S_i^{II}, and S_i^{III}, $i=1,2,3,4$)	Number of Modes			Activity			
			Optic		Acoustic	D_{2h}		C_{2v}	
C_{2v}	D_{2h}		External	Internal		IR	Raman	IR	Raman
A_1	A_g	$\dfrac{xM_1 - xM_2 + xM_3 - xM_4}{yM_1 + yM_2 - yM_3 - yM_4}$	—	10	—	—	$\alpha_{aa}, \alpha_{bb}, \alpha_{cc}$	$E \parallel c$	$\alpha_{aa}, \alpha_{bb}, \alpha_{cc}$
	B_{1u}	$zM_1 + zM_2 + zM_3 + zM_4$	—	4	1	$E \parallel c$	—		
A_2	A_u	$zM_1 - zM_2 - zM_3 + zM_4$	—	5	—	—	—	—	α_{ab}
	B_{1g}	$\dfrac{xM_1 + xM_2 - xM_3 - xM_4}{yM_1 - yM_2 + yM_3 - yM_4}$	2 (rot.)	8	—	—	α_{ab}		
B_2	B_{2g}	$zM_1 + zM_2 - zM_3 - zM_4$	—	5	—	—	α_{bc}	$E \parallel b$	α_{bc}
	B_{3u}	$\dfrac{xM_1 - xM_2 - xM_3 + xM_4}{yM_1 + yM_2 + yM_3 + yM_4}$	1 (transl.)	8	1	$E \parallel b$	—		
B_2	B_{3g}	$zM_1 - zM_2 + zM_3 - zM_4$	—	5	—	—	α_{ac}	$E \parallel a$	α_{ac}
	B_{2u}	$\dfrac{xM_1 + xM_2 + xM_3 + xM_4}{yM_1 - yM_2 - yM_3 + yM_4}$	1 (transl.)	8	1	$E \parallel a$	—		

4.4.3. Dielectric Dispersion

In the presence of an electromagnetic field, some of the optical modes are coupled to this field when the selection rules of energy and momentum conservation are obeyed. The range of crystal lattice vibrations is 0 to 10^{13} Hz. The shortest wavelength of photons which interact with phonons is

$$\lambda_{min} = \frac{c}{v_{max}} = 3 \times 10^{-5} \, m.$$

As the crystal lattice constant, d, is of the order of 10^{-9} to 10^{-10} m (i.e., $d \ll \lambda_{min}$), an electromagnetic field can interact only with the long-wavelength lattice vibrations for which $q \to 0$ or $\lambda \to \infty$. In order to obtain the equations of the coupled system (crystal + electromagnetic field), it is necessary to add to Maxwell's equations the equations describing the phonon field and the polarization:

$$D = \varepsilon_0 E_{int} + P,$$
$$P = N e_T u + \varepsilon_0 \varepsilon_\infty E_{int}$$

(4.10)

and the equation of motion:

$$\ddot{u} + \gamma \dot{u} + \omega_{TO}^2 u = \frac{e_T}{M} E_{int},$$

(4.11)

where M is the oscillator mass, e_T is the transverse effective charge:

$$e_T = e_s \frac{\varepsilon_\infty + 2}{3},$$

and e_s is the Szigeti charge, u is given by Equation 4.3, ε_∞ is the contribution of electronic polarization ($\varepsilon_\infty - 1$) to dielectric permittivity, and E_{int} is the internal electric field.

Equations 4.3 and 4.11 enable deduction of u, and introducing this value of u into Equation 4.10 provides the dynamic polarization and displacement from which the dispersion equation (see also Section 2.4) of the dynamic dielectric permittivity is deduced:

$$\varepsilon^*(\omega) = \varepsilon' - i\varepsilon'' = \varepsilon_\infty + \sum_i \frac{\Delta\varepsilon_i \omega_{TOi}^2}{\omega_{TOi}^2 - \omega^2 - i\gamma_i\omega},$$

(4.12)

where

$$\Delta\varepsilon_i = \varepsilon(0) - \varepsilon_\infty = \frac{Ne_T^2}{\varepsilon_0 M\omega_{TOi}^2} \tag{4.13}$$

is the contribution of the optical mode of the frequency ω_{TOi} to static dielectric permittivity (oscillator strength), which in the limiting case $\omega = 0$, gives the value

$$\varepsilon(0) = \varepsilon_\infty + \sum_i \Delta\varepsilon_i.$$

From Equation 4.13 it follows that mainly low frequency of the optical modes causes high static dielectric permittivity of crystals.

Consider now the third Maxwell equation:

$$sD = \varepsilon(sE_{int}) = 0,$$

where $s = q/|q|$ is a unit vector. As long as the internal electric field $E_{int} \neq 0$, this equation will be satisfied for the two cases: $\varepsilon = 0$ and $(sE_{int}) = 0$. The first case corresponds to $u \parallel P \parallel E_{int}$ (i.e., a longitudinal wave propagation whose frequency, ω_{LO}, is a solution of Equation 4.12):

$$\omega_{LO}^2 = \omega_{TO}^2 + \frac{Ne_T^2}{\varepsilon_0 \varepsilon_\infty M} = \omega_{TO}^2 \frac{\varepsilon(0)}{\varepsilon_\infty}. \tag{4.14}$$

Equation 4.14 is the Lyddane–Sachs–Teller (LST) relationship (see Equation 2.31) for a single undamped oscillator. When a crystal has i dielectrically active undamped modes, the LST relationship can be written in the form

$$\frac{\varepsilon(0)}{\varepsilon_\infty} = \prod_{i=1}^n \frac{\omega_{LOi}^2}{\omega_{TOi}^2}. \tag{4.15}$$

The second case, $s \neq 0$, corresponds to $E_{int} \perp s$. The propagating wave is therefore transversal, and the frequencies ω_{TOi} in Equation 4.12 are transversal optical modes. Their damping constants are $\gamma_i = 1/\tau_{fi}$, and τ_{fi} is the TO phonon lifetime. Indeed, the TO and LO frequencies are complex values of ω in Equation 4.12 when $\varepsilon > 0$ and $\varepsilon = 0$, respectively. The frequency ω_{TO}, a pole of $\varepsilon^*(\omega)$, is given by

$$\omega_{TO} = \omega_r + i\omega^i, \tag{4.16}$$

where

$$\omega_r = \omega_{TO}[1 - (\gamma/2\omega_{TO})^2]^{1/2} \quad \text{and} \quad \omega^i = -\gamma/2.$$

The real part of ω_{TO} (i.e., ω_r), represents the resonant frequency while the imaginary part is the measure of damping. It is necessary for ω^i to be negative in order to solve Equation 4.11 with u in the form of Equation 4.3.

Dielectric dispersion of Sb_2S_3-type crystals in the far-infrared and submillimeter wave range was studied by Petzelt and Grigas (1973).

The reflectivity spectra of the natural Sb_2S_3 crystal of the largest $E \| c$ reflectivity are shown in Figure 4.7. The spectra are unusually complex. There is no pronounced temperature dependence for $E \| c$ polarization, whereas a number of peaks become more pronounced and stronger by lowering the temperature in the $E \perp c$ case. These changes are largest in the 220 to 370 K region. In all other melt-grown and vapor-grown Sb_2S_3 crystals, the $E \| c$ spectra differ from that in Figure 4.7 only by an absolute value of reflectivity, not by the frequencies of peaks. Even the fine structure in the 70 to 200 cm^{-1} region roughly reproduces

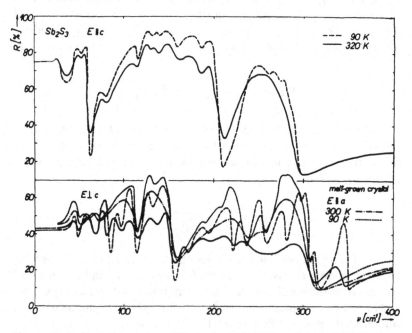

Figure 4.7. Reflectivity spectra of Sb_2S_3. Solid (320 K) and dashed (90 K) lines are for natural crystal. Lines for melt-grown crystal are also shown. From Petzelt and Grigas (1973).

frequencies[1]. Nevertheless, reflectivity values are systematically lower by an amount of 10 to 20 percent of the values shown in Figure 4.7. The variation of the $E \| c$ reflectivity-peak heights in different samples is a result of different long-range lattice perfectness within the defective surface layer of a crystal (the penetration depth of the infrared radiation at reflectivity peaks is only 3×10^{-5} to 10^{-4} cm). The violation of the translational symmetry in such a layer leads to the relaxation of the quasi-impulse-conservation selection rule. This results in a decrease of the one-phonon oscillator strength of a perfect crystal. Dielectric permittivity of Sb_2S_3 also depends on stoichiometry (vide infra).

In $E \perp c$ spectra, the differences between different crystals are greater. All modes become more distinct and stronger at lower temperatures, as observed with the natural crystal.

The reflectivity spectra of the vapor-phase crystals are closely similar to those of natural crystals. The number of peaks exceeds the number of first-order resonances for C_{2v} symmetry. One can suppose that the measured surface of the natural and vapor-grown crystals was (110) or that the extra peaks are second-order bands.

New peaks appear or become stronger at lower temperatures 140, 225 and 300 K in $E \perp c$ spectra. The new peaks at about 300 K appear at 250 and 335 cm^{-1}, whereas the peaks at 135 and 286 cm^{-1} split (Figure 4.8). These changes in the spectrum indicate structural changes in the lattice.

Parameters of the dielectric dispersion are obtained from the reflectivity spectra employing the Kramers–Kronig dispersion relations (Equations 2.4 and 2.5). The following set of equations are determined using Equation 3.121:

$$\varepsilon'(\omega) = \frac{(1 - R_p)^2 - 4R_p \sin^2 \phi}{(1 + R - 2R^{1/2} \cos \phi)^2},$$

$$\varepsilon''(\omega) = \frac{-4(1 - R_p)(R_p)^{1/2} \sin \phi}{[1 + R - 2(R_p)^{1/2} \cos \phi]^2},$$

$$(4.17)$$

where R_p is the experimentally measured reflectance power connecting the incident and reflected intensities,

$$\phi(\omega) = \frac{\omega}{\pi} \int_0^\infty \frac{\ln[R_p(\omega')] - \ln[R(\omega)]}{\omega^2 - \omega'^2} d\omega' \qquad (4.18)$$

---

[1] In this chapter and in Chapter 5 the wave number (in cm^{-1}) is termed by frequency.

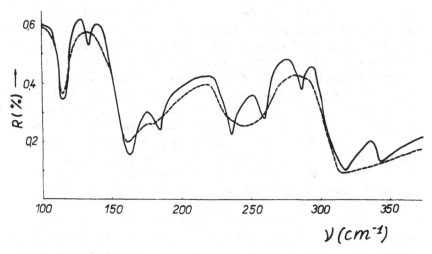

Figure 4.8. Reflectivity spectra of Sb_2S_3 for $E \perp c$ polarization at $T = 260 \, K$ (solid line) and $T = 320 \, K$ (dashed line). From Grigas (1980a).

is the phase difference between the waves. Having $R_p(\omega)$ spectra, $\phi(\omega)$ can be evaluated by using Equation 4.18, and the dielectric parameters follow from Equations 4.17.

In all Sb_2S_3-type crystals, reflectivity above $400 \, cm^{-1}$ is supposed to be constant. Below $25 \, cm^{-1}$, the reflectivity is in rough agreement with the static or microwave dielectric data (Grigas et al. 1976a; Grigas et al. 1978a; Juodviršis et al. 1969). The results of the Kramers–Kronig analysis are shown in Tables 4.5 to 4.7. The absolute errors of reflectivity in the spectral region 50 to $400 \, cm^{-1}$ of ± 0.03 lead to the errors in the oscillator strengths of about 20 percent. At low-frequency modes, the errors are much greater. However, the frequencies v_{Ti} and v_{Li} are determined with an accuracy of $\pm 1 \, cm^{-1}$. The accuracy of calculation of the damping, γ_i, depends on the shape of the peak and varies from about ± 1 to $\pm 5 \, cm^{-1}$.

Figures 4.9 to 4.12 show dielectric spectra of these crystals. Since the effect of free carriers is negligible in the entire spectral region, the absorption may be supposed to result mainly from one-phonon processes.

The number of peaks that can be treated as one-phonon absorption can vary in a rather wide range. When polarization is $E \parallel c$ the number of peaks is 3 to 13 for Sb_2S_3, 3 to 10 for Bi_2S_3, and 3 to 5 for Sb_2Se_3. When polarization is $E \parallel a$ the number of peaks is 9 to 13 for Sb_2S_3, 9 to 12 for Bi_2S_3, and about 8 for Sb_2Se_3. If the symmetry of these crystals is D_{2h}, all

Table 4.5. Mode parameters for Sb₂S₃

Sym.	$\nu_T[\text{cm}^{-1}]$ 90	$\nu_T[\text{cm}^{-1}]$ 320	$\nu_L[\text{cm}^{-1}]$ 90	$\nu_L[\text{cm}^{-1}]$ 320	$\gamma_T[\text{cm}^{-1}]$ 90	$\gamma_T[\text{cm}^{-1}]$ 320	$\Delta\varepsilon$ 90	$\Delta\varepsilon$ 320	ε_∞ 90	ε_∞ 320	$\varepsilon(0)$ 90	$\varepsilon(0)$ 320
$E\parallel c$	<25	<25	32	34	?	?	x	y	9.5	9.5	92 + x	92 + x
	48	48	62	62	6	11	34	33				
	78	78	86	88	7	9	1.4	1.6				
	108	112	114	114	7	13	16	16				
	121	122	140	139	8	8	22	15				
	139	141	159	163	12	15	4	9				
	160	160	185	185	14	24	2	4.5				
	237	233	293	292	15	30	2.8	3.6				
$E\perp c$	47	47	48	48	3	8	0.6	1.5	7.2	7.2	22	22
	63	63	66	66	7	9	1.3	2.0				
	81	77	84	80	4.5	4.5	1.4	1.3				
	95	91	96	95	3.5	8	0.4	1.6				
	107	108	113	113	5	11	2	1.8				
	129	132	132	138	5	12	1.3	1.7				
	139	144	154	153	8	8	2.7	0.6				
	189	190	191	197	7	20	0.4	1.1				
	211	—	220	—	14 ⎱	60	1.8 ⎱	2.6				
	229	—	234	—	8 ⎰	60	0.6 ⎰	2.6				
	248	—	256	—	8 ⎱	35	1 ⎱	1				
	273	—	280	—	12 ⎰	35	0.9 ⎰	1				
	290	286	316	207	10	35	0.7	1				
	341	336	350	343	8	16	0.3	0.2				

Table 4.6. Mode parameters for Bi_2S_3

Sym.	$\nu_T[cm^{-1}]$ 90	$\nu_T[cm^{-1}]$ 300	$\nu_L[cm^{-1}]$ 90	$\nu_L[cm^{-1}]$ 300	$\gamma_T[cm^{-1}]$ 90	$\gamma_T[cm^{-1}]$ 300	$\Delta\varepsilon$ 90	$\Delta\varepsilon$ 300	ε_∞ 90	ε_∞ 300	$\varepsilon(0)$ 90	$\varepsilon(0)$ 300
$E\parallel c$	<20	<20	26	26	?	?	x	y	16	13	71+x	65+x
	36	35	37	36	3	8	2	5.7				
	45	43	47	45	5	10	4.4	4.6				
	67	–	68	–	10	–	0.6	–				
	126	122	128	?	15	17	11	13				
	138	134	149	196}	38	40}	34	27				
	152	–	188	–		–}						
	215	215	254	252	7	26	2.7	2.7				
$E\perp c$	38	–	40	–	5	–	3	–	9	9	42	38
	45	45	47	47	4	11	3	7				
	59	58	64	63	9	10	5	4				
	84	85	91	91	11	15	5	4				
	117	–	119	–	10	–	3	–				
	130	131	139	139	16	40	6.6	10				
	143	–	163	–	12	–	3	–				
	192	194	231	226	10	26	4	3.6				
	240	239	267	267	11	15	0.7	0.6				

Table 4.7. Mode parameters for Sb_2Se_3

Sym.	$\nu_T[cm^{-1}]$ 90	300	$\nu_L[cm^{-1}]$ 90	300	$\gamma_T[cm^{-1}]$ 90	300	$\Delta\varepsilon$ 90	300	ε_∞ 90	300	$\varepsilon(0)$ 90	300
$E\|c$	44	45	63	64	12	23	73	84	15.1	15.1	128	133
	66	–	?	–	14	–	3.3	–				
	93	87	128	138	8	18	25	30				
	112	–	?	–	10	–	4	–				
	131	–	140	–	12	–	2.2	–				
	158	155	198	194	7	12	4.8	4				
$E\perp c$	<25	<25	29	32	?	?	x	y	14.5	13.7	52 + x	53 + y
	35	37	40	41	6	8	14	12				
	46	47	51	53	9	11	9	8.5				
	81	81	86	86	6	11	3.9	5				
	94	95	104	105	5	8	3.5	3.3				
	134	137	140	144	8	18	2.8	3.4				
	165	168	197	202	6	15	4.8	6.8				
	204	201	209	206	9	10	0.3	0.3				

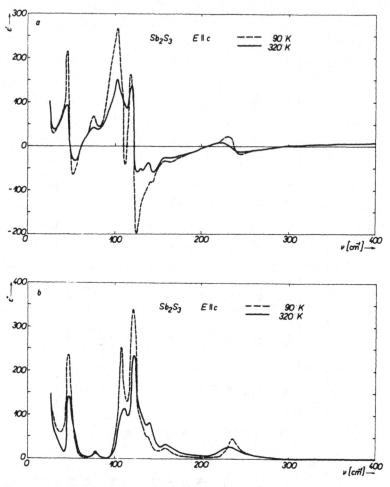

Figure 4.9. Dielectric spectra of Sb_2S_3 for $E \parallel c$ at $T = 320\,K$ (solid line) and $T = 90\,K$ (dashed line). From Petzelt and Grigas (1973).

the preceding peaks in ε'' cannot be first-order transitions. These data for Sb_2S_3 and Bi_2S_3 are compatible with both symmetries, D_{2h} and C_{2v}. C_{2v} symmetry can be produced by slightly displacing the atoms from their positions along the c-axis of the crystal. The small distortion of the lattice required to change the symmetry to C_{2v} introduces only weak additional resonances. Thus, the number of observed peaks supports the notion that Sb_2S_3, as well as Bi_2S_3, might be weakly polar. However, the latest (fourth) refinement of the crystal structure of Sb_2S_3 by McKee and McMullan (1975) indicates that at room temperature, Sb_2S_3 is in the

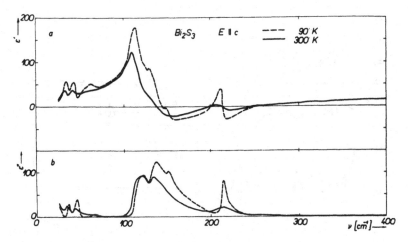

Figure 4.10. Dielectric spectra of Bi_2S_3 for $E \parallel c$ at $T = 300\,K$ (solid line) and $T = 90\,K$ (dashed line). From Petzelt and Grigas (1973).

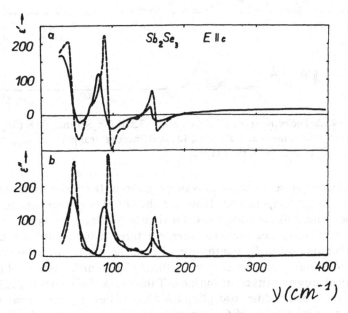

Figure 4.11. Dielectric spectra of Sb_2Se_3 for $E \parallel c$ at $T = 300\,K$ (solid line) and $T = 90\,K$ (dashed line). From Petzelt and Grigas (1973).

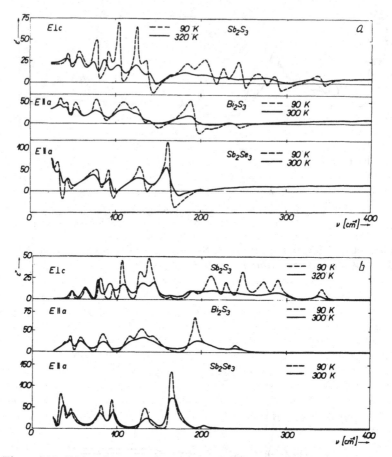

Figure 4.12. Dielectric spectra for $E \| a$ at $T = 320\,K$ (Sb_2S_3) and $300\,K$ (Sb_2Se_3 and Bi_2S_3) (solid lines) and $T = 90\,K$ (dashed lines); (a) real part, (b) imaginary part. From Petzelt and Grigas (1973).

paraelectric phase. Hence, D_{2h}^{16} is supposed to be the appropriate space group at high temperatures. However, the lattice constants depend on temperature (Lukaszewicz, 1994). Obviously, Sb_2Se_3 is nonpolar.

In all three compounds, there are three main modes for $E \| c$ polarization—in Bi_2S_3 the third lies below the low-frequency limit— instead of four as predicted by group theory. The fourth mode would lie within microwaves (its contribution in Tables 4.5 and 4.6 is designated as x), or the broad middle absorption band would be composed of two (or even more in the case of C_{2v} symmetry) one-phonon peaks. In Sb_2S_3 a further mode was found in the microwave region (Figure 4.13).

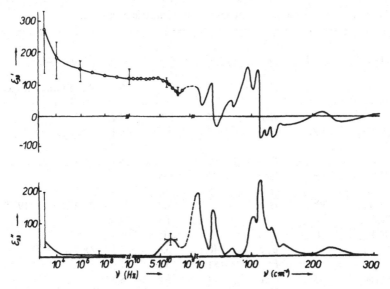

Figure 4.13. Complete dielectric spectrum of Sb_2S_3 along the c-axis at $T = 297\,K$. Limits correspond to various crystals. From Grigas (1980b).

From the comparison of the frequencies of individual modes in all three crystals it follows that in the higher-frequency modes, the motion of S(Se) dominates. The frequency is markedly lowered by the exchange $S \rightarrow Se$ and remains nearly constant by the exchange $Sb \rightarrow Bi$. In the lowest-frequency modes, it is assumed that the Sb_{II} (or Bi_{II}) and S_I (or Se_I) sublattices are vibrating 180 degrees out of phase, just as in SbSI-type compounds (Koutsoudakis et al. 1976). No exact conclusion can be drawn regarding the polarization vector of normal modes because the number of modes is too large. The picture is in agreement with the assumption (Section 4.2.1) that the unusual dielectric properties of Sb_2S_3-type semiconductors are predominantly determined by the weak bonds in $Sb_{II}S_5$-type pyramids.

In contrast to covalent semiconductors or even other layered crystals (Wieting and Schlüter 1979), some of the polar B_{1u} phonons have unusually low frequencies, and the dielectric dispersion, at least in Sb_2S_3 and Bi_2S_3, proceeds to the microwave region. As a result, according to Equation 4.13, the dielectric contribution, $\Delta\varepsilon$, of these phonons to the static permittivity is much larger than the electronic one ($\varepsilon_\infty - 1$) (Table 4.8).

The infrared reflectances of Bi_2Te_3, Bi_2Se_3 and Sb_2Te_3 (Richter et al. 1977; see also Wieting and Schlüter 1979) revealed two optically

Table 4.8. Lowest phonon frequencies and dielectric parameters

Crystal	T(K)	Polari-zation	$\nu_{TO}(cm^{-1})$	$\Delta\varepsilon$	ε_∞	References
Sb_2S_3	320	$E\|c$	10	> 110	9.5	Petzelt and Grigas (1973)
		$E^\perp c$	47	22	7.2	
Sb_2Se_3	300	$E\|c$	45	133	15.1	
		$E\|a$	< 25	> 53	13.7	"
Bi_2S_3	300	$E\|c$	< 20	> 65	13	"
		$E\|a$	38	38		
Sb_2Te_3	300	$E^\perp c$		500	51	Richter et al. (1977)
Bi_2Se_3	15	"	69	83	29	
Bi_2Te_3	15	"	49	205	85	"
As_2S_3	300	$E\|c$	140	3.6	7.3	Zallen (1975)
		$E\|a$	181	3.3	11	
As_2Se_3	300	$E\|c$	94	3.6	8.8	
		$E\|a$	106	3.4	10.5	
SbSI	295	$E\|c$	10	2700	10	Koutsoudakis et al. (1976)
SbSeI	300	"	35	220		"
	40		19	650		
BiSI	292	"	39	150		"
BiSeI	300	"	40	140		"
BiSBr	300	"	40	180		"
BiSeBr	300	"	47	110		"
SbSBr	22	"	15	9500		Inushima et al. (1982)

active infrared modes: E_u^2 mode (in which the Bi(Sb) and chalcogen $(C_{3v} + D_{3d})$ sublattices vibrate out of phase), and E_u^3 mode (in which the chalcogen (C_{3v}) and chalcogen (D_{3d}) sublattices vibrate out of phase). In all three crystals the oscillator strength of E_u^2 mode is similar to that in the Sb_2S_3 family.

From all of these low-dimensional $A_2^V B_3^{VI}$ and $A^V B^{VI} C^{VII}$ semiconductors, only covalent As_2S_3 and As_2Se_3 crystals exhibit predominantly electronic polarization. In all the other crystals, as a result of the *stibnite paradox*, the phonon frequencies are low enough. Their contribution to the static dielectric permittivity is much larger than the electronic one, and, therefore, they cause high values of the static dielectric permittivity $\varepsilon(0) = \Delta\varepsilon + \varepsilon_\infty$ for these semiconductors.

Inelastic neutron scattering studies of Sb_2S_3 (Rao et al. 1982) have revealed disorder in the *ab* plane. This can be associated with static (polytypism in layered structure, or incommensurable structure) or dynamic (very low frequency modes that cannot be resolved from the elastic intensity, overdamped modes, and such) causes, and also rather flat low frequency TA-TO modes along the Σ and Λ directions. These modes along the Σ direction (belonging to Σ_2 and Σ_3 representations) and the TO mode along the Λ direction of Λ_2 representation, soften as a function of the radius parameter of Sb_{II}. The polarization vector of this branch indicates that the two $(Sb_4S_6)_n$ chain units take part in the rigid-layer vibrations along the c-axis—in phase at $q = 0$ for the TA mode, progressively changing over to out-of-phase at $q = 0$ TO branch.

A complete dielectric spectrum of Sb_2S_3 shows (Figure 4.13) that, in addition to the B_{1u} modes, there is a damped excitation at microwaves that can be associated with a disorder in the crystal. The coupling of the disorder mode, Ω, with the low-frequency optic B_{1u} mode, ω_1, results in enhancement of the dielectric permittivity (Burns 1976) and dielectric dispersion at microwaves and high frequencies:

$$\varepsilon^*(\omega) = \varepsilon_\infty + \Delta\varepsilon + \frac{\Omega^2[(\Delta\varepsilon + 2)/3]^2}{\omega_1^2 - \frac{1}{3}\Omega^2[(\Delta\varepsilon + 2)/3] - \omega^2 - i\omega\Gamma}, \qquad (4.19)$$

where $\Omega = 4\pi ne^2/m$, Γ is the damping of the disorder mode ($\Gamma/\Omega = 1.6$), and $\Delta\varepsilon$ is the phonon contribution to $\varepsilon(0)$. Layered crystal structures of the Sb_2S_3-type are favorable for impurities or deviation from stoichiometry (Grigas et al. 1973). Even for a small number of impurities, coupling of these disorder modes (see also Reinecke and Ngai 1977) to the optic mode, due to the high value of $\Delta\varepsilon$, gives sizable effects on the static dielectric permittivity:

$$\varepsilon(0) = \varepsilon_\infty + \Delta\varepsilon + \frac{[(\Delta\varepsilon + 2)/3]^2}{\omega_1^2/\Omega^2 - 1/3[(\Delta\varepsilon + 2)/3]}, \qquad (4.20)$$

which can reach values up to 1400 (Grigas et al. 1973). A similar situation results in other compounds in question.

The high dielectric permittivity and small loss at microwaves make these semiconductors useful for some microwave applications.

4.4.4. Phase Transitions

Experimental findings which might indicate structural phase transitions, for which all preconditions exist will now be discussed for these soft,

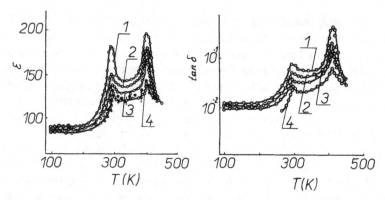

Figure 4.14. Temperature dependence of dielectric permittivity and $\tan\delta$ of Sb_2S_3 along the c-axis at frequencies: 1) 1 MHz, 2) 10 MHz, 3) 100 MHz, 4) 10 to 42 GHz. From Grigas (1980a).

low-dimensional semiconductors. Indeed, according to the vibronic theory (Konsin 1978), electron-phonon interaction in semiconductors can induce a spontaneous lattice distortion and substantial changes in electron and phonon spectra (soft mode at an arbitrary point of the Brillouin zone, dielectric anomalies, and such). Low-dimensional semiconductors reveal the strongest tendency towards structural phase transitions. Softness of the crystal lattice (low phonon frequencies), a relatively narrow gap, and an anisotropy of electronic spectra favor the structural phase transitions. Ionic-covalent interactions and high polarizability of chalcogenide ions are also the driving mechanism for ferroelectricity.

Figure 4.14 shows the temperature dependence of dielectric permittivity and $\tan\delta$ of a natural stoichiometric Sb_2S_3 crystal. Dielectric anomalies at $T_{c1} = 300$ K and $T_{c2} = 420$ K indicate the second order phase transitions. Dielectric anomalies are supposed to be caused by a small softening of the low-frequency B_{1u} mode and are increased by the disorder mode. When the dielectric contribution of this mode is small in comparison with the contribution of hard modes and electronic polarization (Table 4.5), then the dielectric anomalies at microwaves are small. The frequency dependence of ε and $\tan\delta$ can be accounted for by Equation 4.19. Below 180 K the dielectric permittivity along the c-axis increases and shows one more anomaly at $T = 13$ K (Grigas 1980a). The magnitude of dielectric parameters and the temperatures T_{c1} and T_{c2} depend on the concentration of impurities (Grigas et al. 1973) which changes disorder mode parameters. Impurities predominantly occupy the tetrahedral space between the chains, bonding them into a more

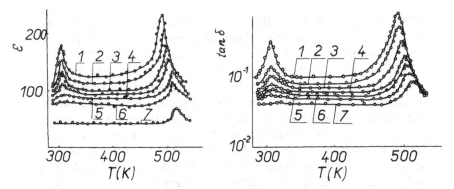

Figure 4.15. Temperature dependence of dielectric permittivity and $\tan\delta$ of Sb_2S_3 crystal along the c-axis on hydrostatic pressure: 1) 1 bar, 2) 600 bar, 3) 1.5 kbar, 4) 2 kbar, 5) 3 kbar, 6) 4 kbar, 7) 5 kbar. From Kachalov et al. (1975).

stable lattice, and destroy very sensitive to any changes of the local field $sp_z d_z$ (σ and π) bonds in $Sb_{II}S_5$ groups. Therefore, dielectric anisotropy is decreased. The double chains are bonded in a similar manner in complex sulfides. Deviation from stoichiometry has a similar effect upon dielectric properties. With the increase of the impurity concentration, the Lorentz field in $Sb_{II}S$ polyhedra increases. But at higher than $3 \times 10^{24}\,m^{-3}$ concentrations, it decreases again (Gavelis 1977). Impurities cause the potential screening of the free charges in Sb_2S_3 and leads to an increase or decrease of the phase transition temperatures T_{c1} and T_{c2}. With the change of stoichiometry from $Sb_2S_{2.18}$ to $Sb_2S_{3.4}$, and concentration of the impurities of Ag, Hg, Cu or Zn, the temperatures T_{c1} and T_{c2} change from 293 to 308 K and 420 to 492 K, respectively.

Due to the lone pairs of electrons, the $Sb_{II}S_5$ polyhedra are active centers of the donor-acceptor interactions with the impurities. That is why these semiconductors are so selective to impurities, which are difficult to exclude from the crystal and strongly exert their influence on electric properties.

Hydrostatic pressure decreases dielectric permittivity and $\tan\delta$ (Figure 4.15). This in turn strongly suppresses polarization fluctuations and dielectric anomalies at T_{c1} and T_{c2}. At pressures of 3 kbar, the dielectric anomaly at T_{c1} disappears, while the temperature T_{c2} increases according to $T_{c2}^p = T_{c2}^o + Kp$, where $K = 4.8 \times 10^{-3}\,K/kbar$ (Kachalov et al. 1975). The positive coefficient K shows that the phase transition at T_{c2} is of second order and the coefficient of the bulk electrostriction is negative. If the polarization fluctuations in isostructural SbSI crystals are large even at $T = 4\,K$ and hydrostatic pressure 7.8 kbar (Samara

1975), they vanish more rapidly in Sb_2S_3 even at high temperatures. This indicates the great sensitivity of dielectric properties and the phase transitions to small structural distortions in the coordination of Sb_{II} atoms.

Discovery of spontaneous polarization in the perfect single crystals, which appears along the c-axis above T_{c1} and gradually increases below T_{c1} (Orliukas and Grigas 1974), and visualization of domain structure by the etching technique (Grigas et al. 1976b) indicate that Sb_2S_3 is *weakly polar*. Anomalies of attenuation of the longitudinal and transversal ultrasonic waves at T_{c1} and T_{c2} (Samulionis et al. 1974) and the second acoustic and microwave harmonics generation (Meškauskas et al. 1977), indicating the acoustic and dielectric nonlinearities, confirmed the existence of structural phase transitions at temperatures T_{c1} and T_{c2}. No second microwave harmonic generation was obtained in Sb_2Se_3 crystals.

NQR investigations of local electric field distribution and the dynamics of Sb atoms in both coordination polyhedra (Abdulin et al. 1977) show that the asymmetry parameter, η, of the Sb_I nucleus is 0.8 percent, and that of the Sb_{II} nucleus is 38 percent (i.e., the $Sb_{II}S_5$ polyhedra are more distorted). The spin-lattice relaxation time, which indicates a mobility of atoms in a lattice, is 3 to 5 ms for Sb_I atoms and 12 to 18 ms for Sb_{II} atoms.

Figure 4.16 shows the temperature dependence for two of the ten transition frequencies and the asymmetry parameter. Changes of frequencies and changes in asymmetry parameters for the polyhedra $Sb_{II}S_5$ were found at 140, 225, 300 and 420 K. For Sb_IS_3, they are found at 355 and 420 K. At these temperatures, some phonon modes disappear upon heating. Besides, at 300 K, one frequency splits into two frequencies of transitions $1/2 \rightarrow 3/2$ $^{121}Sb_{II}$ and $3/2 \rightarrow 5/2$ $^{123}Sb_{II}$.

Analysis of the splitting of NQR lines in a magnetic field shows that the polar distortion of the lattice due to Sb_{II} atoms displacement is small. Upon cooling, this distortion starts at the temperature T_{c2} in both polyhedra, Sb_IS_3 and $S_{II}S_5$, and increases at temperatures of 300, 225 and 140 K. The changes at the last three temperatures are most probably connected with the weakening of the donor-acceptor interactions in $Sb_{II}S_5$ polyhedra. Due to the low energy of these interactions, the change of internal energy of the crystal is small. Therefore, anomalies in physical properties should also be small.

The influence of impurities on the relaxational processes shows that the impurities are mainly localized in the vicinity of $Sb_{II}S_5$ polyhedra, or enter into the coordination sphere of Sb_{II} atoms. The increase in concentration of impurities, decreases anomalies in η and NMR frequencies, and dielectric parameters.

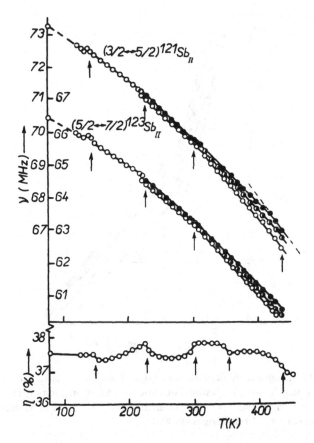

Figure 4.16. Temperature dependence of quadrupole frequencies and asymmetry parameter in Sb_2S_3. From Abdulin et al. (1977).

The NQR spectrum in Sb_2Se_3 is shifted to lower frequencies. The difference in the asymmetry parameters for Sb_I and Sb_{II} nuclei is smaller than in Bi_2S_3 (Penkov et al. 1970). The asymmetry parameter for Bi_I is $\eta = 0$, and $\eta = 90$ percent for Bi_{II}. The lines for Bi_{II} are much broader than in Sb_2S_3.

Even small structural distortions of the lattice cause the electronic spectrum of these crystals to change. As a result, the temperature coefficient of the energy gap at all temperatures of the NQR anomalies changes in both vapor-grown (Grigas et al. 1977) and melt-grown (Gaumann et al. 1977) Sb_2S_3 crystals. These changes indicate the successive phase transitions in Sb_2S_3 crystals.

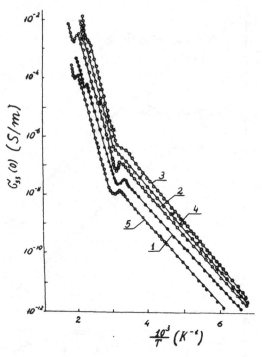

Figure 4.17. Temperature dependence of d.c. electric conductivity σ_c of Sb_2S_3 crystals: (1) stoichiometric, (2) stoichiometric with additional impurities of 10^{-2} at. % Pb and Cu; (3) and (4) nonstoichiometric with 10^{-1} at. % of As impurities, (5) conductivity $\sigma_{\perp c}$. From Grigas (1980a).

The stoichiometric Sb_2S_3 crystals containing a concentration of impurities of 10^{-3} percent or less show jumps in electric conductivity at the temperatures T_{c1} and T_{c2} (Figure 4.17) (Orliukas et al. 1975). Anisotropy of the conductivity $\sigma_c/\sigma_{\perp c} = 10$ corresponds to the anisotropy of permittivity. Deviation from stoichiometry and increases of impurities abolish the jumps in conductivity. The change in the activation energy of acceptor levels is related to the change of the electronic permittivity, $\varepsilon_e = \varepsilon_\infty - 1$, at the phase transition temperatures. This results from the fact that the energy of the impurity levels is defined by the ε_e (i.e., $E_a \sim \varepsilon_e^{-2}$). Supposing that the effective mass of the holes does not change substantially at the phase transitions, the ratio of the activation energies below E_a^b and above E_a^a is given by

$$\frac{E_a^b}{E_a^a} = \frac{(\varepsilon_e^a)^2}{(\varepsilon_e^b)^2}. \tag{4.21}$$

The activation energies below T_{c1} for various crystals are $E_a^b = 0.45 - 0.52\,\text{eV}$. Above T_{c1}, the activation energies are $E_a^a = 1.58 - 1.78\,\text{eV}$. And above T_{c2}, the activation energy is $E_a = 2.2\,\text{eV}$. The change of ε_e at T_{c1} is $\varepsilon_e^a/\varepsilon_e^b = 0.53$ (Petzelt and Grigas 1973), which explains the change of the activation energy well.

Due to the high value of the electron-phonon interaction constant:

$$\alpha_{ef} = \left(\frac{R}{\hbar\omega_{LO}}\right)^{1/2}\left(\frac{m^*}{m_e}\right)^{1/2}\left(\frac{1}{\varepsilon_e} - \frac{1}{\varepsilon(0)}\right), \tag{4.22}$$

where R is the Rydberg constant, m^* is the effective mass, m_e is the electron (hole) mass (for Sb_2S_3: $\alpha_{ef} = 5.6(m^*/m_e)^{1/2}$ for $E\,\|\,c$; $\alpha_{ef} = 27(m^*/m_e)^{1/2}$ for $E\perp c$ polarization), and the charge carriers are polarons of small radius. The jump in conductivity at the temperatures of the phase transitions can also be related to the decrease in the activation energy of the polarons, which is defined by a change of dielectric parameters. The polarons determine small mobility of the charge carriers (Lipskis et al. 1971), since conductivity is caused by a hopping process of the polarons. It depends on the electric field frequency according to $\sigma(\omega) \sim \omega^{0.8}$ (see Chapter 8).

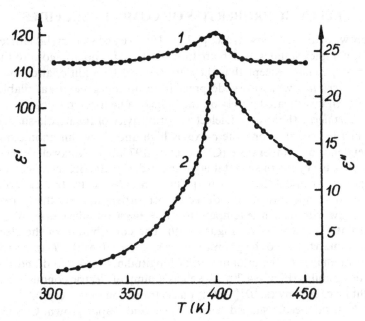

Figure 4.18. Temperature dependence of ε' (1) and ε'' (2) of Sb_2Se_3 crystal along the c-axis at the frequency of 9.3 GHz. From Grigas (1978).

In Sb_2Se_3 crystals, the dielectric anomalies are weaker (Figure 4.18) than in Sb_2S_3. Below 100 K, the dielectric permittivity increases with the decrease of temperature and, at $T = 36$ K, shows one more anomaly (Grigas 1980a). Also, small changes in the temperature coefficient of the energy gap and birefringence (Grigas et al. 1978b) probably reflect the weakening of the donor-acceptor interactions. The sequence of anomalies of various physical parameters, which can be interpreted as successive phase transitions, have also been found in Bi_2S_3 (Grigas 1980a).

Thus, dielectric spectroscopy has revealed high and anisotropic dielectric permittivity ($\varepsilon(0) \gg \varepsilon_e$), as well as the sequence of anomalies in dielectric properties for essentially covalently bonded semiconductors. Investigations of other physical parameters have confirmed that the anomalies might be regarded as being due to successive phase transitions. As the small structural changes affect only a part of the chains formed by $Sb_{II}S_5$-pyramids (Figure 4.1), while another part of the covalently bonded chains are relatively stable, the anomalies at these *weak phase transitions* are smaller than at ordinary ferroelectric phase transitions.

4.5. DIELECTRIC PROPERTIES OF COMPLEX SULFIDES

Microwave dielectric spectroscopy has also revealed soft crystal lattices in the complex sulfides, which are formed by chains homologous to the Sb_2S_3-type chain except that they are bonded by additional cations. *Natural crystals*, which are widespread in nature and generally available, have been investigated purely on the basis of a heuristic purpose.

Berthierite ($FeSb_2S_4$). Dielectric permittivity of the needle-shaped berthierite crystals along the c-axis is high and shows an anomalous dependence on temperature (Grigas et al. 1975a). As Maxwell's relaxation frequency for this crystal is on the order of 10^7 Hz, the dielectric properties at and below this frequency are affected by free charges. However, at microwaves, the dielectric permittivity follows the Curie–Weiss law with the Curie temperature $T_c = 348$ K and the Curie–Weiss constant $C = 1.2 \times 10^3$ K (Figure 4.19). The contribution of the electronic polarization to the permittivity is $\varepsilon_e - 1 = 10$. At the T_c temperature, anomalies of the attenuation of longitudinal ultrasound and the elastic modulus (Figure 4.20), as well as a jump of electric conductivity, might be caused by the $D_{2h}^{16} \rightarrow C_{2v}^9$ phase transition in this crystal.

Natural needle-shaped, melt-grown and vapor-grown $CuSbS_2$ crystals also show high dielectric permittivity along the chain axis, as well as the Curie–Weiss type anomaly at $T_c = 366$ K (Figure 4.21) with

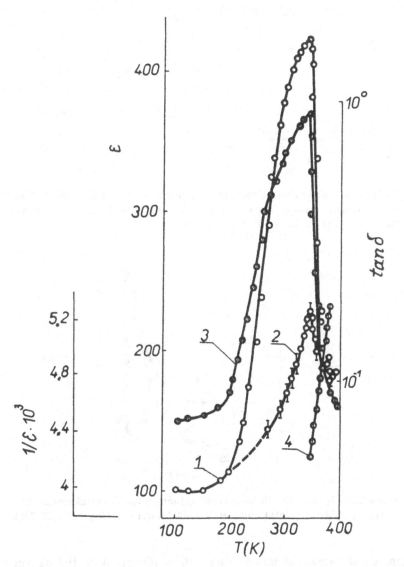

Figure 4.19. Temperature dependence of ε (curves 1 and 2), $\tan \delta$ (curve 3) and $1/\varepsilon$ along the c-axis of $FeSb_2S_4$ at frequencies: (1 and 3) 100 MHz, (2 and 4) 10 GHz. From Grigas (1978).

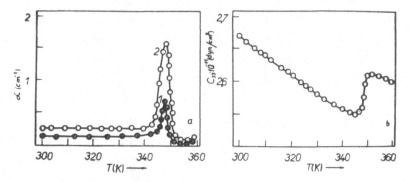

Figure 4.20. Temperature dependence of attenuation of longitudinal ultrasound along the c-axis at frequencies 10 MHz (1) and 15 MHz (2) and modulus of elasticity C_{33} in berthierite. From Grigas (1978).

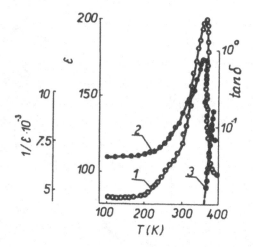

Figure 4.21. Temperature dependence of ε (1), tan δ (2) and $1/\varepsilon$ (3) along the chain axis of the $CuSbS_2$ crystal at a frequency of 100 MHz. From Grigas et al. (1975b).

the Curie–Weiss constant $C = 1.1 \times 10^3$ K (Grigas et al. 1975b). The contribution of phonons in this crystal considerably exceeds the contribution of the electronic polarization, $\varepsilon_e - 1 = 9$. A jump in electric conductivity, an anomaly of attenuation of longitudinal ultrasound of the velocity 3.13×10^3 m/s (Samulionis et al. 1975), and an anomaly of the quadrupole interaction constant have been obtained at the temperature T_c. These anomalies suggest similar structural phase transition as in berthierite.

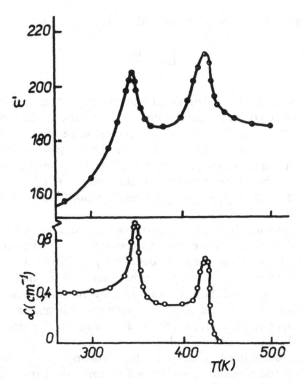

Figure 4.22. Temperature dependence of ε at a frequency of 9.15 GHz and a coefficient of absorption of longitudinal acoustic waves in aikinite. From Grigas et al. (1975c).

Aikinite ($CuPbBiS_3$)—needle-shaped crystals—is one of the complex bismuth sulfides which along the chain axis, possesses high dielectric permittivity (Grigas et al. 1975c). At the temperatures $T_{c1} = 345$ K and $T_{c2} = 426$ K it exhibits anomalies (Figure 4.22). The permittivity follows the Curie–Weiss law with the constant $C = 1.3 \times 10^3$ K at $T > T_{c2}$. Also, the contribution of phonons is much larger than the electronic one, $\varepsilon_e = 9$. Anomalies of electric conductivity, velocity, and attenuation of longitudinal ultrasound at temperatures T_{c1} and T_{c2} indicate the structural phase transitions. The adiabatic elastic moduli at T_{c1} and T_{c2} are $C_{22} = 3.48 \times 10^2$ N/m^2 and 3.3×10^2 N/m, respectively.

In all three high-conductivity semiconductors, the electromechanical coupling has been detected by measuring the acousto-motive force (Samulionis et al. 1975). The acoustic wave carries along free charge carriers and, due to the piezoelectric effect, creates an electric field following the wave. The acousto-motive force is given by

$$E = \frac{\alpha_a \mu I}{\sigma v_a},$$

where α_a is the electronic damping, I is the intensity of the acoustic wave, μ is the mobility of the charge carriers, σ is the electric conductivity of the crystal, and v_a is the velocity of the acoustic wave. By measuring the acousto-motive force, it was found that in the low-temperature phases these semiconductors are piezoelectrics. One can infer that the phase transition in these three complex sulfides might be of D_{2h}^{16} to C_{2v}^9 type. These crystals, as well as Sb_2S_3, might be *weak ferroelectrics* at low temperatures.

Livingstonite ($HgSb_4S_8$) is also a semiconductor and possesses a high value of microwave dielectric permittivity ε_c (Brilingas et al. 1977a). In various specimens the dielectric permittivity at $T = 295\,K$ varies between 90 and 125, tan δ is on the order of 10^{-2}, and $\varepsilon_{\perp c} = 27$. At temperatures of 15 and 280 K, ε_c and tan δ show anomalies characteristic of nonferroelectric structural phase transitions. Anomalies of attenuation of longitudinal ultrasound and a jump in the energy gap, $\Delta E_g = 0.03\,eV$, have also been obtained.

Thus, the complex sulfides, homologues of Sb_2S_3, also possess high lattice dielectric permittivity and show structural phase transitions.

Other complex sulfides, formed of more complicated chains, also have high dielectric permittivity at microwaves (Table 4.9). *Bournonite*

Table 4.9. Dielectric parameters of the complex sulphides at $T = 295\,K$

Crystal	Dielectric permittivity	Frequency (GHz)	Electronic contributions, ε_e	Temperature of supposed phase transition (K)
$FeSB_2Se_4$	200	10	10	348
$CuSbS_2$	100	10	9	366
$HgSb_4S_8$	105	10	7	280
$CuPbBiS_3$	166	10	8	345 and 426
$CuPbSbS_3$	40	10	–	755
$CuPbAsS_3$	40	2	–	330–370
$FePb_4Sb_6S_{14}$	50	24	–	605–635
$CuPbBi_5S_9$—Bi_2S_3	130	10	9	–
$5PbSx4Sb_2S_3$	180	10	6.2	–
$6PbSx7Sb_2S_3$	80	10	6	–

($CuPbSbS_3$) is also an efficient semiconductive piezoelectric with a high coefficient of electromechanical coupling for the transverse acoustic waves (Brilingas et al. 1976a).

In *seligmanite* ($CuPbAsS_3$), a considerable phonon contribution to the dielectric permittivity and anomalies of electric properties (Orliukas et al. 1979) are probably caused by weak bonds and small structural changes in Cu and Pb coordination spheres.

The dielectric permittivity of *jamesonite* ($FePb_4Sb_6S_{14}$) reaches its maximum value $\varepsilon = 60$ and $\tan \delta = 0.25$ at 400 K (Brilingas et al. 1977a).

Soft lattices have some superstructures of homologues of Sb_2S_3 and Bi_2S_3. One such superstructure is Bi_2S_3-$CuPbBi_5S_9$ (Brilingas et al. 1977b), where the phonon contribution to the dielectric permittivity considerably exceeds the electronic one.

There are also compounds with the general formula $mPbS \times nSb_2S_3$, where m and n are ordinary numbers. Because both Sb_2S_3 and PbS possess high dielectric permittivity (for PbS, $\varepsilon = 160$ at 25 GHz), these compounds also possess high permittivity (Brilingas et al. 1976b; 1977c).

In Ag_5SbS_4 a superionic (Chapter 8) phase transition in the Ag_2S sublattice occurs with a jump in permittivity from 20 to 800 at $T_c = 452$ K and a frequency of 9.2 GHz (Brilingas et al. 1978; Orliukas et al. 1980).

Thus, the semiconductive antimony (and bismuth) complex sulfides also possess a high value of dielectric permittivity and show a variety of anomalies in physical properties due to structural phase transitions. According to the results of spectral and chemical analysis, these crystals are rather defective. Growing crystals of good quality is still matter of chance. Therefore, it is not clear yet to what degree the phase transitions in these compounds are intrinsic or induced (for instance, by off-center impurities). Matters are further complicated because the phase transitions show both order-disorder and displacive features.

4.6. PHASE TRANSITIONS IN COMPLEX HALOGENIDES

Fluorides are more transparent than sulfides. Therefore, they are of interest for optoelectronics. Some of them are promising laser materials or superionic conductors. A large number of fluorides with a perovskite-type structure reveal structural phase transitions (Aleksandrov et al. 1981) caused by rotational distortions of the crystal lattice. A sizable subclass of complex fluorides is also formed by antimony and bismuth. The crystal lattices of fluorides, like those of the antimony and bismuth sulfides, are anisotropic, low-symmetry, and "friable", which makes these compounds good candidates for the structural phase transitions as well.

Phase transitions in crystals of the $MBiF_4$ family (M = K, Rb, Tl) were discussed by Lucat et al. (1977). Even though fluorides do not belong to semiconductors, a brief discussion will be included in this chapter pertaining to the antimony fluorides. The phase transitions and stereochemical properties of these complex fluorides—despite their ionic dielectric character—are similar to semiconductive sulfides. Crystallographic data suggest that the "driving" mechanisms of the phase transitions in fluorides and sulfides might be similar.

Dielectric investigations of the $M_6Sb_4(SO_4)_3F_{12}$ single crystals (Grigas et al. 1978b), where M = NH_4 or Rb, have revealed isostructural phase transitions of the first order $P3 \leftrightarrow P3$ at the temperatures $T_c = 256$ K and 240 K, respectively. The peak value of the static dielectric permittivity along the c-axis in the NH_4-compound is $\varepsilon_c = 50$. Perpendicular to the c-axis it is $\varepsilon_a = 7.6$ and is independent of temperature. In the $Rb_6Sb_4(SO_4)_3F_{12}$ crystal, $\varepsilon_c = 90$ and $\varepsilon_a = 8.7$. The crystal structure consists of infinite columns $[Sb_4(SO_4)_3F_{12}]_n^{6n-}$ running along the c-axis and NH_4 or Rb cations. The unit cell contains three crystallographically independent columns and six types of antimony sites differing in respect to their environment. The main coordination polyhedra of the antimony sites are $SbEF_3$ tetrahedra, where E is the lone pair of electrons. At the phase transition temperature, both distances Sb—Sb and Sb—F change (Udovenko et al. 1981), and the columns move along the 3-fold-axis. But the main distortion in the lattice occurs due to the displacement of the polar tetrahedra $SbEF_3$, a lone pair of which is directed along the 3-fold-axis. The phase transition in $(NH_4)_6Sb_4(SO_4)F_{12}$ was confirmed later (Agarwal and Chand 1985) by EPR study.

At temperatures of 293, 258, 168 and 142 K, microwave dielectric, NQR, NMR, calorimetrical, and acoustic studies (Avkhutskij et al. 1983) have revealed the successive phase transitions $Cmcm \leftrightarrow C2/c \leftrightarrow C2/c \leftrightarrow P2_1/c \leftrightarrow P2_1/c$ in $(NH_4)_2SbF_5$ crystals caused by the rotational motion of NH_4^+ and the axial reorientations of the square-pyramidal $[SbEF_5]^{2-}$ anions. The dynamics of the ammonium ions cause an order-disorder transition at 168 K, while angular displacement of the square-pyramidal $[SbEF_5]^{2-}$ anions causes a displacive structural transition at 293 K (Kobayashi et al. 1988). Upon cooling, at the phase transitions temperatures, the Sb—F distance shortens. The largest change (0.43 Å) is in the surroundings of the E-pair, and the least change (0.12 Å) is in the $[SbEF_5]^{2-}$ group. Due to the displacement of atoms, the intermolecular interactions Sb...Sb and Sb...F become stronger upon lowering the temperature. As a result, changes occur in the symmetry of the crystal, the position of the E-pair, and in all coordination spheres of the $[SbEF_5]^{2-}$ polyhedra (Volkova and Udovenko 1988).

Similar successive phase transitions associated with the reorientational movements, the rotation of $[SbEF_5]^{2-}$ anions, and the partial disordering of the lattice above room temperature have been revealed in K_2SbF_5 (Zemnukhova et al. 1983), $KSbF_4$ (Davidovich et al. 1984), other complex fluorides (Grigas et al. 1980; Urbonavičius et al. 1982a),

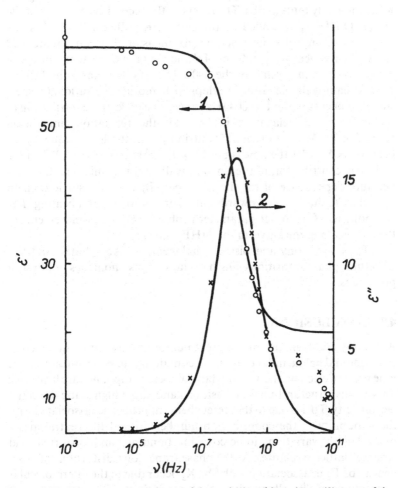

Figure 4.23. Frequency dependence of the real (1) and imaginary (2) parts of the dielectric permittivity of $K_3BiCl_6 \cdot 2KCl \cdot KH_3F_4$ crystal at T_c. The solid lines represent domain-wall dispersion. From Brilingas et al. (1986).

and antimony halogenides (Volkova and Udovenko 1988). Some of them show structural, ferroelastic, and superionic phase transitions.

When fluoride compounds are crystallized in solutions containing chloride, bismuth (in contrast to antimony) is surrounded by chlorine atoms and forms saturated chloride complexes.

The $K_3BiCl_6 \cdot 2KCl \cdot KH_3F_4$ crystal synthesized in this way at room temperature belongs to the space group $D_{6h}^4 (P6_3/mmc)$. It has a layered structure consisting of two types of layers alternating parallel to the (001) plane. The layers of the first type are formed by the octahedra $BiCl_6$ and KCl_6 joined by mutual ribs. The layers of the second type consist of the anions $[F(HF_3)]^-$, bonded by the cations K^+. Below $T_c = 130\,K$ the crystal is an improper ferroelectric (Brilingas et al. 1986). Figure 4.23 shows the frequency dependence of ε' and ε'' at the Curie temperature of the multi-domain crystal. In the 100 kHz to 1 GHz range, the Debye-type domain-wall relaxation overlaps with another relaxational dispersion that occurs in the 1 to 80 GHz range. The dielectric contribution of the domain-wall relaxation is $\Delta\varepsilon_d = 41$, the relaxation time being $\tau_d = 1.3 \times 10^{-9}\,s$. The dielectric contribution of the high-frequency relaxation is $\Delta\varepsilon = 10$, the time is $\tau = 1.5 \times 10^{-11}\,s$, and $\varepsilon_\infty = 10$. When the dielectric contribution of the domain walls $\Delta\varepsilon_d$ is eliminated, the temperature dependence of the dielectric permittivity of a single-domain crystal at T_c shows a downward jump from 20 to 10 upon heating. The components of the order parameter probably are the displacements of the atoms along the c-axis in the $F(HF_3)$ layers.

Thus, antimony and bismuth halogenides, like sulfides, undergo structural phase transitions which excite various anomalies of physical properties.

4.7. CONCLUSIONS

Antimony and bismuth form a great number of low-dimensional chalcogenide and halogenide compounds containing stereochemically active lone pairs of electrons (E-pairs). Many of these compounds are favorable for various structural phase transitions and show high static dielectric permittivity. This is due to the stereochemical properties associated with the variability of the coordination number around the central atom, bond length variation, ionic-covalent bonding, and distorted and "friable" lattice structures. At the phase transitions, distortions of polyhedra $SbEF_5$ in association with $SbEX_3$ occur due to the intermolecular interactions of Sb...Sb and Sb...X atoms with or without a change of symmetry (X is a chalcogen or halogen). At the phase transitions, crystals containing $SbEX_3$ polyhedra show only the displacement of these groups.

It appears that the presence of phase transitions is a common property of antimony and bismuth chalcogenides and halogenides which contain coordination polyhedra with stereochemically active lone pairs of electrons. The lone pairs can easily change their position giving a possibility of atomic displacements under the influence of external forces (see also Section 5.5). For instance, under the influence of the d.c. or microwave electric field, the stereochemically active lone pairs of electrons can decrease their stereochemical activity up to the inertial state and cause a kind of dielectric-metal phase transition with the S-type of the current versus voltage characteristic, which has been revealed in Sb_2S_3 (Audzijonis et al. 1970) and Cs_2SbF_5 compounds (Grigas et al. 1979).

The ionic-covalent nature and high anisotropic chalcogenide or halogenide ion polarizabilities (as described by the microscopic polarizability model mentioned in Section 4.4.1) can be decisive factors for ferroelectricity in these crystals. High static dielectric permittivity and its anomalies lead to the conclusion that at least some of these crystals are either real ferroelectrics or incipient ones.

5 SEMICONDUCTIVE FERROELECTRICS

5.1. INTRODUCTION

Following the discussion of lattice instability in chalcogenides and halogenides, this chapter considers the semiconductive ferroelectrics which unambiguously show temperature-dependent (i.e., soft) mode ferroelectric instability. Microwave dielectric spectroscopy has played a substantial role in finding out peculiarities of the soft mode behavior in the vicinity of ferroelectric phase transitions (PT) temperature.

The IV–VI semiconductive ferroelectrics are excluded from consideration. These ferroelectrics exhibit the simplest soft mode lattice instability. Thus, the static values of the dielectric permittivity have been determined by microwave techniques. The reader is referred to the books by Jantsch and Bussman-Holder et al. (1983) for details on PT and the soft modes in these semiconductors.

In displacive ferroelectrics, in accordance with the soft mode concept (first introduced by Cochran 1960), the lowest TO mode frequency (see Section 4.4.3) is related to PT temperature T_c by the mean-field relation:

$$\omega_{TO}^2 = \omega_s^2 = A(T - T_c)^{\gamma'}, \qquad (5.1)$$

where $\gamma' = 1$ is called the mean-field critical exponent. Generally, γ' deviates from the value one in the quantum limit of $T = 0$ and above the mean-field regime. However, there are now enough data to show that proper ferroelectrics display a more enigmatic behavior than is suggested by the simplest concept of the soft mode. In particular, the presence of a central peak in the neutron scattering and Raman spectra in displacive crystals signifies that the frequency dispersion of the susceptibility, which corresponds to the order parameter, occurs at much lower frequencies than that of the soft phonon mode.

The soft mode or displacive PT signifies that there is not only the soft mode behavior given by Equation 5.1 (with the Curie–Weiss law obeyed in the paraelectric phase), but also that the temperature dependence of the oscillator strength of the soft mode (Equation 4.13) for proper ferroelectrics is responsible alone for the behavior of the low-frequency clamped dielectric permittivity $\varepsilon(0)$ (i.e., $\Delta\varepsilon_s \approx \varepsilon(0)$, and, therefore, $\Delta\varepsilon_s \times \omega_s^2 \approx$ constant). The LST relation (Equation 4.14) allows examination of this important property.

There are many examples of proper ferroelectrics to show that the soft mode oscillator strength does not explain the magnitude and behavior of $\varepsilon(0)$. On approaching the Curie temperature, T_c, the soft mode—even in ferroelectric perovskites—becomes overdamped and the ratio γ/ω_s (γ is the damping in Equation 4.11) strongly increases with the increase of the $\varepsilon(0)/\Delta\varepsilon_s$ ratio. It shows an anharmonic motion in a local potential—even relaxation—and a crossover from the displacive to order-disorder behavior (Gervais 1984). The soft mode in $BaTiO_3$ becomes temperature independent at temperatures greater than 100 K above T_c. Ti ions are shifted from the centers of the oxygen octahedra along direction [111] and linearly correlated along [100] (Itoh et al. 1985). This crossover can be viewed as a formation of TiO_6 dipoles. A relaxation at 10^{10} to 10^{11} Hz gives evidence in favor of dipolar contribution to $\varepsilon(0)$. A progressive change of the dominant regime from a displacive-like to order-disorder-like mechanism on cooling in the paraelectric phase emerges also from computer simulations (Bruce 1978). It is now generally accepted that both microscopic mechanisms mentioned represent only limiting cases, and are not very common in nature. Displacive-like and order-disorder-like behaviors are, in fact, regimes dominant at different temperatures.

Phase transitions in $KTa_{1-x}Nb_xO_3$ (KTN) system, as well as in ATN and LTN, are also neither simply of displacive, nor of order-disorder type. Raman studies (Bouziane et al. 1992; DiAntonio et al. 1994) and dielectric studies in the frequency range 10^6 to 10^{12} Hz (Volkov et al. 1995) have revealed a phonon softening, as well as the relaxator, a precursor of order in the form of polar nanoregions. These regions are formed by off-center Nb or Li ions in the highly polarizable lattice and cause an intense central peak in Raman spectra.

It is instructive to deduce the response of the dynamical susceptibility for a simple double-well local potential. Neither the mean-field theory nor the correlated effective-field approximation is restricted to the response of damped-harmonic form. Long ago, Onodera (1970) gave a unified mean-field treatment of the dynamic dielectric properties of both displacive and order-disorder phase transitions. The potential energy of the soft mode was assumed to have a simple anharmonic form:

$$V(q_s) = Aq_s^2 + Bq_s^4, \tag{5.2}$$

where q_s is the soft mode normal coordinate. Exact classical calculation of the dynamic susceptibility for mean-field, ferroelectrically interacting oscillators has shown that when $A > 0$, the case corresponds to displacive ferroelectrics, and the collective oscillation has a frequency given by Equation 5.1. When $A < 0$ (double-minimum potential), the dynamic response depends upon the temperature. If the well depth between the minima is small ($\Delta U \ll kT_c$), the system again behaves very much as if it were of the displacive type, with the response of a damped oscillator and a soft mode frequency given by Equation 5.1 with $T'_c < T_c$. For cases where kT_c is less than but in the order of the well depth, the phase transition looks like a transient between the displacive and order-disorder types. The dielectric permittivity consists of two resonant modes which, roughly speaking, correspond to oscillating motion in one potential minimum for the low excited states and, in the whole double well, for the highly excited states. This theory predicts the nonexistence of soft phonon modes on increasing $\Delta U/kT_c$ to about 5 where the phase transition passes into order-disorder type. In that case, the critical slowing down of the relaxational process is responsible for the Curie–Weiss behavior of the static permittivity. The soft mode concept has been extended to such relaxational systems (Blinc and Žekš 1979) where, in a mean-field approximation, the soft relaxational mode frequency is given by

$$\omega_s = 2\pi/\tau = A(T - T_c). \tag{5.3}$$

Recent studies have shown that the double-well structures in ferroelectrics dominate. As a result, the concept of a soft mode driven PT has altered. Instead of thinking of the high temperature phase as becoming unstable due to a soft mode, the more correct interpretation would be that the low temperature phase becomes unstable when enough energy is available to traverse the barrier ΔU between energy minima. The soft mode response, when it exists, is a symptom of the presence of the double-well potential.

The soft ferroelectric mode behavior near T_c can be examined directly by microwave dielectric spectroscopy. In the following sections, several peculiar examples of strongly anharmonic, primarily displacive semiconductive ferroelectrics are considered. These include incommensurate crystals for which, on approaching the Curie temperature, the frequency of the overdamped soft mode drops into the microwave region, becomes strongly overdamped, and shows the crossover from the displacive to order-disorder behavior.

5.2. SbSI-TYPE CRYSTALS

5.2.1. Dielectric Dispersion and Soft Mode

From the semiconductive $A^V B^{VI} C^{VII}$ compounds (Section 4.2.3), SbSI and SbSBr exhibit a number of outstanding strongly coupled semiconductive and ferroelectric properties (Fridkin 1980). Other compounds are not ferroelectrics. Since the discovery of the first-order ferroelectric PT in SbSI at $T_c = 295\,K$ by Fatuzo et al. (1962), the unusual lattice dynamics of this material have attracted much attention. X-ray studies by Kikuchi et al. (1967) showed that the PT is connected with shifts along the c-axis direction of Sb and S atoms relative to I atoms by an amount of 0.20 and 0.05 Å, respectively. This causes the symmetry change D_{2h}^{16} $(Pnam) \rightarrow C_{2v}^9$ $(Pna2_1)$, the spontaneous polarization $P_s = 25\ \mu C/cm^2$ along the c-axis, and an anomaly of the static dielectric permittivity $\varepsilon(0)$ following the Curie–Weiss law with the maximum value of $\varepsilon(0) = 5 \times 10^4$. The phase transition is attributed to a displacive type with a Cochran-type zone-center soft mode.

The group analysis of the long wavelength optical phonons in the SbSI structure (Petzelt 1969) revealed 2 and 8 active modes for $E \| c$ polarization in the paraelectric and ferroelectric phase, respectively (B_{1u} and A_1 modes). Similarly, it revealed 10 and 16 modes for $E \| (001)$ polarization ($B_{2u} + B_{3u}$ and $B_1 + B_2$ modes).

For the soft B_{1u} mode eigenvector, Sb and S ions vibrate against I ions (Riede and Sobota 1978).

Soft mode behavior was actually observed in early optical experiments (Petzelt 1969; Agrawal and Perry 1971; Sugawara and Nakamura 1972). However, further analysis of the low-frequency part of Raman and infrared (IR) reflectivity spectra has shown that the order parameter behaves in a more complex manner than is suggested by the simplest concept of a soft mode. In particular, Raman spectra in the ferroelectric phase reveal that, on approaching the PT, the soft mode frequency varies according to $\omega_s = 8.7(T_c - T)^{0.33}$ cm^{-1} and crosses one or even two (Teng et al. 1972) other optical modes of the same symmetry interacting with them strongly. Steigmeier et al. (1975) reported that close to T_c the central mode acts as a strongly overdamped oscillator of the frequency 3–4 cm^{-1}. This has been attributed to the phonon density fluctuations. Neutron scattering data (Pouget et al. 1979) present evidence that the simple soft phonon picture is inadequate to describe the phase transition in SbSI.

In the paraelectric phase, only IR and dielectric spectra provide information on the behavior of a soft mode and its contribution to the static permittivity $\varepsilon(0)$. IR spectra show that, on approaching T_c, the soft mode frequency varies according to Equation 5.1 and stabilizes near 10 cm^{-1} (Koutsoudakis et al. 1976). The contribution of the soft mode to the static permittivity (Equation 4.13) obtained from the IR spectra varies from 500 (Petzelt 1969; 1973) to 5,000 (Bartzokas and Siapkas 1980), and is less temperature dependent than the Curie–Weiss law requires. The contribution to $\varepsilon(0)$ does not account at all for the high value of permittivity at microwaves, which is about 30,000 (Hosoya and Nakamura 1970). It was concluded (Irie 1978) that the LST relation is not obeyed for lattice vibrations of the SbSI crystal. The above results show that the dispersion of the permittivity corresponding to the order parameter occurs at lower frequencies than 10 cm^{-1} = 300 GHz (i.e., occurs at microwaves).

Direct dielectric investigations of the needle-shaped single crystals in the frequency range of 10^3 to 10^{11} Hz (Grigas and Beliackas 1978) have shown that the value of permittivity in different vapor-phase grown crystals varies in a wide range. The main dielectric dispersion in all of them occurs at microwaves. In some crystals, this dispersion is similar to a resonant and can be described by the dispersion relation for a damped oscillator (Equation 4.12). In other crystals, this dispersion is more relaxational, with the microwave dielectric dispersion contributing more than 85 percent of $\varepsilon(0)$.

The soft mode parameters deduced from IR spectra are also strongly sample-dependent; the soft mode frequency near T_c is given from 20 cm^{-1} (Petzelt 1973) to 4 cm^{-1} (Sugawara and Nakamura 1972). This is possibly due to the variable quality of the samples, influence of the defective surface layer (see Section 5.3), and use of the $\varepsilon(0)$ value instead of ε at microwaves when the reflection spectra are analyzed by means of Kramers–Kronig relations (Equations 4.17 and 4.18).

Long discussion on the mechanism of the phase transition in SbSI (see e.g., Petzelt et al. 1987) stimulated reexamination of these crystals

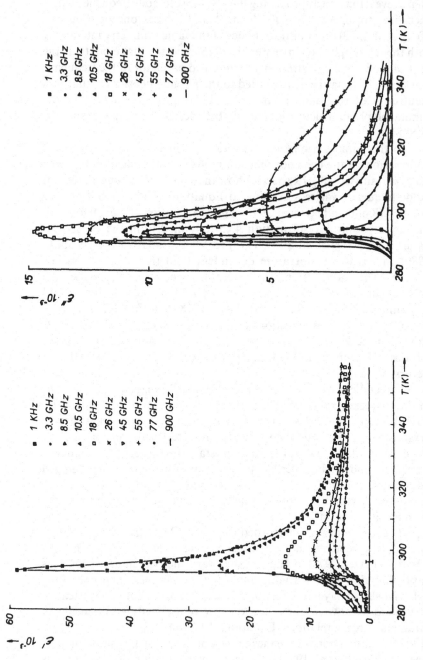

Figure 5.1. Temperature dependence of ε' and ε'' of SbSI at various frequencies. From Grigas (1990). The curve at 900 GHz is from Petzelt (1969).

(Grigas 1990). Figure 5.1 shows the latest results which confirm that the main dielectric dispersion in the paraelectric phase near T_c occurs at microwaves. At $v > 50\,\text{GHz}$, the permittivity depends slightly on temperature. The dielectric dispersion in the 10^7 to $10^{12}\,\text{Hz}$ region can be described by Equation 5.4, allowing for the Curie–Weiss law:

$$\varepsilon^*(v, T) = \varepsilon_\infty + \frac{C}{T - T_0} \cdot \frac{v_s^2(T)}{v_s^2(T) - v^2 + iv\gamma(T)}, \qquad (5.4)$$

with the one overdamped soft mode of the frequency $v_s \approx 75\,\text{GHz}$, the damping constant $\gamma = 265\,\text{GHz}$ at $T_c = 296\,\text{K}$, and $v_s = 109\,\text{GHz}$ at $315\,\text{K}$. In this case, however, as has been shown in Chapter 2, v_s and γ lose the original meanings separately, and the damped harmonic oscillator becomes difficult to interpret physically.

The main question whether the dielectric dispersion in microwave and IR ranges is caused by the same soft B_{1u} mode or by two coupled excitations triggering the phase transition required detailed microwave investigations, especially in the short millimeter wave range. The answer to the problem is based on the investigation of the dielectric dispersion in various SbSI crystals as well as in solid solutions (Table 5.1). By replacing Sb $\rightarrow$ Bi, S $\rightarrow$ Se, and I $\rightarrow$ Br, the phase transition temperature and a maximum value of the permittivity decrease (this favors microwave investigations), but the replacement I $\rightarrow$ Cl increase the T_c. In such a way, the modified SbSI with $T_c = 334\,\text{K}$ was grown from the melt.

In the SbSI family, only $\text{SbSI}_{1-x}\text{Br}_x$ mixed crystals are ferroelectric throughout the full range of compositions. Their Curie temperature decreases from 295 to 22.8 K (Inushima et al. 1981).

Table 5.1. Parameters of SbSI-type crystals

Compound	$T_c(\text{K})$	$C(\text{K})$	$T_0(\text{K})$	$\varepsilon'_{\text{max}}(v = 9\,\text{GHz})$
SbSI	295	2.3×10^5	287	35000
$\text{Sb}_{0.95}\text{B}_{0.05}\text{SI}$	254	1.7×10^5	230	12200
$\text{Sb}_{0.91}\text{Bi}_{0.09}\text{SI}$	210	2.1×10^5	207	12000
$\text{Sb}_{0.85}\text{Bi}_{0.15}\text{SI}$	150	2×10^5	143	10000
$\text{SbS}_{0.7}\text{Se}_{0.3}\text{I}$	194	1.4×10^5	189	17200
$\text{SbS}_{0.5}\text{Se}_{0.5}\text{I}$	122	0.9×10^5	100	3300
$\text{SbSI}_{0.91}\text{Br}_{0.09}$	304	3.7×10^5	290	22500
$\text{SbSI}_{0.77}\text{Br}_{0.23}$	270	1.7×10^5	263	15000
$\text{SbSI}_{0.51}\text{Br}_{0.49}$	188	0.9×10^5	167	6000
SbSI (modified)	334	2×10^5		10000

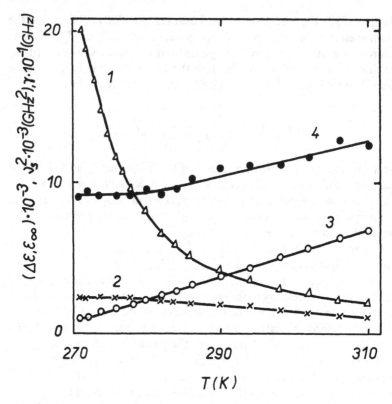

Figure 5.2. Temperature dependences of the parameters of a soft mode of $SbSI_{0.77}Br_{0.23}$ crystal in the paraelectric phase: 1) $\Delta\varepsilon$; 2) ε_∞; 3) v_s; 4) γ. From Kalesinskas et al. (1983a).

In the paraelectric phase of these crystals, the investigated mode is soft (Figure 5.2), and its frequency varies in accordance with Equation 5.1. This mode makes the main contribution to the static permittivity $\varepsilon(0) = \Delta\varepsilon + \varepsilon_\infty$ near T_c. The value of ε_∞ depends much less on temperature than one would expect on the basis of the Curie–Weiss law. The soft mode is strongly overdamped. At $T \to T_c$, the ratio γ/v_s increases as expected on the basis of the soft mode concept. In the case of $SbSI_{0.77}Br_{0.23}$, at $T = T_c$, the ratio in question is $\gamma/v_s = 2.9$. At $T - T_c = 45\,K$, it is $\gamma/v_s = 1.43$.

Figure 5.3 shows the temperature and frequency dependences of ε' and ε'' in the paraelectric phase of $SbSI_{0.77}Br_{0.23}$ crystals. They are deduced from the dispersive parameters of the model oscillator plotted in Figure 5.2. The IR spectra analysis of this composition shows that the mode softens only up to $16\,cm^{-1}$ (Bartzokas and Siapkas 1980).

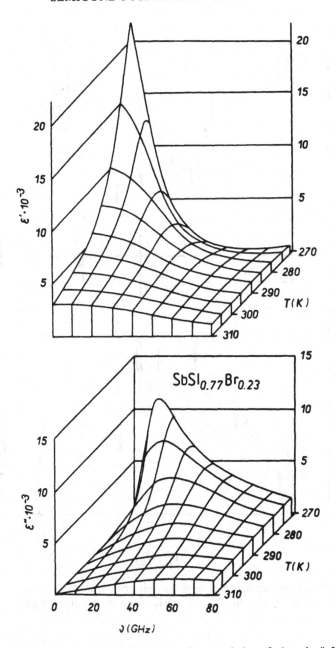

Figure 5.3. Temperature and frequency characteristics of ε' and ε'' for the paraelectric phase. From Kalesinskas et al. (1983a).

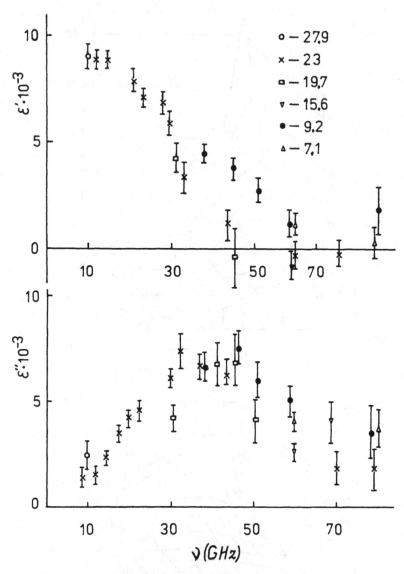

Figure 5.4. Frequency dependence of the dielectric permittivity of $SbS_{0.7}Se_{0.3}I$ specimens of different radii at $T - T_c = 9\,K$. The radii are given in μm.

The similar dielectric dispersion at microwaves has been found in mixed $SbSI_{1-x}Br_x$ crystals and in the other composition crystals (Kalesinskas et al. 1983a), $Sb_{1-x}Bi_xSI$ (Kalesinskas et al. 1983b) and $Sb_{0.7}Se_{0.3}I$ (Kalesinskas et al. 1982).

Figure 5.4 shows microwave dielectric dispersion for $SbS_{0.7}Se_{0.3}I$ specimens of different radii in the paraelectric phase. One can see that the dispersion frequency within the limits of the measurement accuracy does not depend considerably on the radius of the sample (i.e., it is not caused by the electrodynamic resonance (see Equation 3.71)).

The dielectric permittivity of the nonferroelectric $A^V B^{VI} C^{VII}$ crystals at microwaves (Figure 5.5) is much smaller and less temperature dependent than of that mentioned above. It is also almost frequency independent. Microwave measurements (Kalesinskas et al. 1981) do not confirm the existence of a soft mode and ferroelectric phase transition at $T_c = 113\,K$ in BiSI as concluded by Siapkas (1974) from IR measurements.

Figure 5.6 shows the temperature dependence of the soft mode parameters of modified SbSI crystals with the Curie temperature $T_c = 334\,K$ (Paprotny et al. 1984). In the vicinity of the Curie temperature, the frequency of the soft mode varies according to Equation 5.1 with $A = 9\,GHz/K^{1/2}$ and explains the temperature dependence of $\varepsilon(0)$. The mode is strongly overdamped. The contribution of the soft mode to $\varepsilon(0)$ near T_c is $\Delta\varepsilon = 26{,}000$ and considerably exceeds the contribution of IR modes and electronic polarization for SbSI.

Figure 5.7 presents the frequency dependences of ε'' and ε' for various temperatures which shows beyond doubt the presence of the main dielectric dispersion at microwaves. In the paraelectric phase, assuming there are no nuclei of the new phase due to impurities, defects and such, the microwave dispersion is not due to phonon density fluctuations as has been supposed by Steigmeier et al. (1975). In the paraelectric phase, unlike in the ferroelectric phase, the oscillations of the polarization do not cause oscillations of temperature. The defect content of various SbSI-type crystals varies considerably, but the dielectric dispersion at microwaves is observed in all of them and is the main contribution to $\varepsilon(0)$. The greatest dispersion is found in the thinnest whiskers, which are regarded as being the most perfect. Therefore, it seems unlikely that the microwave dispersion in SbSI-type crystals is due to defects.

Due to its large dielectric permittivity, SbSI is a nonlinear material at microwaves, since the microwave nonlinear coefficient, d_m, is $d_m \sim \varepsilon^3$. Large nonlinearity in the paraelectric phase can be used for electrically controlled switches, filters and phase shifters in microstrip lines

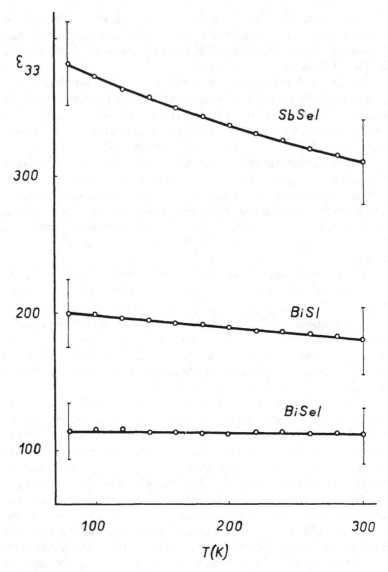

Figure 5.5. Dielectric permittivity of nonferroelectric $A^VB^{VI}C^{VII}$ crystals at 9 GHz. From Kalesinkas et al. (1981).

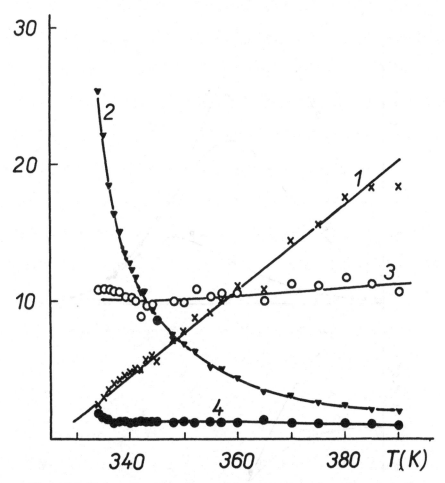

Figure 5.6. Temperature dependence of the parameters of a soft mode in the paraelectric phase of modified SbSI: 1) $v_s^2 \times 4 \times 10^{-2}$ (GHz^2); 2) $\Delta\varepsilon \times 10^{-3}$; 3) γ (GHz); 4) $\varepsilon_\infty \times 10^{-3}$. From Paprotny et al. (1984b).

(Beliackas et al. 1975), as well as for frequency mixing or harmonic generation applications (Meškauskas and Grigas 1976). However, the absorption loss limits the interaction length in practical devices, and the elements require thermal stabilization.

The soft mode of unusually low frequency in SbSI-type crystals is a manifestation of a large anharmonicity of the lattice. This points to an intermediate between the displacive and order-disorder types phase transition. In this case, as Onodera (1970) has shown, the dielectric dispersion may behave as if it has two soft modes (see Section 5.1).

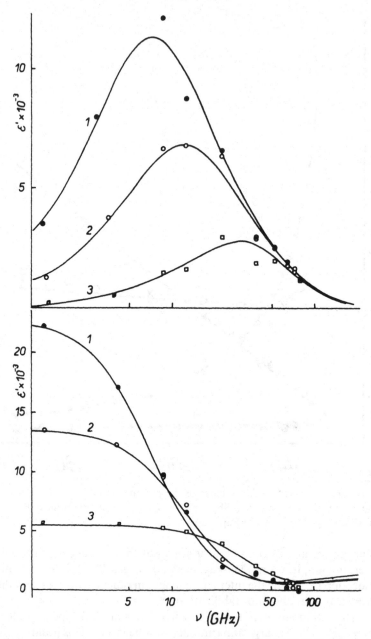

Figure 5.7. Frequency dependence of ε' and ε'' of the modified SbSI at temperatures (K): 1) 335; 2) 340; 3) 360. Curves are the oscillator fits with the parameters given in Figure 5.6. Points are experimental. From Paprotny et al. (1984b).

Even in a case of a slightly anharmonic crystal, the dielectric dispersion may be quite complex. It is possible that, when necessary allowance is made for the interaction of different branches of vibrations, the dispersion $\varepsilon^*(v)$ is close to what is observed, especially since SbSI shows strong electrostrictive interaction (Samara 1975).

Crystal structure analysis of SbSI (Itoh et al. 1980) shows that Sb ions have thermal motion along the c-axis of anomalously large amplitude. This suggests that the potential is rather flat or even has the form of a small double-well. In the next Section, a model of the phase transition is proposed (Stasyuk et al. 1983; Grigas et al. 1984a) taking into account the large lattice anharmonicity.

5.2.2. Model of Phase Transition

The phonon Hamiltonian of the crystal can be written as

$$H = \sum_{mf} H_{mf}^{Sb} - \frac{1}{2} \sum_{mm'} \sum_{ff'} V_{ff'}^{\alpha\beta}(mm') u_{mf}^{\alpha} u_{m'f'}^{\beta}$$

$$- \frac{1}{N^{1/2}} \sum_{\substack{qj \\ mf\beta}} e^{iqR_{mf}} \tau_{jf}^{\beta}(q)(b_{qj}^- + b_{-qj}^+) u_{mf}^{\beta} + \sum_{qj} \hbar\omega_j(q) b_{qj}^+ b_{qj}^-, \quad (5.5)$$

where H_{mf}^{Sb} is the Hamiltonian describing the motion of the Sb ion in the potential well produced by the nearest S and I ions, u_{mf} are the displacements of Sb ions, and b_{qj}^- and b_{qj}^+ are Bose-operators of the "bare" harmonic phonons introduced on the reduced set of the normal coordinates, from which the local normal coordinates of the relative motion of Sb, S and I ions are excluded (m is the cell number, and f is the ion index). The second term in Equation 5.5 describes the interaction between Sb ions, while the third one describes the interaction with the remaining ions of the lattice arising at their relative displacements. At small values of the wave vector q

$$\tau_{j_{ac}}^{\beta}(q) \sim q^{1/2}; \quad \tau_{j_{opt}}^{\beta}(q) \sim \text{constant}. \quad (5.6)$$

The spectrum of the frequency of the lattice vibrations can be determined by means of the two-time Green functions. Only the random phase approximation (RPA), which gives the possibility of finding the frequency of vibrations (the damping can be accounted for phenomenologically) will be considered. Calculations will be confined to the paraelectric phase.

In RPA, the use of the Green function

$$\langle\langle u^{\gamma}_{m_1 f_1} | u^{\gamma'}_{m_2 f_2} \rangle\rangle$$

leads to

$$\langle\langle u^{\gamma}_{f_1} | u^{\gamma'}_{f_2} \rangle\rangle_{q,\omega} = \frac{\hbar}{2\pi} \delta_{f_1,f_2} Z^{\gamma\gamma'}_{f_1}(\omega) - \sum_{\alpha f' \beta} Z^{\gamma\alpha}_{f_1} [V^{\alpha\beta}_{f_1 f'}(q)$$

$$- \chi^{\alpha\beta}_{f_1 f'}(\omega_1 q)] \langle\langle u^{\beta}_{f'} | u^{\gamma'}_{f_2} \rangle\rangle_{q_1\omega}, \qquad (5.7)$$

where

$$Z^{\gamma\gamma'}_{f_1}(\omega) = \frac{2\pi}{\hbar} \langle\langle u^{\gamma}_{f_1} | u^{\gamma'}_{f_1} \rangle\rangle^0_\omega$$

is the single-ion Green function of the displacements

$$X^{\alpha\beta}_{f_1 f}(\omega, q) = \sum_j \tau^{\alpha}_{jf_1}(q) \frac{2\hbar\omega_j(q)}{\hbar^2\omega^2 - \hbar^2\omega_j^2(q)} \tau^{\beta}_{jf}(-q). \qquad (5.8)$$

The solution to Equation 5.7 in matrix form is

$$\hat{G} = \frac{1}{\hbar} \langle\langle u|u \rangle\rangle = \frac{1}{2\pi} [\hat{Z}^{-1} + \hat{V} - \hat{\chi}]^{-1}. \qquad (5.9)$$

Therefore, the frequency spectrum of the phonon vibrations is determined from the equation

$$\text{Det} \| \hat{Z}^{-1} + \hat{V} - \hat{\chi} \| = 0.$$

The case of the long wave vibrations ($q \approx 0$) will be considered in greater detail. The explicit form of the interaction matrices can be established by proceeding from the transformation properties of the ($f\alpha$) ($f = 1\ldots 4, \alpha = x, y, z$) basis at the transformation from the point group of the wave vector, and also by using the expressions for the $\tau^{\beta}_{jf}(q)$ matrices (Stasyuk and Kaminskaya 1974). After the transition to the symmetrized basis, transformed according to the irreducible representations of the point group D_{2h}, the condition of equality to zero of the determinant in the long wave limit leads to the equation

$$Z^{-1}(\omega) + V_{ii} + \sum_{j(i)} \frac{L_{j(i)}}{\hbar^2\omega^2 - \hbar^2\omega_{j(i)}^2(q)} = 0, \qquad (5.10)$$

where $\omega_{j(i)}$ are the frequencies of the "bare" optical vibrations of the symmetry "i" ($i = B_{1u}, B_{3g}, A_u, B_{2g}$); $Z(\omega) = Z_f^{zz}(\omega)$. If the Sb ion were in the harmonic potential well, the function $Z(\omega)$ would have one pole at $\omega^2 = \omega_0^2$, where ω_0 is the frequency of the corresponding Einstein oscillator. In this case, Equation 5.10 would determine the frequencies of the two possible optical vibrations of B_{1u}-type. If the single-ionic potential for Sb is anharmonic, the spectrum (determined by the $Z(\omega)$ function) will be more complex. The anharmonicity can cause the splitting of the frequency ω_0.

A model potential describing the motion of the Sb ion along the c-axis with respect to the surrounding S and I ions is

$$W(\omega) = \frac{1}{2}m\omega_0^2 u^2 + Ve^{-\tilde{\beta}u^2}. \tag{5.11}$$

The anharmonic addend, which is significant at small displacements u, is included alongside the usual harmonic term. The bottom of the potential well is almost flat at

$$\tilde{\beta}V \sim \frac{m\omega_0^2}{2},$$

and there are two minima divided by the barrier at

$$\tilde{\beta}V > \frac{m\omega_0^2}{2}.$$

One of the reasons for such a potential in SbSI might be the strong electron-phonon interaction.

The single-ionic Hamiltonian with the use of the dimensionless variable $\zeta = u(m\omega_0/\hbar)^{1/2}$ can be written as

$$H^{Sb} = \frac{\hbar\omega_0}{2}\left[\zeta^2 - \frac{d^2}{d\zeta^2}\right] + Ve^{-2\tau\zeta^2}, \tag{5.12}$$

where

$$\tau = \frac{\tilde{\beta}\hbar}{2m\omega_0}.$$

The calculated Green function of displacements are

$$Z(\omega) = \frac{2\pi}{m\omega_0}\langle\langle\zeta|\zeta\rangle\rangle_\omega = \frac{2\pi\hbar}{m\omega_0}g(\omega). \tag{5.13}$$

Using the equations of motion for the operators ζ and $d/d\zeta$, it follows

$$\hbar^2(\omega^2 - \omega_0^2)\langle\langle\zeta|\zeta\rangle\rangle_\omega = \frac{\hbar}{2\pi}\hbar\omega_0 - 4\tau V\hbar\omega_0\langle\langle\zeta e^{-2\tau\xi^2}\xi^2|\zeta\rangle\rangle_\omega. \quad (5.14)$$

The function

$$\langle\langle\zeta e^{-2\tau\zeta^2}|\zeta\rangle\rangle_\omega$$

can be expressed through the irreducible Green functions:

$$\langle\langle\zeta e^{-2\tau\zeta^2}|\zeta\rangle\rangle_\omega = \sum_{m=1}^{\infty} K_m^{ir}\langle\langle\zeta^m|\zeta\rangle\rangle_\omega, \quad (5.15)$$

$$K_m = \left\langle \frac{1}{m!}\frac{d^m}{d\zeta^m}\zeta e^{-2\tau\zeta^2}\right\rangle. \quad (5.16)$$

The most simple approximation is to neglect the irreducible functions with $m > 2$. This case leads to the so-called self-consistent Einstein phonons. Then

$$g_0(\omega) = \frac{\hbar\omega_0}{2\pi}(\hbar^2\omega^2 - \hbar^2\tilde{\omega}_0^2)^{-1}, \quad (5.17)$$

where

$$\tilde{\omega}_0 = \left[\omega_0^2 - \frac{4\tau V\omega_0}{\hbar}K_1\right]^{1/2}. \quad (5.18)$$

In this approximation

$$K_1 \equiv \langle e^{-2\tau\zeta^2}(1 - 4\tau\zeta^2)\rangle = \left[1 + 2\tau\frac{\omega_0}{\tilde{\omega}_0}(1 + 2\bar{n})\right]^{-3/2}, \quad (5.19)$$

where

$$\bar{n} = [e^{\beta\tilde{\omega}_0} - 1]^{-1}.$$

The frequency, $\tilde{\omega}_0$, is determined in a self-consistent way from Equations 5.18 and 5.19. At high temperatures, it approaches the frequency ω_0 of the harmonic vibrations, while at low temperatures, it decreases, and in the case of the two-minima potential ($4\tau V > \hbar\omega_0$), the local normal vibration of the Sb ion with respect to S and I ions becomes unstable.

The equation for the irreducible Green function $^{ir}\langle\langle\zeta^m|\zeta\rangle\rangle_\omega$ can be obtained by differentiation over the second time argument:

$$\hbar^2(\omega^2 - \omega_0^2)^{ir}\langle\langle\zeta^m|\zeta\rangle\rangle_\omega = -4\tau V\hbar\omega_0 \,^{ir}\langle\langle\zeta^m|\zeta e^{-2\tau\zeta^2}\rangle\rangle_\omega. \quad (5.20)$$

Using a procedure analogous to Equation 5.15 and substituting the function $^{ir}\langle\langle\zeta^m|\zeta\rangle\rangle_\omega$ from Equation 5.14 into Equation 5.15 gives

$$g(\omega) = g_0(\omega) + g_0(\omega)\Pi(\omega)g(\omega), \quad (5.21)$$

where

$$\Pi(\omega) = (2\pi)^2 \cdot 16\tau^2 V^2 \sum_{m,p=2}^\infty K_m K_p \frac{1}{\hbar} \,^{ir}\langle\langle\zeta^m|\zeta^p\rangle\rangle_c^{ir}. \quad (5.22)$$

(Here, only connected parts of Green functions are considered). The approximation corresponding to the second order of the self-consistent perturbation theory gives:

$$\Pi(\omega) = 2\pi \frac{16\tau^2 V^2}{\hbar}\left(\frac{\tau'}{\tau}\right)^3 \int_{-\infty}^{+\infty} \frac{d\omega'}{\omega - \omega'}(e^{\beta\omega'} - 1)$$

$$\times \int_{-\infty}^{+\infty} \frac{dt}{2\pi}e^{-i\omega't}\langle\zeta(t)\zeta\rangle[(1 - 16\tau'^2\langle\zeta(t)\zeta\rangle^2)^{-3/2} -]. \quad (5.23)$$

Here, the parameter

$$\tau' = \tau\left[1 + 2\tau\frac{\omega_0}{\tilde{\omega}_0}(1 + 2\bar{n})\right]^{-1},$$

characterizing the renormalized anharmonic part of the single-ionic potential, is introduced. At

$$\tau \ll \left[2\frac{\omega_0}{\tilde{\omega}_0}(1 + 2\bar{n})\right]^{-1}, \quad \tau' \approx \tau,$$

and at

$$\tau \gg \left[2\frac{\omega_0}{\tilde{\omega}_0}(1 + 2\bar{n})\right]^{-1}, \quad \tau' \approx \frac{\tilde{\omega}_0}{2\omega_0(1 + 2\bar{n})}.$$

The second case is realized at sufficiently high temperatures when $1 + 2\bar{n} \gg 1$. Then $\tau' \ll 1$, and the effective single-ionic potential cannot be distinguished from the harmonic one. The anharmonicity of the potential reveals itself only at low temperatures.

The inequality $\tau'^2 \bar{n}(1 + \bar{n}) \ll 1$ is valid both at high and low temperatures. Because of this the corresponding expansion in Equation 5.23 can be carried out, and the polarization operator can be presented as

$$\Pi(\omega) = 2\pi \cdot 288 \frac{\tau'^5}{\tau} \frac{\omega_0^3}{\tilde{\omega}_0^3} \hbar \tilde{\omega}_0 V^2 \left[\frac{1 + 3\bar{n}(1 + \bar{n})}{\hbar^2(\omega^2 - 9\tilde{\omega}_0^2)} + \frac{\bar{n}(1 + \bar{n})}{\hbar^2(\omega^2 - \tilde{\omega}_0^2)} \right]. \quad (5.24)$$

For the pair correlator $\langle \dot{\zeta}(t)\zeta \rangle$, the expression obtained in the self-consistent phonons is used:

$$\langle \zeta(t)\zeta \rangle = \frac{\omega_0}{2\tilde{\omega}_0} [e^{i\tilde{\omega}_0 t} \bar{n} + e^{-i\tilde{\omega}_0 t}(1 + \bar{n})]. \quad (5.25)$$

Substitution of Equation 5.25 into Equation 5.21 leads to

$$g(\omega) = \frac{\omega_0}{2\pi h} \sum_i \frac{f_i}{\omega^2 - \omega_i^2},$$

where

$$\omega_1 - 3\tilde{\omega}_0 \left(1 + \frac{8}{9} A \right)^{1/2}, \quad \omega_{2,3} = \tilde{\omega}_0 [1 \pm 4A(B \pm 1)]^{1/2},$$

$$f_1 = A, \quad f_{2,3} = \tfrac{1}{2}(1 - A)(1 \pm 1/B), \quad (5.26)$$

and

$$A = \phi[1 + 3\bar{n}(1 + \bar{n})],$$

$$\phi = \frac{9\tau'^5 V^2 \omega_0^4}{2\tau \hbar^2 \tilde{\omega}_0^6}, \quad B = \left[1 + \frac{4\bar{n}(1 + \bar{n})}{\phi[1 + 3\bar{n}(1 + \bar{n})]^2} \right]^{1/2}. \quad (5.27)$$

Under the influence of the anharmonicity, the threefold splitting of the frequency of the local normal vibration of the Sb ion takes place. These frequencies shift with the change of temperature, and their oscillator strengths change as well.

At high temperatures (i.e., when $kT \gg \hbar\omega_0$ and $\bar{n} \gg 1$), $\tau' \sim 1/\bar{n}$, $\phi \sim (1/\bar{n})^5$, and $B \sim (\bar{n})^{3/2}$. By this means, $f_1 \sim (1/\bar{n})^3$ and decreases with increasing temperature, while f_2 and f_3 approach 1/2. Correspondingly, $\omega_1 \rightarrow 3\tilde{\omega}_0$, and $\omega_{2,3} \rightarrow \tilde{\omega}_0$. At low temperatures ($kT \ll \hbar\omega_0$), it follows that

$$\omega_1 \rightarrow 3\tilde{\omega}_0 \left(1 + \frac{8}{9} \phi \right)^{1/2}, \quad \omega_2 \rightarrow \tilde{\omega}_0, \quad \omega_3 \rightarrow \tilde{\omega}_0(1 - 8\phi)^{1/2},$$

$$f_1 \rightarrow \phi, \quad f_2 \rightarrow 0, \quad f_3 \rightarrow 1 - \phi, \quad (5.28)$$

and the vibration of the frequency ω_3 predominates. With the increase of temperature, the statistical weight of the vibrations ω_1 and ω_2 increases.

The determination of the optical lattice vibrational B_{1u}-type modes can be reexamined using Equation 5.10. Furthermore, taking into account Equations 5.13 and 5.16, the following expression can be obtained:

$$\frac{1}{m}\sum_{i=1}^{3}\frac{f_i}{\omega^2-\omega_i^2}+\left[V_{11}+\frac{L}{\hbar^2\omega^2-\hbar^2\omega_{j(1)}}\right]=0. \quad (5.29)$$

This equation has four solutions that correspond to four different B_{1u} modes. One of them is a high frequency mode, while three others, resulting from the splitting of the frequency ω_0, are of low frequency. Their contribution to the dielectric permittivity $\varepsilon_{33}=1+\chi_{33}$ is different because

$$\chi_{33}=-\frac{2\pi}{V}q_{ef}^2\,G=-\frac{q_{ef}^2}{V}\left[Z^{-1}(\omega)+V_{11}+\frac{L}{\hbar^2(\omega^2-\omega_{j(1)}^2)}\right]^{-1/2}, \quad (5.30)$$

where q_{ef} is the effective charge. At low frequencies, this decomposes to fractions and becomes

$$\varepsilon_{33}=\varepsilon_\infty+\sum_{i=1}^{3}\frac{F_i}{\omega_i^2-\omega^2}. \quad (5.31)$$

In the case of a small statistical weight of the local vibration of the frequency ω_1 (high temperatures or $\phi\ll 1$), the following approximated expressions are obtained for the frequencies and oscillator strengths:

$$\tilde{\omega}_1=3\tilde{\omega}_0\left[1+\frac{8}{9}A-\frac{\tilde{\alpha}_1}{9\tilde{\omega}_0^2}A\right]^{1/2},$$

$$\tilde{\omega}_2=\tilde{\omega}_0\left[1+16A^2B^2\frac{\tilde{\omega}_0^2}{\tilde{\alpha}_1}\right]^{1/2},\quad \tilde{\omega}_3=\tilde{\omega}_0\left[1-8A-\frac{\tilde{\alpha}_1}{\tilde{\omega}_0^2}(1-A)\right]^{1/2},$$

$$(5.32)$$

$$F_1=\frac{q_{ef}^2}{mV}A,\quad F_2=\frac{q_{ef}^2}{mV}(1-A)\,16A^2B^2\frac{\tilde{\omega}_0^4}{\tilde{\alpha}_1^2},$$

$$(5.33)$$

$$F_3=\frac{q_{ef}^2}{mV}(1-A)\left[1-16A^2B^2\frac{\tilde{\omega}_0^4}{\tilde{\alpha}_1^2}\right],$$

$$\tilde{\alpha}_1=\frac{1}{m}\left(V_{11}-\frac{L}{\hbar^2\omega_{j(1)}^2}\right), \quad (5.34)$$

A is a parameter that decreases with the increase of temperature. When $\tilde{\alpha}_1 > 0$, the vibration with the frequency $\tilde{\omega}_3$ is the true soft mode. The given frequency at $T \to T_c$ decreases. This is caused both by the decrease of $\tilde{\omega}_0$ and by the increase of A. The tending of $\tilde{\omega}_3$ to zero corresponds to the temperature instability of the paraelectric phase at T_0.

The frequency $\tilde{\omega}_2$ slightly differs from $\tilde{\omega}_0$ and also decreases with the decrease of temperature, remaining finite at T_c. Its oscillator strength and the contribution to $\varepsilon(0)$ are considerably smaller than these of the ω_3 mode (the ratio of the contributions $(F_2/\tilde{\omega}_2^2)/(F_3/\tilde{\omega}_3^2) = \zeta$ at T_c is approximately 0.1).

The soft mode at microwaves in SbSI-type crystals apparently corresponds to the $\tilde{\omega}_3$ mode. The pseudosoft mode, which appears in the IR spectrum, can be an $\tilde{\omega}_2$ mode. Due to the insignificant oscillator strength, the $\tilde{\omega}_1$ is not seen in the infrared spectrum. The high frequency $\omega_{j(1)}$ mode apparently corresponds to the phonon with the frequency of $180\,cm^{-1}$ observed in IR experiments.

The comparison with the experimental data for SbSI allows some numerical estimations of the parameters predicted by the theory. The ratio $\Delta_0 = \tilde{\omega}_0^2/\omega_0^2$ will be used instead of the frequency of the self-consistent phonons $\tilde{\omega}_0$. At temperatures satisfying the inequality $\theta \gg \hbar\tilde{\omega}_0$, the parameter Δ_0 is defined by the equation

$$\Delta_0 = 1 - \frac{4\tau V}{\hbar\omega_0}\left[1 + \frac{4\tau}{\hbar\omega_0}\frac{\theta}{\Delta_0}\right]^{-3/2}. \tag{5.35}$$

At the temperature T_0 the condition

$$\omega_0^2(1 - 8A_0)\Delta_0^0 - \tilde{\alpha}_1(1 - A_0) = 0 \tag{5.36}$$

is satisfied, and at T_c:

$$\omega_0^2(1 - 8A_c)\Delta_0^c - \tilde{\alpha}_1(1 - A_c) = \tilde{\omega}_{3c}^2,$$

$$\omega_0^2\Delta_0^2 + 16A_c^2B^2\frac{\omega_0^4}{\tilde{\alpha}_1}(\Delta_0^c)^2 = \tilde{\omega}_{2c}^2. \tag{5.37}$$

From Equations 5.33 and 5.37 the relation follows

$$16A_c^2\left(1 + \frac{4}{3A_c}\right)\frac{(1 - A_c)^2}{(1 - 8A_c)^2} \approx \frac{\zeta\tilde{\omega}_{2c}^2}{\tilde{\omega}_{3c}^2 + \zeta\tilde{\omega}_{2c}^2}, \tag{5.38}$$

from which the magnitude of the A_c parameter can be estimated using the experimental values of the frequencies $\tilde{\omega}_2$ and $\tilde{\omega}_3$ and also the ratio ζ of the mode's contribution. The calculated value of $A_c = 0.03$. Equation 5.37 gives

$$\omega_0^2 \Delta_0^c = 141 \times 10^{22} \, s^{-2}, \quad \tilde{\alpha}_1 = 116 \times 10^{22} \, s^{-2}. \tag{5.39}$$

Some more parameters of the theory can be established from Equation 5.27. Using Equation 5.35, the parameter τ' can be expressed as

$$\tau' = \tau a^{-2/3} (1 + \Delta_0)^{2/3}, \tag{5.40}$$

where $a = (4\tau/\hbar\omega_0)V$. Elimination of the quantity ϕ from Equation 5.27, yields

$$A = \frac{27}{512} \left(\frac{4\tau\theta}{\hbar\omega_0} \right)^2 \frac{(1 - \Delta_0)^{10/3}}{a^{4/3} \Delta_0^4}. \tag{5.41}$$

From Equations 5.35, 5.36 and 5.41 the following is obtained:

$$A_0 \approx A_c, \quad \Delta_0^0 = 0.260, \quad \Delta_0^c = 0.262,$$

$$\zeta = 4.1 \times 10^{-4} \, K^{-1}, \quad a^{2/3} = 1.0094.$$

Here $4\tau\langle \zeta^2 \rangle = \zeta\theta/\Delta_0 = 0.45$. The given product enters the parameter K_1, which determines the change of the frequency spectrum of the Einstein phonons when the Gaussian addend in the single-particle potential is available. At $4\tau\langle \zeta^2 \rangle \ll 1$, the situation corresponds to the phase transition of the displacive type when the anharmonicity of the fourth order in the potential is restricted. In this case, the transition in SbSI-type crystals is intermediate between the displacive and order-disorder type transitions.

When $\tilde{\alpha}_1 < 0$, the situation is different. Only the $\tilde{\omega}_2$ mode can be soft. The phase transition takes place when $\tilde{\omega}_0 \to 0$ and is connected with the instability of local modes. Unlike the preceding case, the oscillator strength of the soft mode is small. The Sb_2S_3-crystal may be an example of such a situation. The low frequency mode in this crystal makes a small contribution to $\varepsilon(0)$ (Table 4.5).

Confining the discussion to the $\tilde{\omega}_3$ mode and formally taking into account its damping, Equation 5.31 can be written in form coinciding with Equation 5.4:

$$\varepsilon_{33} = \varepsilon_\infty + \frac{1}{\tilde{\omega}_3^2} \cdot \frac{F_3 \tilde{\omega}_3^2}{\tilde{\omega}_3^2 - \omega^2 + i\gamma\omega} = \varepsilon_\infty + \frac{C}{T - T_0} \cdot \frac{\tilde{\omega}_3^2}{\tilde{\omega}_3^2 - \omega^2 + i\gamma\omega}. \tag{5.42}$$

Near T_c, $\tilde{\omega}_3 \cong (F_3/C)(T - T_0)$. Large values of the constant C (Table 5.1) indicate a rather slow increase of the $\tilde{\omega}_3$ frequency moving away from the temperature T_c.

In microwave and far IR spectra, the frequencies ω_3 and ω_2 can reveal themselves separately or show one large absorption peak in the frequency range of 10^{10} to 10^{12} Hz accompanied by deep dielectric dispersion (Grigas 1990). It looks like the resonant mode in the IR range is coupled to the relaxational mode at microwaves.

Thus, the considered model explains the observed splitting of the low frequency B_{1u} mode and the appearance of the "additional" excitation at microwaves, which do not enter the set of the harmonic phonon modes and make the main contribution to the static dielectric permittivity. Unlike the transition of the displacive type in which the contribution from the anharmonicities to the frequency of the soft mode increases with the increase of temperature, in the present case, the anharmonicity is "low temperature" (i.e., the change of the frequency spectrum caused by it occurs with the decrease of temperature). In the case of strong anharmonicity, the phase transition in the model considered is near the transition of the order-disorder type. In the case of mean anharmonicity, it is of intermediate type.

Calculation of the dependence of electronic potential of the B_{1u} mode in both para- and ferroelectric phases at Sb, S and I sites (Kvedaravičius et al. 1993) shows that in the paraelectric phase in the vicinity of T_c the electronic potential at Sb site becomes of the form of a small double well, while at S and I sites it has the form of a single well in both phases.

Flocken et al. (1992) have also shown that the experimental dielectric response in SbSI can be reproduced quite closely by a model of a damped double well oscillator. They used a single Hamiltonian of the form

$$H = \frac{1}{2}\frac{\lambda^2 \partial^2}{\partial x^2} - Ax^2 + Bx^4 - \chi\langle x \rangle x,$$

where $\lambda = \hbar/(m)^{1/2}$, M is the oscillator mass, $A = 2E_b/x_0^2$ and $B = E_b/x_0^4$, E_b is the height of the energy barrier of the well, x_0 is the position of the energy minima. The constant χ provides a linear coupling between a given oscillator and all others in the lattice, which have an average position $\langle x \rangle$, and is regarded as the order parameter. Above T_c, the order parameter is zero and the well is symmetric. The expression for the dielectric response is

$$\varepsilon = 1 + q_{ef}^2/\varepsilon_0 V\{\pi(\omega, T)/(1 - \chi(0)\pi(\omega, T))\},$$

where $\chi_0(0)$ is the response function of the oscillator at $q = 0$, and $\pi(\omega, T)$ is a sum over states given by

$$\pi(\omega, T) = \sum_{\alpha\beta} \{(f_\alpha - f_\beta) | \langle \phi_\alpha | x | \phi_\beta \rangle |^2 \rho(\omega_{\alpha\beta})\}/(\omega - \omega_\alpha - \omega_\beta - i\gamma\omega),$$

in which ω_α and ϕ_α are the eigenfrequency and eigenvector of the αth state, respectively. f_α is the thermal weighting factor of the αth state, calculated using

$$f_\alpha = \exp(-h\omega_\alpha/kT) \bigg/ \sum_\beta \exp(-h\omega_\beta/kT).$$

The factor ρ is the relaxation factor, given by $\rho = 1 + i/\omega_{\alpha\beta}\tau$, where $\omega_{\alpha\beta}$ is the transition frequency between the αth and βth state, and τ is the relaxation factor. The relaxation term accounts for the damping effect of other phonons in the lattice, with τ being a relaxation time associated with these interactions near T_c.

By using an oscillator mass of 100 a.u. and the effective charge of three electronic charges (i.e., close to the mass and the ionic charge of Sb), and $E_b = 100$ K, 50 K, and 150 K, Flocken et al. (1992) described remarkably well dielectric dispersion of SbSI (Grigas and Beliackas 1978) in microwave and submillimeter wave region. It appears that this simple double-well model reproduces the general features of dielectric function using a variety of physically realistic well depths and widths.

5.2.3. Influence of Microwave Radiation on Phase Transition

Microwave radiation exerts a specific influence on the phase transition in SbSI, which is not related to the usual dielectric nonlinearity. The reflection and transmission coefficients depend nonlinearly on the incident average microwave power P_{av}. When P_{av} is increased until it reaches a threshold value P_{th}, which depends on the ambient temperature T_a [$P_{th} \sim (T_c - T_a)$], a crystal switches abruptly from the ferroelectric to the paraelectric phase. There is also a corresponding change of reflection or transmission coefficients (Figure 5.8). Such a crystal in a waveguide can function as a power limiter (Mizeris et al. 1985) or relay-type microwave power transducer. The switching time is in the order of 10 ms (Grigas et al. 1982). The threshold power could also be altered by a static electric field, by optical excitation, etc. In the vicinity of T_c, a crystal switches to the paraelectric phase in very weak microwave fields ($E_m < 1$ V/cm). Consequently, ε' is dependent on E_m. In the region of

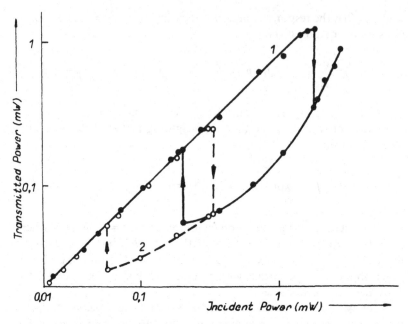

Figure 5.8. Transmitted vs. incident power for SbSI crystal ($r = 61\,\mu$m) at 9 GHz: $T - T_c$(K): 1) 1.67; 2) 0.26. From Mizeris et al. (1985).

dielectric dispersion, when ε'' increases strongly, the crystal is heated even when the microwave power is low. As a result, the temperature dependence of ε' for the same needle-like crystal with an effective radius $r = 27\,\mu$m at two levels of the microwave power is different (Figure 5.9). When the power is low, the crystal is not heated and curve 1 represents the usually observed temperature dependence of ε' in weak microwave fields. Curve 2 indicates a characteristic dependence $\varepsilon'(T)$ in stronger microwave fields.

Under the action of a continuous microwave field, the heat evolved in the crystal in one second is

$$Q_1 = 2\pi\varepsilon''\varepsilon_0 E_m^2,$$

It increases the crystal temperature T_e. The heat transferred to the ambient medium of temperature T_a is

$$Q_2 = A(T_e - T_a),$$

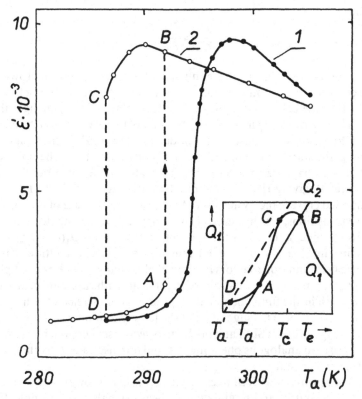

Figure 5.9. Temperature dependence of permittivity of SbSI crystal for micro-wave powers of 3 μw (curve 1) and 30 mW (curve 2). The insert explains curve 2. From Grigas et al. (1982).

where A is a constant proportional to the rate of heat exchange between the crystal and the ambient medium. A steady state is established when

$$Q_1 = Q_2 \quad \text{and} \quad \frac{\partial Q_1}{\partial T_e} < \frac{\partial Q_2}{\partial T_e}.$$

Curve 2 in Figure 5.9 should be considered together with the inset. In the inset, the plotted curve with a maximum represents the dependence of ε'' and, therefore, the dependence of Q_1 on the crystal temperature T_e. Also, the parallel straight lines correspond to Q_2 for a given value of A but different ambient temperatures T_a. An increase in T_a at the point A, corresponding to the condition $Q_1 = Q_2$, results in a loss of the stability because

$$\frac{\partial Q_1}{\partial T_e} > \frac{\partial Q_2}{\partial T_e}.$$

Then the crystal, due to the nonlinear absorption of the microwave power, becomes abruptly heated to a temperature corresponding to the point B in the paraelectric phase. A further increase in T_a changes the crystal temperature T_e more slowly because of temperature self-stabiliz-ation (Fousek 1965). Consequently, ε' depends less on T_a. In the course of cooling, the stability is retained up to T_e corresponding to the point C, and then the temperature decreases abruptly to a value corresponding to the point D, giving rise to a thermal hysteresis $\Delta T = T_a - T'_a$. For the mentioned crystal, the hysteresis is $\Delta T = 5.5$ K. In the case of first order phase transitions, this hysteresis is not related to a metastable phase in the usual understanding of the concept. It depends on $Q_1(\varepsilon'', v, E_m^2)$ and the constant A, and can have different values. Optical excitation, static electric field, and other factors altering ε'' also change ΔT. For example, for $P_{av} = 10$ mW, illumination of a crystal with a laser beam of wavelength in the fundamental absorption region increases the hyster-esis to $\Delta T = 10$ K. This suggests that the large photoinduced shift of the Curie temperature in SbSI and other ferroelectric semiconductors (Frid-kin 1980) is an analogous phenomenon, and it is also caused by the local heating of a crystal.

The sensitivity of SbSI to external factors in microwave fields could be used to control and modulate microwave signals. For example, for $P_{av} = 6$ mW, $E = 0.5$ kV/cm, and $T = 299$ K, the illumination of the crystal with a laser beam results in 50 percent optical modulation of R for a microwave. Similar shifts of T_c and an increase in the hysteresis, varying with the laser radiation power, have been attributed by Uchinokura et al. (1981) to the optical bistability of SbSI.

Nonlinear heating of ferroelectrics by radiation and self-stabiliz-ation of its temperature are rather common features and ought to be revealed in any dispersive ($\varepsilon'' > 0$) region (e.g., IR, submillimeter, and such), when radiating power is of milliwatts or more and can distort results near the PT temperature.

5.3. $Sn_2P_2S_6$ AND $Sn_2P_2Se_6$

$Sn_2P_2S_6$ and $Sn_2P_2Se_6$ are one-axial semiconductive ferroelectrics (Car-penter and Nitsche 1974). At the Curie temperature $T_c = 339$ K, $Sn_2P_2S_6$ undergoes the second order proper ferroelectric phase transition $P2_1/c \rightarrow P_c$ of the displacive type without changing the unit cell volume

(Dittmar and Schafer 1974). The vibrational spectra of $Sn_2P_2S_6$ have been investigated by the Raman and IR techniques. All the Raman and IR modes predicted by the lattice vibration group analysis have been found in the ferroelectric and paraelectric phases.

In the ferroelectric phase, the frequency of the soft mode A', active in the Raman spectra, decreases from $40\,cm^{-1}$ at 77 K to $12\,cm^{-1}$ at T_c according to the law, $v_s = 4.5(T_c - T)^{1/2}\,cm^{-1}$ (Gomonnai et al. 1983). The mode becomes overdamped ($\gamma/v_s \geq 2$) in the vicinity of T_c (Slivka et al. 1978) where a central peak with a half width of $1\,cm^{-1}$ is found in the Brillouin spectrum (Ritus et al. 1985) appearing several degrees above T_c and disappearing below T_c. A model of the phase transition in $Sn_2P_2S_6$ has been proposed on the basis of the normal vibration group analysis and the structural data (Vysochansky et al. 1979). According to the model, the main contribution to the eigenvector of the B_u soft mode is due to Sn^{2+} translations along the [100] and [010] axes (Figure 5.10). This causes the spontaneous polarization along the [100] axis and an

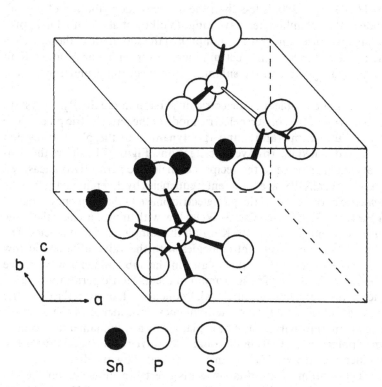

Figure 5.10. Unit cell of $Sn_2P_2S_6$ crystal. From Dittmar and Schafer (1974).

anomaly of the static dielectric permittivity following the Curie–Weiss law. However, the soft mode frequency, deduced from the submillimeter spectra (Volkov et al. 1983b), decreases only to $12\,cm^{-1}$ at T_c. Its temperature dependence is extremely weak and the dielectric contribution, $\Delta\varepsilon = 1000$, is much smaller than the static dielectric permittivity, which is estimated to be 25,000 (Mayor et al. 1984a). Since the temperature dependence of any phonon modes in the proper displacive ferroelectric $Sn_2P_2S_6$ is not responsible for the behavior of the static dielectric permittivity $\varepsilon(0, T)$, and since the entropy of the phase transition is unusually large—six times larger than the value for one ordering ion— (Mayor et al. 1983), the additional excitation was expected to be in the microwave dielectric spectrum of $Sn_2P_2S_6$ with a high dielectric contribution. However, dielectric measurements (Mayor et al. 1984a) in the frequency range of 10^2 to 4×10^{10} Hz did not provide a clear picture. The dielectric dispersion found in the frequency range of 10^7 to 4×10^{10} Hz is not of the Debye shape and is hardly related to any fundamental mechanism of polarization. Moreover, the dielectric permittivity at $v = 43\,GHz$ ($\varepsilon' < 390$) is less than the dielectric contribution of the soft mode in the submillimeter wave range (Volkov et al. 1983b). On approaching the Curie temperature, damping of the soft mode in the ferroelectric phase in the Raman spectrum increases from $2.3\,cm^{-1}$ at 77 K to $30\,cm^{-1}$ at T_c, while in the submillimeter spectrum it undergoes a decrease.

Thus, the problem of dielectric dispersion in the $Sn_2P_2S_6$ crystals, the behavior of the soft ferroelectric mode in the paraelectric phase close to the Curie temperature, and the dynamics of the phase transition required detailed microwave investigations. Figure 5.11 shows the temperature dependence of the reciprocal dielectric permittivity measured by Grigas et al. (1988a) at different frequencies by different methods. The dielectric permittivity of the paraelectric phase in the frequency range of 1 kHz to 27 GHz fits the Curie–Weiss law with approximately the same Curie constant $C = 5.7 \times 10^4$ K for all the samples and frequencies. The value of the Curie constant coincides with the value obtained at low frequencies (Mayor et al. 1984a) and in the submillimeter wave range (Volkov et al. 1983b). The maximum of the dielectric permittivity takes place at the Curie temperature at all frequencies. However, the dielectric loss is small ($tg\delta < 0.1$), and the frequency dependence of the complex dielectric permittivity cannot be explained by any fundamental mechanism of polarization. Only in the millimeter wave range does the dielectric loss increase (at $v = 27$ GHz, $\tan\delta = 0.5$) and show the dispersion.

The motion of the domain walls exerts an influence on the dependence $\varepsilon(T)$ in the ferroelectric phase up to $v = 4\,GHz$. At higher

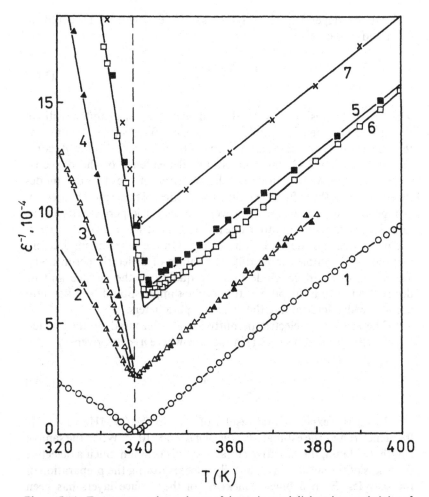

Figure 5.11. Temperature dependence of the reciprocal dielectric permittivity of $Sn_2P_2S_6$ at the frequencies: 1) 1 kHz, 2) 20 MHz, 3) 600 MHz, 4) 4 GHz, 5) 8.9 GHz, 6) 27 GHz, 7) soft mode contribution, deduced from submillimeter spectra. From Grigas et al. (1988a).

frequencies, the dependence $\varepsilon(T)$ fits the Curie–Weiss law with $C^f = 7.8 \times 10^3$ K.

It follows from the calorimetric investigations (Mayor et al. 1983) that the phase transition in $Sn_2P_2S_6$ is of the second order. The soft mode behavior in the Raman spectra (Vysochansky et al. 1979) supports this idea. However, the reciprocal dielectric permittivity does not become zero at higher frequencies upon approaching T_c.

All the dependences of $1/\varepsilon(T)$ shown in Figure 5.11 could be described by the formula

$$\frac{1}{\varepsilon} = \frac{T - T_c}{C} + n, \qquad (5.43)$$

where n is a dimensionless value depending on frequency and the ratio of the surface volume of a sample. The value $n \neq 0$ reflects the fact that at the Curie point, the dielectric permittivity of the sample is finite. According to Equation 5.43, the temperature dependence of the dielectric permittivity could be caused by the coupling of the disorder modes (Burns 1976) to the soft mode if the contribution of these modes does not change the Curie constant. However, the dielectric permittivity of the same sample measured by the same method at $v < 4$ GHz in the paraelectric phase does not depend on frequency. On the other hand, different methods and samples show different values of dielectric permittivity. Therefore, one can conclude that the quality of the surface and its destruction during the preparation of the sample should be taken into account while determining the dielectric parameters.

The complex dielectric permittivity, ε_m^*, measured by the capacitance technique (Section 3.2.1) in the case where $n \ll 1$ is given by

$$\frac{1}{\varepsilon_m^*} = \frac{1}{\varepsilon^*} + n. \qquad (5.44)$$

From the experimental results of $\varepsilon(T)$ in the 20 MHz to 4 GHz frequency range, the estimated value of n is 2.6×10^{-4}. When the sample length $d = 1$ mm, the defective surface layer is 0.26 μm. Such a defective thickness of the surface layer usually appears during the preparation of the samples. Even a bigger thickness of the surface layers has been known, for example, in TGS (from 4 to 13 μm).

In the centimeter and millimeter wave ranges, the electric field vector of the incident wave is parallel to the sample surface. When the simplified model of the surface consists of the crystal regions separated by air spaces perpendicular to the electric field vector, and when the higher spatial harmonics of the electric field near the crystal surface are neglected, then the measured dielectric permittivity is given by the same Equation 5.44.

At low frequencies, the influence of the surface layer on the dielectric parameters can be significantly decreased by using large samples or special electrodes. For microwaves and IR experiments, the surface layer cannot be eliminated. However, in a second-order phase transition, when $\varepsilon(T_c) \to \infty$, there is a criterion for the estimations of n, and one can eliminate the influence of the defective surface layer from Equation 5.44:

$$\varepsilon^* = \frac{\varepsilon_m^*}{1 - n\varepsilon_m^*}. \tag{5.45}$$

In the dielectric dispersion region, the defective surface layers of the small permittivity complicate the calculation of intrinsic dispersion parameters and give higher dispersion frequency. In the case of a resonant dispersion, when $\varepsilon^*(v)$ is described by the dispersion relation for a damped soft mode (Equation 5.4) and when $n\varepsilon_\infty \ll 1$, it follows that

$$v_s(T) = A[(T - T_c) + nC]^{1/2}. \tag{5.46}$$

The second term in Equation 5.46 causes the "saturation" of the soft mode frequency upon approaching the Curie point. Such a "saturation", as in SbSI-type crystals, was found by Volkov et al. (1983b) in $Sn_2P_2S_6$, and it is known in perovskites (see Section 5.1). The soft mode in this case contributes less than the value of $\varepsilon(0)$ and is unable to explain the temperature dependence of $\varepsilon(0)$ in the vicinity of the Curie point.

By using the procedure described above and Equation 5.45, one can eliminate the influence of the defective surface layers for every sample and every method of measurement.

Figure 5.12 shows the temperature dependence of the real and imaginary parts of the complex dielectric permittivity at 78.5 GHz for a $Sn_2P_2S_6$ crystal before and after elimination of the surface layer influence. The parameter n is found to be 3.8×10^{-4}. The minimum in $\varepsilon'(T)$ and the high dielectric loss show that the 78.5 GHz frequency lies in the dispersion region. The temperature dependence of the dispersion parameters (Figure 5.13) of the soft ferroelectric mode, calculated from the temperature dependence of ε' and ε'', shows that in the ferroelectric phase close to T_c, the values of the soft mode frequency and the damping constant are similar to the values calculated from the Raman spectra (Vysochansky et al. 1979). In the paraelectric phase, the frequency of the soft mode depends on temperature according to the law:

$$v_s = 35(T - T_c)^{1/2} \text{ GHz}.$$

The soft mode is strongly overdamped. At $T = 375\,\text{K}$, the damping constant $\gamma = 550\,\text{GHz}$, while the relative damping is $\gamma/v_s = 2.5$. As the temperature approaches the Curie point, the damping increases still more. At $T = 340\,\text{K}$, $\gamma = 750\,\text{GHz}$ and $\gamma/v_s = 14$. Such a mode in the

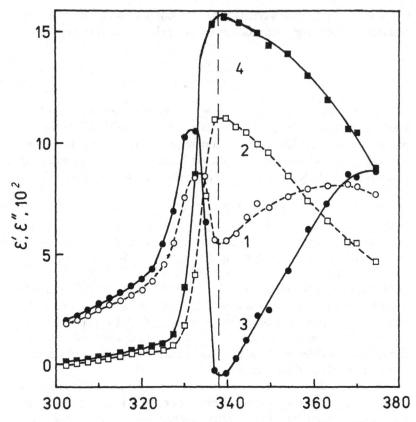

Figure 5.12. Temperature dependence of the complex dielectric permittivity of $Sn_2P_2S_6$ at 78.5 GHz: 1) and 2) ε' and ε'' respectively without elimination of the surface influence, 3) and 4) ε' and ε'' after elimination of the surface influence. From Grigas et al. (1988a).

vicinity of T_c corresponds to the relaxational motion of the Sn^{2+} ions in the strongly anharmonic potential. As in SbSI-type crystals, v_s and γ lose the original meanings. Such a "mode" does not show the saturation in the vicinity of T_c. Thus, the frequency of the same B_u ferroelectric mode softens to the millimeter range. It turns into a relaxational mode. Its dielectric contribution exhibits the Curie–Weiss law and explains the value of the static dielectric permittivity.

Sn$_2$P$_2$Se$_6$ undergoes two phase transitions: the second-order transition to the incommensurate (IC) phase at $T_i = 220$ K, and the first-order lock-in transition to the commensurate ferroelectric phase at $T_c = 193$ K with the peak value of $\varepsilon(0) = 3.5 \times 10^3$ (Mayor et al. 1984b).

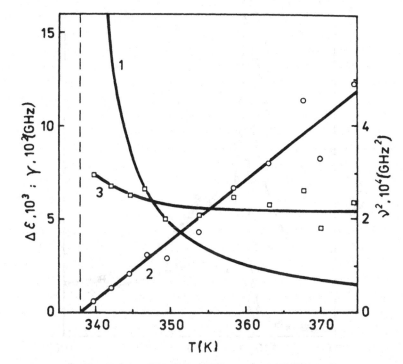

Figure 5.13. Temperature dependence of the soft mode parameters: 1) $\Delta\varepsilon$, 2) v_s^2 and 3) γ. From Grigas et al. (1988a).

Figure 5.14 shows the temperature dependences of the real and imaginary parts of the complex dielectric permittivity of $Sn_2P_2Se_6$ crystals. In the paraelectric phase, $\varepsilon(0)$ obeys the Curie–Weiss law with the Curie constant $C = 6.6 \times 10^4$ K. In this and in the incommensurate phases, there is no dispersion in the frequency range 1 kHz to 4 GHz. In the ferroelectric phase, there is dielectric dispersion in the 10 MHz to 1 GHz range, caused by the dynamics of the domain walls. Its characteristic relaxation frequency is $v_d \simeq 100$ MHz.

The ferroelectric dispersion near the PTs temperatures T_i and T_c occurs at microwaves (Grigas et al. 1988b). In the paraelectric phase close to T_i, as in the $Sn_2P_2S_6$ crystals, the dielectric dispersion is caused by the soft B_u mode. Within the incommensurate phase, the soft mode splits into a phason and amplitudon (see Section 5.4). Close to T_i, the main contribution to ε' is caused by the amplitudon, the frequency of which exceeds 78 GHz and increases with the decrease of temperature.

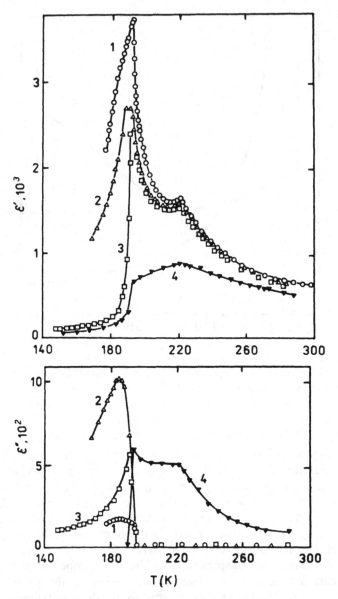

Figure 5.14. Temperature dependence of real and imaginary parts of dielectric permittivity of $Sn_2P_2Se_6$ at frequencies: 1) 1 kHz; 2) 80 MHz; 3) 1.2 GHz; 4) 78.5 GHz. From Grigas et al. (1988b).

As the lock-in transition is approached, the phason contribution to $\varepsilon(0)$ rapidly increases (see Dvořak 1980). The Debye-shape dielectric dispersion caused by the phason lies between 4 and 78 GHz. At the frequency of 78.5 GHz the dielectric dispersion close to T_c is almost over. The dielectric contribution $\Delta\varepsilon = 370$, and is mainly caused by the amplitudon branch. The relaxation frequency of the phason (i.e., the phason gap at T_c), is $\Delta_{ph} \approx 20$ GHz. This frequency is two orders higher than the relaxation frequency of the domains. It means that the lock-in transition from the soliton lattice to the domain structure is discontinuous with the essential change of the dimensions of a homogeneous polarization regions. Apparently, the lock-in transition occurs before the wave number of the polarization wave goes to zero and is of the first order. Therefore, the gaps on the amplitudon and phason branches at $q = 0$ are close to each other, and the dielectric dispersion regions exhibit a partial overlapping.

The above results indicate that the soft modes of extremely low frequency and the change of displacive-like to order-disorder like regimes on approaching the Curie point are also characteristic features of $Sn_2P_2S_6$-type ferroelectrics. To a certain extent, the model described in the previous Section could be as equally well applied for the phase transition in $Sn_2P_2S_6$.

It should be mentioned that $Sn_2P_2S_6$ is one of the best piezoelectric materials for ultrasonic devices. $Sn_2P_2S_6$ and $Sn_2P_2Se_6$ in the vicinity of PT exhibit large acoustic nonlinearity, as detected by the second ultrasonic harmonic generation (Samulionis et al. 1990). The amplitude of the second longitudinal ultrasonic harmonic is given by

$$u'' = \frac{\beta k^2 u_0^2}{16\alpha v_s^2}(e^{-2\alpha 1} - e^{-4\alpha 1}), \qquad (5.47)$$

where β is the nonlinear elastic parameter, which can be described by elastic coefficients of the second and third order, k is the ultrasonic wave number, v_s is the ultrasonic velocity, l is the length of the sample, α is the attenuation coefficient on the frequency of the first harmonic, and u_0 is the amplitude of the displacement at the input of the sample. This equation and the measurements of v_s, α and u'' enable the calculation of the temperature dependence of the nonlinear parameter β, which at 20 MHz is shown in Figure 5.15. Similarly, as in SbSI (Samulionis et al. 1971), the nonlinear elastic parameter in the vicinity of the PT of $Sn_2P_2S_6$ shows the anomaly that is related to the soft ferroelectric mode. Nonlinearity in the ferroelectric phase is connected with the real domain structure.

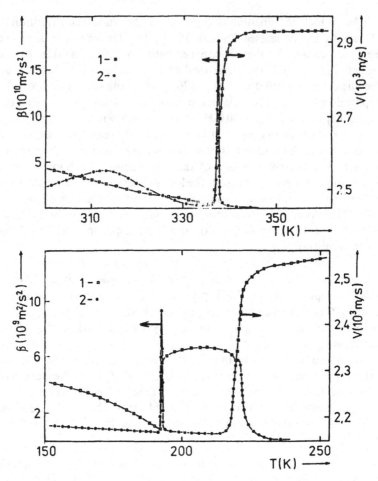

Figure 5.15. Temperature dependence of ultrasonic velocity (1) in [100] direction and nonlinear elastic parameter (2) for $Sn_2P_2S_6$ and $Sn_2P_2Se_6$ crystals. From Samulionis et al. (1990).

For $Sn_2P_2Se_6$ crystals, the acoustic nonlinearity is also considerable in the IC phase due to the domain-like structure and reaches its maximum near the T_i temperature. The domain-like structure also causes anomalous temperature hysteresis of u'' in the IC phase. Therefore, $Sn_2P_2Se_6$ crystals can be used for nonlinear acoustic signal processing devices.

Other properties of these crystals are discussed in detail by Vysochansky and Slivka (1994).

5.4. INCOMMENSURATE TlGaSe₂ AND TlInS₂

The ternary layer-structured semiconductive ferroelectrics $TlGaSe_2$ and $TlInS_2$ are typical examples of crystals with an extremely low soft mode frequency.

At room temperature, both materials are reported (Henkel et al. 1982) to have a monoclinic structure, $C2/c$ (C_{2h}^6) with two layers per unit cell, and upon cooling at the temperature T_i, they undergo a second-order phase transition from a paraelectric to an incommensurate phase. At the Curie temperature, T_c, they undergo a first-order lock-in transition to a ferroelectric phase. The temperatures are $T_i = 214$ K and $T_c = 202$ K in $TlInS_2$, and $T_i = 119$ K and $T_c = 107$ K in $TlGaSe_2$ (Aliev et al. 1984). The phase between T_i and T_c has been investigated by neutron diffraction (Vakhrushev et al. 1984) for $TlInS_2$, and by triple-crystal x-ray scattering technique for $TlGaSe_2$ (McMorrow et al. 1990). This is a modulated phase with a wavevector $q_s = (\delta, \delta, 1/4)$, where $\delta = 0.02$ reciprocal lattice units. Below T_c, the distorted structure has $q_s = (0, 0, 1/4)$.

Though the crystals have large temperature dependent static dielectric permittivity and are isotropic in the (001) plane (Aliev et al. 1984), investigations of their lattice dynamics by IR reflection and Raman scattering spectroscopy (Henkel et al. 1982; Gasanly et al. 1983) have not revealed any soft ferroelectric modes. The soft ferroelectric modes of very low frequency in the paraelectric phase, far from T_i, have been revealed in the submillimeter frequency range by Volkov et al. (1983a, 1984a). Their frequency at $T \to T_i$ moves into the $v < 120$ GHz range.

Microwave dielectric spectroscopy of the $TlGaSe_2$ (Paprotny and Grigas 1985; Banys et al. 1987) and $TlInS_2$ (Banys et al. 1988; 1989) crystals, as well as the pinning effect on microwave dielectric properties and soft mode in these crystals (Banys et al. 1990), shows that the ferroelectric phase transitions are related to a strongly overdamped soft mode, the frequency of which lies at microwaves.

The paraelectric-IC phase transition is the result of the condensation of a soft mode at the point q_s in the Brillouin zone. Within the IC phase, the soft mode splits into an acoustic-like phason with the linear dispersion and optic-like amplitudon (Figure 5.16). One of the most interesting questions is whether the phason branch extrapolates to zero at $q = q_s$ (as, for instance, in $ThBr_4$ and biphenyl), or whether there is a gap Δ_{ph} (as in K_2SeO_4). The gap in real crystals could be produced by discrete lattice effects or impurities, which can pin the phason. (For details related to the IC structures see, for example, Blinc and Levanyuk (1986)). Inelastic neutron scattering has succeeded in reveal-

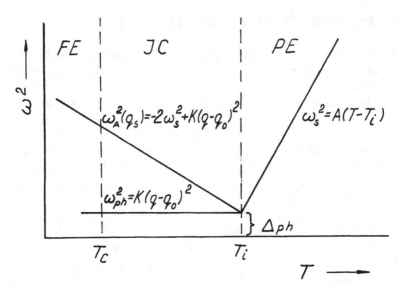

Figure 5.16. Schematic representation of the temperature dependence of the incommensurate soft mode (ω_s), the amplitudon(ω_A) and the phason(ω_{ph}) frequencies.

ing unambiguously the existence of phasons in the above mentioned crystals.

TlInS$_2$ and TlGaSe$_2$ have a wide IC phase, soft mode of extremely low frequency, and $q_s \approx 0$. They are unique, proper, primarily displacive crystals for a separate determination of parameters of both amplitudon and phason excitations at $q = 0$ by microwave dielectric spectroscopy. The phason and amplitudon are active in dielectric spectra due to their coupling to homogeneous polarization (Dvořak 1980).

The crystals are isotropic in the (001) plane. Therefore, the temperature and frequency characteristics of the complex dielectric permittivity of the component, ε_{11}^* will be considered. The component $\varepsilon_{33} \approx 16$ and is weakly dependent on temperature and frequency.

Figure 5.17 shows the temperature dependence of the real and imaginary parts of the complex dielectric permittivity of TlInS$_2$ crystals at several microwave frequencies. A strong dielectric dispersion at microwaves is observed. As in these proper ferroelectrics the role of the order parameter plays the polarization components P_x and P_y, the static permittivity should have a larger anomaly at T_c than at T_i (Dvořak 1980). The divergence of $\varepsilon(0)$ at T_c is due to the fact that, when q approaches zero, a homogeneous polarization $(P_s = 0.2\mu\,C/cm^2)$ in the crystal occurs

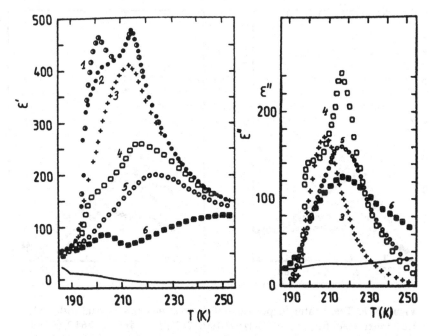

Figure 5.17. Temperature dependence of ε' and ε'' of $TlInS_2$ crystal for various frequencies (GHz): 1) 0.001; 2) 0.1; 3) 10; 4) 25; 5) 38; 6) 68. The continuous curves for 353 GHz. From Banys et al. (1988).

spontaneously. Near T_i the main contribution to $\varepsilon(0)$ gives the amplitudon branch while, near T_c the amplitudon contribution is smaller due to a large gap on the amplitudon branch. The main contribution to $\varepsilon(0)$ near T_c gives the low frequency phason since the phason frequency approaches zero as does q_s when the temperature approaches T_c. However, at 0.1 GHz, ε' has the highest value at T_i. At T_c it is suppressed by frequency. At the frequency of $v = 10$ GHz, the maximum of ε'' lies at T_c. At the frequencies of $v > 20$ GHz, it lies near T_i. This means that, at T_c the dispersion is caused by a lower frequency excitation than at T_i.

Any sort of imperfections cause pinning of the phason. Owing to the pinning, the phason frequency becomes higher and the anomalies of ε' and ε'' at a given frequency become clearly expressed (Figure 5.18).

Similar behavior is found in $TlGaSe_2$. In pure crystals, anomalies of ε' and ε'' are suppressed at T_c at frequencies of several hundred MHz (i.e., the phason frequency is very low (Banys et al. 1990)). Impurities (e.g., of Fe) smear the phase transitions and decrease the permittivity. However, owing to phason pinning on the defects, the phason frequency becomes

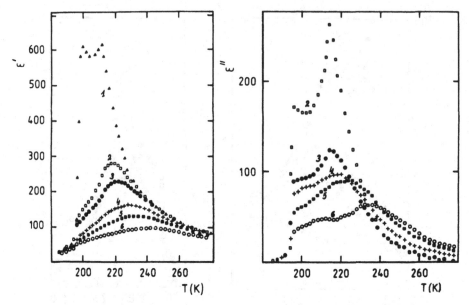

Figure 5.18. Temperature dependence of ε' and ε'' of $TlInS_2$ crystal with 1.5% Cu admixture for frequencies (GHz): 1) 0.001; 2) 27; 3) 39; 4) 51.4; 5) 64.2; 6) 76.4. From Banys et al. (1990).

higher, and the anomaly of ε' at 200 MHz becomes clearly expressed at T_c (Figure 5.19).

Figure 5.20 shows the temperature dependence of ε' and ε'' of $TlGaSe_2$ at different frequencies. The dielectric dispersion near T_c at frequencies of $v < 1\,GHz$ is caused mainly by the phason. The loss caused by the phason at T_c is higher than the loss caused by the soft mode at $T > T_i$. The main contribution to the dielectric loss at frequencies of $v > 10\,GHz$ is due to the soft mode (amplitudon).

The dielectric dispersion occurs (Figure 5.21) in the frequency region of 10^8 to 10^{12} Hz and is caused by a ferroelectric soft mode that can be described by the dispersion relation for a single damped oscillator (Equation 5.4). Loss shows one broad maximum. There are no additional excitations (central modes) in these crystals nor in $Sn_2P_2S_6$ as suggested by Petzelt et al. (1987). The temperature dependence of the parameters of the soft mode, as calculated from the experimental results according to Equation 5.4 by the least-squares method, are shown in Figure 5.22. The frequency of the paraelectric soft mode decreases according to Equation 5.1 to the millimeter wave region. However, the

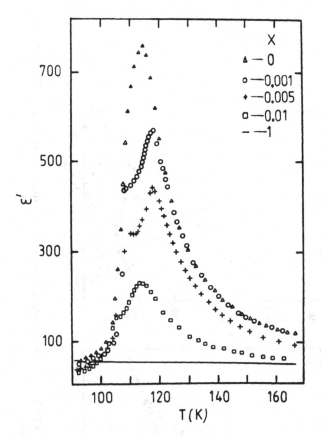

Figure 5.19. Influence of Fe impurities on the temperature dependence of ε' of $TlGa_{1-x}Fe_xSe_2$ crystal at a frequency of 200 MHz. From Banys et al. (1990).

soft mode is strongly overdamped and becomes relaxational near T_i. As $\varepsilon(0) \sim 1/v_s^2$, it follows that $\varepsilon(T_c) > (T_i)$.

Table 5.2 shows the soft mode parameters for investigated crystals of this family (first column). In the second and third columns, the Curie temperature, T_c, and the temperature of the paraelectric-IC phase transition, T_i, are given. The fourth column shows the width of the IC phase. In the following two columns, the Curie–Weiss constant is given. The next two columns show the constant A calculated from Equation 5.1 for the paraelectric and ferroelectric phases. The next two columns give the values of the soft mode frequency at T_i and T_c temperatures, which lie in the millimeter wave region. Impurities shift the dispersion region to the lower frequencies, decreasing the soft mode frequency and increasing its

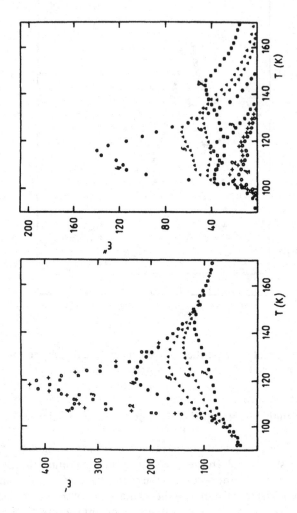

Figure 5.20. Temperature dependence of ε' and ε'' of TlGaSe$_2$ crystal with 0.5% Fe admixture for frequencies (GHz): 1) 0.02; 2) 0.14; 4) 1; 4) 39; 5) 54; 6) 64.2; 7) 76.4. From Banys et al. (1990).

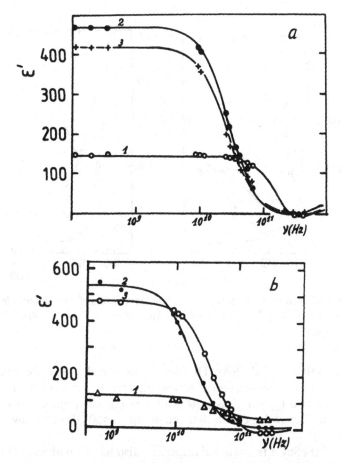

Figure 5.21. Frequency dependence of ε' for various temperatures (K): a) TlInS$_2$: 1) 251; 2) 214; 3) 206; b) TlGaSe$_2$: 1) 106; 2) 113; 3) 123. The continuous curves are the oscillator fits with the parameters given in Figures 5.22 and 5.23. The points are experimental values. From Banys et al. (1990).

relative damping constant, which is shown in the last column. The values of the relative damping correspond to the relaxational motion of the particles in an anharmonic potential, as in SbSI. The poles that represent the soft mode (see Chapter 2) move along the imaginary v-axis, and the modulus of the complex v_s^* fits the law (5.1).

The substitution in TaGaSe$_2$ 50 percent S → Se changes the structure of the crystal from monoclinic to tetragonal (Henkel et al. 1982). The incommensurate phase vanishes, and the value of dielectric permittivity

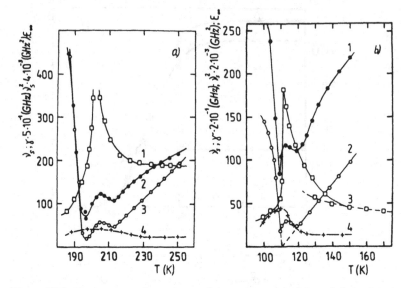

Figure 5.22. Temperature dependence of the parameters of the soft ferroelectric mode of TlInS$_2$: 1) γ; 2) v_s; 3) v_s^2; 4); ε_∞; b) TlGaSe$_2$: 1) v_s; 2) v_s^2; 3) γ; 4) ε_∞. From Banys et al. (1988).

decreases (Banys et al. 1990). The phase transition does not occur in the region of positive temperatures. An incompletely frozen ferroelectric soft mode forms a typical glass state with a broad absorption and relaxational dielectric dispersion that ranges up to 450 GHz (Volkov et al. 1992).

In TlFeSe$_2$, a relaxational dispersion also has been observed in the microwave region. The dielectric contribution of the relaxational mode fits the Curie–Weiss law with the parameters given in Table 5.2. In the 60–80 GHz frequency range, $\varepsilon = \varepsilon_\infty = 9.5$.

The effective values of the soft mode parameters in the IC phase are shown in Figures 5.22. Due to the coupling of the phason and amplitudon to the polarization and strain components, they both contribute to the static dielectric permittivity. When T approaches T_c the phason frequency decreases and is always overdamped. This excitation is a relaxational one. Cole–Cole plots (Banys et al. 1988) have shown that the dielectric dispersion in the IC phase can be described by the Debye equation. Similar relaxation has been found by Deguchi et al. (1986) in the IC phase of NaNO$_2$.

In the IC phase, the dynamical dielectric permittivity of the coupled soft mode (amplitudon) and the relaxational mode (phason) is given

Table 5.2. Incommensurate phase and soft-mode characteristics

Crystal	T_c (K)	T_i (K)	$T_i - T_c$ (K)	C(K)		A $(GHz/K)^{1/2}$		ν_s(GHz)		γ/ν_s $(T = T_i)$
				$T > T_i$	$T < T_c$	$T > T_i$	$T < T_c$	$T = T_i$	$T = T_c$	
TlInS$_2$	202	214	12	6.5×10^3	2.2×10^2	29	120	106	60	3.9
TlInS$_2$	196	205	9	3.3×10^3	3.8×10^2	27	106	40	50	9.7
1% Fe TlInS$_2$ 1,5% Cu	196	214	18	4.1×10^3	10^2	29	130	40	40	6
TlInSSe	170	–	–	26×10^3	4.5×10^3	–	–	–	–	–
TlGaSe$_2$	107	119	12	5.5×10^3	300	33	109	110	80	5
TlGaSe$_2$ 0,1% Fe	108.5	118.5	10	5.5×10^3	400	–	–	–	–	–
TlGaSe$_2$ 0,5% Fe	108	117	9	5.1×10^3	370	34	75	40	40	10.5
TlGaSe$_2$ 1% Fe	105	114	9	2.8×10^3	200	–	–	–	–	–
TlFeSe$_2$	220	–	–	31×10^3	–	–	–	–	–	–

by

$$\varepsilon^*(v) = \varepsilon_\infty + \frac{S_s(1 + iv\tau) + 2S_sS_r\alpha + S_r(v_s^2 - v^2 + i\gamma v)}{(v_s^2 - v^2 + i\gamma v)(1 + iv\tau) - \alpha^2 S_s S_r}. \tag{5.48}$$

Here

$$S_s = \frac{n_1 e_1^2}{\varepsilon_0 m_1} = \Delta\varepsilon_s v_s^2, \quad S_r = \frac{n_2 e_2^2}{\varepsilon_0 m_2} = \Delta\varepsilon_r,$$

where S_s, v_s and γ are the oscillator strength, frequency and damping constant of the soft mode (amplitudon), respectively. The parameters of the phason are $\Delta\varepsilon_r$ and τ. It follows from Equation 5.48 that the static dielectric permittivity caused by the contribution of the phason and amplitudon is

$$\varepsilon(0) = \varepsilon_\infty + \frac{\Delta\varepsilon_s + 2\alpha\Delta\varepsilon_s\Delta\varepsilon_r + \Delta\varepsilon_r}{1 + \alpha^2\Delta\varepsilon_s\Delta\varepsilon_r}. \tag{5.49}$$

The coupling constant, α, calculated from this equation shows maximum value within the IC phase (Banys et al. 1990). The coupling of the amplitudon and phason considerably increases the effective damping constant. It is much smaller for the bare soft mode as calculated from Equation 5.48.

From the experimental results and Equation 5.48 the phason relaxation time or frequency can be obtained. The relaxational time for $TlInS_2$ is given by

$$\tau = \frac{4.9 \times 10^{-9}}{T - 200} \, (s).$$

The frequency of the phason at $T = T_c$ is equal to $v_{ph} = 1/2\pi\tau = 76\,MHz$. Impurities of Cu increase the frequency of the phason to $v_{ph} = 1.34\,GHz$. As the result, the coupling constant, α, increases.

The frequency of the phason for $TlGaSe_2$ crystals at $T = T_c$ is equal to $v_{ph} = 60\,MHz$. Because of pinning by the impurities of 0.5 percent Fe, the frequency increases to $v_{ph} = 1.4\,GHz$.

An external electric field can play the same role as defects and change the coupling constant. Simple modelling shows that changing the coupling constant, α, by only a few percent, considerably changes the static permittivity (given by Equation 5.49) in the vicinity of T_c. Also, the coupled amplitudon-phason can produce even a new maximum in the IC phase (Banys and Grigas 1993). The height of the maximum depends on the value of the coupling constant.

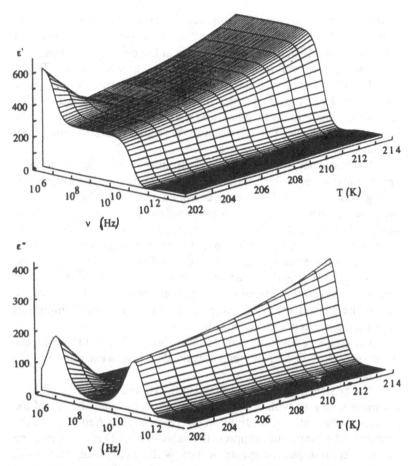

Figure 5.23. Contribution of the phason and amplitudon to ε' and ε'' within the IC phase of TlInS$_2$ crystal. From Banys et al. (1990).

Figure 5.23 shows the temperature dependence of the dielectric permittivity and loss in TlInS$_2$ crystals within the IC phase in the dispersion region. This is calculated from Equation 5.48 using the experimentally found parameters of the phason and amplitudon. The existence of a phason and the shift of its wavevector to zero on approaching T_c remarkably increase the dielectric contribution at T_c and cause the dielectric dispersion at and below microwave frequencies. The main contribution at $T = T_i$ causes a strongly overdamped soft mode (amplitudon) whose frequency lies in the millimeter wave region and increases with a decrease of temperature. Thus, the gaps in the amplitudon and phason branches at $q = 0$ are close to each other, and their coupling causes high loss and an overlap of the dielectric regions.

The experimental results lead to the final conclusion that $TlInS_2$ and $TlGaSe_2$ are proper semiconductive ferroelectrics, which, upon cooling, undergo a second-order phase transition from a paraelectric monoclinic phase to an IC phase and a first-order lock-in transition to a ferroelectric phase. The IC structure is related to a soft mode from a polarization branch with q_s near zero.

The frequency of the soft mode is extremely low in the vicinity of T_i and T_c. This causes the high static dielectric permittivity. The soft mode is strongly overdamped. The crossover from a displacive-like to an order-disorder-like mechanism occurs upon cooling in the paraelectric phase, as with SbSI. At T_i, the soft mode splits into an acoustic-like phason and an optic mode (amplitudon). The phason in real crystals is pinned, and is active in the dielectric spectra. It reveals itself as a relaxator. In the less defective crystals, the frequency of the phason is about 10^7 Hz. Crystal imperfections result in the pinning of the phason and an increase of the gap in the phason spectrum. Even small concentrations of impurities increase the frequency of the phason to 10^9 Hz.

The coupling constant in most cases shows a maximum within the IC phase at $T_c < T < T_i$, and increases the damping of excitations in the IC phase.

The existence of a phason branch in the IC phase and the shift of its wavevector to zero with temperature increases the contribution of the phason to the static permittivity at T_c. However, the frequency of the electric field or impurities suppress the value of $\varepsilon(T_c)$. Due to the pinning effect, different impurities spread or narrow the incommensurate phase, and express the anomalies of permittivity at T_i and T_c temperatures.

Thallium monosulphide is also classified to this family. It is ferroelectric at room temperature, and at $T_i = 341.1$ K and $T_c = 318.6$ K it undergoes similar successive phase transitions to the above discussed compounds. The intermediate phase between the T_i and T_c is incommensurate ($q_s = 0.04, 0, 1/4$) (Nakamura and Kashida 1993).

5.5. INCIPIENT FERROELECTRIC $TlSbSe_2$

The remaining part of this chapter deals with the semiconductive $TlSbSe_2$, which at room temperature is also monoclinic (space group $P2_1$ (C_2^2)). The value of the energy gap of this crystal is $E_g = 0.82$ eV. X-ray, DTA, and dilatometric investigations (Salk et al. 1990; Kiosse et al. 1990a,b) reveal successive phase transitions at about 553, 663 and 668 K.

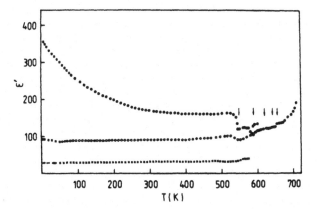

Figure 5.24. Temperature dependence of dielectric permittivity of TlSbSe$_2$ crystals along the b- (●), a- (○) and c-axis (x) at the frequency 3.9 GHz. Arrows indicate the phase transition temperatures. From Banys et al. (1992b).

The high temperature (orthorombic) phase is supposed to be superionic with statistically distributed Tl and Sb atoms at the cation positions. The phase is partly ordered between room temperature and 653 K upon heating, or at 553 K upon cooling. The IC phase is also suggested in the 553 to 633 K temperature range.

Microwave investigations (Banys et al. 1992b) of these conductive crystals ($\sigma_c = 0.1$ S/m) have shown that the lattice dielectric permittivity is high and strongly anisotropic (Figure 5.24) as in the Sb$_2$S$_3$-type semiconductors. Along the c-axis, the permittivity follows the Curie–Weiss law with the Curie constant $C = 8.4 \times 10$ K, the temperature $T_c = -150$ K and $\varepsilon_\infty = 10$. Thus, TlSbSe$_2$ (like TlGaS$_2$ with $T_c = -65$ K (Volkov et al. 1984b)) can be considered as an incipient ferroelectric. High dielectric permittivity is caused by a soft lattice. One of the lattice modes softens according to Equation 5.1 as $T_c \rightarrow 0$ K. However, in the frequency range up to 110 GHz, there is no dielectric dispersion or noticeable loss caused by the soft mode. The soft mode frequency lies at much higher frequencies.

The jumps of the permittivity, accompanied by the anomalies of d.c. conductivity, show a sequence of structural phase transitions that can be described in the framework of Landau theory.

6 QUASI-ONE-DIMENSIONAL H-BONDED FERROELECTRICS

6.1. INTRODUCTION

At present, there is a great number of papers on the microwave dielectric dispersion of hydrogen-bonded ferroelectrics. The dielectric dispersion has been investigated mostly below 10 GHz or above 100 GHz. The experimental results below 10 GHz enable analysis of a critical slowing down process in the close vicinity of the Curie temperature, T_c. Submillimeter or IR spectra give information about only a high frequency tail of fundamental dielectric dispersion. Neither a beginning nor a tail of the dielectric dispersion allows for an unambiguous distinction between various theoretical models. There are practically no complete dielectric spectra of the fundamental dielectric dispersion of H-bonded ferroelectrics in a wide temperature range and at frequencies of 10^6 to 10^{11} Hz. As a result, almost 60 years after the discovery of ferroelectricity in KDP-type crystals, the dynamical models of phase transition of KDP (order-disorder, tunneling, coupled proton-phonon or nonlinear anionic polarizability) are still under discussion (Matsubara 1985; Bussman-Holder and Bilz 1984). The phase transition, in accordance with recent Raman studies, suggests an order-disorder behavior of locally distorted PO_4 dipoles without proton tunneling (Tominaga et al. 1985). On the other hand, the overdamped soft mode in the IR spectra (Simon et al. 1988) indicates a displacive mechanism.

The complex dielectric permittivity at 138.6 GHz has been explained by Gauss et al. (1975) using the model of mode-mode coupling,

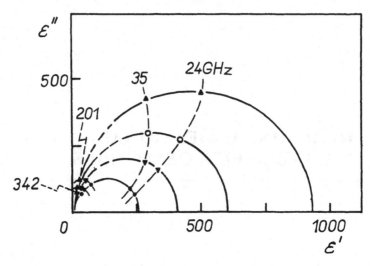

Figure 6.1. Cole–Cole diagrams of KDP. From Horioka and Abe (1990).

while the Volkov et al. (1990) data may be fitted by a simple relaxation model.

According to Takagi (1987), the KDP crystal has a crossover from displacive to order-disorder type, similar to the crystals discussed in Chapter 5.

Cole–Cole diagrams in the paraelectric phase of KDP (Figure 6.1) show that the dielectric dispersion near T_c is of Debye type with a single relaxation time. The relaxation frequency decreases upon approaching T_c, and becomes about 30 GHz at T_c. The value of the oscillator strength of the overdamped soft mode obtained by Simon et al. (1988) is fairly small in comparison to that of $\varepsilon(0)$, and is not consistent with a displacive character of the phase transition. A complete theory of phase transition, which would give a consistent account of all known experimental facts, has yet to be developed.

This chapter will not give equal attention to all the crystals nor review all the papers. The aim of this chapter is to show the examples of complete dielectric spectra of fundamental dielectric dispersion of some low-dimensional representatives of KDP-family crystals in the micro-wave range. Furthermore, it will discuss dynamical models of phase transitions, which, by treating them as the result of instability in a proton subsystem, give a quantitative description of both dynamical and static properties.

6.2. RELAXATIONAL DYNAMICS OF CsH₂PO₄-TYPE FERROELECTRICS

Monoclinic cesium dihydrogen phosphate, CsH_2PO_4 (CDP), and its deuterated isomorph, CsD_2PO_4 (DCDP), with two formula units per primitive unit cell in the high temperature paraelectric phase, undergo a ferroelectric phase transition of the second order $P2_1/m \to P2_1$ at $T_c^{H,D}$, with spontaneous polarization along the monoclinic b-axis (Frazer et al. 1979).

The CsH_2PO_4-type crystals are characterized by the presence of two kinds of hydrogen bonds linking PO_4 tetrahedra (Iwata et al. 1980). The O—H(1)...O bonds link the PO_4 groups along the a-axis, and at room temperature they are ordered. The shorter O—H(2)...O bonds form zigzag quasi-one-dimensional chains along the ferroelectric b-axis (Figure 6.2). They are disordered in the paraelectric phase, while in the

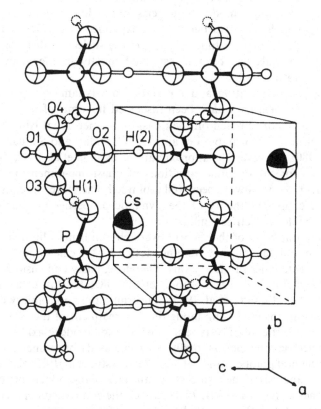

Figure 6.2. The structure of CsH_2PO_4 crystals.

ferroelectric phase they become ordered. CDP-type crystals are one-dimensional versions of KH_2PO_4-type ferroelectrics (Itoh et al. 1983).

When CDP is heated, the monoclinic phase goes to a cubic phase at 504 K. The conductivity of the cubic phase is superionic (Baranov et al. 1989). In the paraelectric phase, the disorder is due to a statistical distribution of the protons between two potential minima of the short $O-H(2)-O$ bond. Raman studies by Romain and Novak (1991) show that there is an additional dynamical disorder in the superionic phase, resulting from a rapid reorientation of the $H_2PO_4^-$ anion.

There are numerous experimental data on the static and dynamic properties of $Cs(H_{1-x}D_x)_2PO_4$ ferroelectrics: such as temperature dependencies of the static and dynamic dielectric permittivity and spontaneous polarization P_s (Deguchi et al. 1982; Kanda et al. 1982; Levstik et al. 1983); specific heat (Kanda et al. 1982); and ultrasonic measurements (Kanda et al. 1983; Aksenov et al. 1984). All the dynamic measurements show a pronounced critical slowing down effect near $T_c^{H,D}$, with the soft relaxational mode frequencies much lower than those in KH_2PO_4 crystals. The dispersion $\varepsilon^*(v)$ is explained by the Debye model with a single relaxation time. However, the dielectric dispersion of CDP-type crystals has been investigated mostly in the vicinity of $T_c^{H,D}$ temperature.

The phase transition and the static and dynamic properties of CDP-type ferroelectrics have been discussed in terms of the existing microscopic theories based on the quasi-one-dimensional Ising model (Kanda et al. 1982; Scalapino et al. 1975; Carvalko and Salinas 1978; Žumer 1980). However, the approximations (Kanda et al. 1982; Žumer 1980) used for the dynamic equations of quasi-one-dimensional ferroelectrics have not always been well grounded. Therefore, the construction of a consistent theory of the dynamic properties of quasi-one-dimensional ferroelectrics remains a problem.

The parameters of the theories were chosen from a limited set of experimental data, and they were not tested by the calculations of the other characteristics of the crystals. Therefore, hasty conclusions were made (Deguchi et al. 1982; Kanda et al. 1982; Kozlov et al. 1984; Kriukova 1984; Kanda et al. 1983) concerning the applicability of the quasi-one-dimensional Ising model in the treatment of the static and dynamic properties of CDP-type crystals. The soft mode relaxation time, τ, and dielectric contribution to $\varepsilon(0)$ were calculated from a small number of experimental points (Kanda et al. 1982; Levstik et al. 1983; Kriukova 1984). The analysis of the results shows the lack of reproducibility of the dependencies $\varepsilon'(v, T)$ and $\varepsilon''(v, T)$. However, the comparison of theoretical results and the values of τ and $\Delta\varepsilon$ obtained from experiments (Kanda et al. 1983; Levstik et al. 1983) carried out in the vicinity of the phase transition temperature show that different parameters of the theory give similar results. Therefore, some of the conclusions about the validity of the quasi-one-dimensional Ising model to the description of the experimental data of CDP-type crystals in a large temperature range are of questionable value.

In addition to the theory of quasi-one-dimensional ferroelectrics (Aksenov et al. 1984; Scalapino et al. 1975; Žumer 1980; Suzuki and Kubo 1968)—which is based on the exact calculations of the one-dimensional Ising model—there is also the cluster approximation for the treatment of the relaxational dynamics (Havlin and Sompolinsky 1980) of the Ising model (Glauber 1963)—which is based on the Glauber one-dimensional version—taking long range interactions into account (Levitsky et al. 1979). Dynamic properties of KDP in the cluster approximation have been discussed by Zachek and Levitsky (1980). The soft relaxational mode dynamics of CDP-type crystals in a cluster approximation is discussed below, taking into account the short-range interactions of deuterons near the PO_4 tetrahedra in the chains of hydrogen bonds, and the long-range interactions between those bonds. Proton tunneling, which has raised strong doubts (Ichikawa and Motida 1988), is omitted. Therefore, the deuteron system in CsD_2PO_4 moving along zigzag chains is considered. The unit cell consists of two neighboring tetrahedra of PO_4 with two short hydrogen bonds belonging to one of them (tetrahedra of A-type). Hydrogen bonds of the other tetrahedra (B-type) belong to the two nearest structure elements that surround it.

The Hamiltonian of such a system in an external electric field, taking into account short- and long-range interactions, has the form (Levitsky et al. 1986)

$$\hat{H} = -\frac{1}{2} w \sum_{q_i q_j} \sigma_1^z(R_{q_i}) \sigma_2^z(R_{q_j})(\delta_{R_{q_i}, R_{q_j}} + \delta_{R_{q_i} + r_2, R_{q_j}})$$

$$-\frac{1}{8} \sum_{\substack{q_i f_i \\ q_j f_j}} J_{f_i f_j}(R_{q_i} R_{q_j}) \sigma_{f_i}^z(R_{q_i}) \sigma_{f_j}^z(R_{q_j}) - \frac{1}{2}(\mu E) \sum_{f_i q_i} \sigma_{f_i}^z(R_{q_i}). \tag{6.1}$$

The first term describes the short-range configuration interactions of deuterons near the tetrahedra of the A-type (first Kronecker symbol) and those of the B-type (second Kronecker symbol). The second term describes the effective long-range interaction through the lattice vibrations. The last term describes the interaction of deuterons with longitudinal electric field. The quasi-spin z-component operator of the deuteron is

$$\sigma_{f_j}^z(R_{q_j}),$$

which is in the f_j-th bond of the q_j-th cell ($f_j = 1, 2$), and

$$\sigma^z_{f_j}(R^z_{q_j}) = \pm 1.$$

The unit translation along the b-axis is r_2, and μ is the effective dipole moment of the unit cell along the b-axis. The parameter, w, describes the short-range correlations of deuterons in the chains of hydrogen bonds and is the parameter of the two-particle cluster.

The discussion will be limited to the two-particle cluster approximation, taking into account the quasi-one-dimensional structure.

In order to treat the dynamic properties of the Hamiltonian (Equation 6.1) for DCDP-type crystals, it is necessary to know the probability

$$P\{\ldots, \sigma^z_{f_j}(R_{q_j}), \ldots, t\}$$

that the deuteron system is in the state

$$\{\ldots, \sigma^z_{f_j}(R_{q_j}), \ldots\}$$

at the moment t. The time-dependent function of that probability is given by

$$\frac{d}{dt} P\{\ldots, \sigma^z_{f_j}(R_{q_j}), \ldots, t\}$$

$$= -\sum_{q_j f_j} [W_{q_j f_j}\{\ldots, \sigma^z_{f_j}(R_{q_j}), \ldots\} \times P\{\ldots, \sigma^z_{f_j}(R_{q_j}), \ldots, t\}] \qquad (6.2)$$

$$+ \sum_{q_j f_j} [W_{f_j q_j}\{\ldots, -\sigma^z_{f_j}(R_{q_j}), \ldots\} \times P\{\ldots, -\sigma^z_{f_j}(R_{q_j}), \ldots, t\}],$$

where

$$W_{q_j f_j}\{\ldots, \sigma^z_{f_j}(R_{q_j}), \ldots\}$$

is the probability per unit time that the f_j-th deuteron in the q_j-th cell goes from the state with quasi-spin

$$\sigma^z_{f_j}(R_{q_j})$$

to the state with quasi-spin

$$-\sigma^z_{f_j}(R_{q_j})$$

due to the interaction with a heat bath. It is assumed that the heat bath is always in the equilibrium state.

Following Suzuki and Kubo (1968) in the cluster approximation, it follows that

$$W_{q_j f_j}\{\ldots, \sigma_{f_j}^z(R_{q_j}), \ldots\} = \frac{1}{2\tau_0}\left[1 - \sigma_{f_j}^z(R_{q_j})\tanh\frac{\beta E_{f_j}(R_{q_j}, t)}{2}\right], \quad \beta = \frac{1}{kT}.$$

(6.3)

Here, τ_0 effectively defines the interaction of the system with the heat bath. The local field is

$$E_{f_j}(R_{q_j}, t),$$

which acts on the f_j-th deuteron in the q_j-th cell:

$$E_1(R_{q_j}, t) = w\sigma_2^z(R_{q_j}) + \frac{z}{\beta},$$

$$E_2(R_{q_j}, t) = w\sigma_1^z(R_{q_j}) + \frac{z}{\beta},$$

(6.4)

where

$$z = \beta[\Delta + \gamma + \mu E], \quad \gamma = \frac{v_1 \eta}{2},$$

(6.5)

$$\Delta = \Delta_{f_j}(R_{q_j}), \quad \eta = \langle \sigma_{f_j}^z(R_{q_j}) \rangle,$$

$$v_1 = \sum_{q_j f_j} J_{f_i f_j}(R_{q_j} R_{q_i}).$$

(6.6)

Here

$$\Delta_{f_j}(R_{q_j})$$

are effective fields created by neighboring bonds outside the cluster boundaries, and γ are the corresponding molecular fields created by long-range correlations. The parameters Δ, η and z are time-dependent.

It is difficult to determine the function

$$P\{\ldots, \sigma_{f_j}^z(R_{q_j}), \ldots, t\}$$

from the solution of Equation 6.2 (when Equation 6.3 is taken into account). Rather, it is advisable to calculate the time-dependent deuteron distribution functions for which it is possible to obtain the system of equations in the cluster approximation on the basis of Equation 6.2 and Equation 6.3:

$$
-\tau_0 \frac{d}{dt} \left\langle \left\{ \prod_{f_i=1}^{2} \sigma^z_{f_i}(\boldsymbol{R}_{q_i}) \right\} \right\rangle
$$

$$
= \sum_{f_j=1}^{2} \left\langle \left\{ \prod_{f} \sigma^z_{f_i}(\boldsymbol{R}_{q_i}) \right\} \times \left[1 - \sigma^z_{f_j}(\boldsymbol{R}_{q_j}) \tanh \frac{\beta E_{f_j}(\boldsymbol{R}_{q_j}, t)}{2} \right] \right\rangle. \quad (6.7)
$$

Using Equation 6.7, the two-particle cluster approximation allows the equation for the one-particle distribution function to be obtained:

$$
-\tau_0 \frac{d}{dt} \eta = (1 - P)\eta - L, \quad (6.8)
$$

where

$$
P = \frac{1 - a^2}{1 + a^2 + 2a \cosh z}, \quad L = \frac{2a \sinh z}{1 + a^2 + 2a \cosh z}, \quad a = e^{-\beta w}. \quad (6.9)
$$

In an analogous way, the corresponding equation for η at a one-particle approximation can be obtained:

$$
-\tau_0 \frac{d}{dt} \eta = \eta - \tanh\left(\frac{\bar{z}}{2}\right), \quad \bar{z} = \beta[2\Delta + \gamma + \mu E]. \quad (6.10)
$$

Thus, in the cluster approximation, the closed system of Equations 6.8 and 6.10 for η and Δ are obtained. On the other hand, attempts to calculate exactly the intrachain and long-range interactions in the mean-field approximation lead (on the basis of the Glauber model) to the chain of coupled equations for the function of deuteron distribution, which cannot be solved without an approximate decoupling of the chain. One of the simplest approximations in decoupling the chain is to take the quasi-static distribution function instead of the time-dependent function. However, the methods (Kanda et al. 1982; Žumer 1980) of decoupling the chain of equations are far from being consistent.

The solutions of Equations 6.8 and 6.10 will be limited to the case of small deviations from the equilibrium; η, Δ, and effective fields will be represented as the sum of the equilibrium term and their deviations from the equilibrium:

$$\eta = \tilde{\eta} + \eta_t, \quad \Delta = \tilde{\Delta} + \Delta_t; \quad t > 0, \quad E = 0,$$

$$z = \tilde{x} + x_t = \beta \left[\tilde{\Delta} + \frac{v_1 \tilde{\eta}}{2} \right] + \beta \left[\Delta_t + \frac{v_1 \eta_t}{2} \right], \tag{6.11}$$

$$\bar{z} = \tilde{\bar{x}} + \bar{x}_t = \beta \left[2\tilde{\Delta} + \frac{v_1 \tilde{\eta}}{2} \right] + \beta \left[2\Delta_t + \frac{v_1 \eta_t}{2} \right].$$

By substituting Equation 6.11 into Equations 6.8 and 6.10, expanding the functions P, L and tanh $(\bar{z}/2)$ into series, excluding Δ and Δ_t, and solving the obtained equations for $\tilde{\eta}$ and η_t, the following equations are yielded:

$$\tilde{\eta} = \frac{\sinh x}{a + \cosh x}, \quad x = \frac{1}{2} \ln \frac{1 + \tilde{\eta}}{1 - \tilde{\eta}} + \frac{v_1 \tilde{\eta}}{4kT}, \tag{6.12}$$

$$\eta_t = (\tilde{\eta}_E - \tilde{\eta})e^{-t/\tau}, \quad \eta = \tilde{\eta} + (\tilde{\eta}_E - \tilde{\eta})e^{-t/\tau}, \tag{6.13}$$

where the relaxation time is expressed by

$$\tau = \tau_0 \frac{3a^2 \cosh x - 2a^2 \cosh^3 x + \cosh x + a^3 + a}{2a \left[2a \cosh x - a \cosh^3 x + a^2 - (1 + a \cosh x)\dfrac{v_1}{4kT} \right]}, \quad T < T_c,$$

$$\tau = \tau_0 \frac{\cosh(w/2kT)}{a - \dfrac{v_1}{4kT}}, \quad T > T_c. \tag{6.14}$$

Equation 6.14 in the paraelectric phase agrees with the results of Kanda et al. (1982). Note that $\tilde{\eta}_E - \tilde{\eta}$ is used as the initial condition when solving the equation for η_t.

The spontaneous polarization is defined as

$$P_s = \frac{\mu}{V} \sum_{q_j f_j} \langle \sigma_{f_j}^z (R_{q_j}) \rangle = \frac{\mu}{V} \tilde{\eta}_E, \tag{6.15}$$

where V is the volume of the system.

The static susceptibility of the deuterated quasi-one-dimensional hydrogen-bonded ferroelectrics is given by

$$\chi_b = \lim_{E \to 0} \frac{dP_s}{dE} = \frac{f_b}{2T} \frac{1}{F_1 - v_1/4kT}, \quad f_b = \frac{\mu^2}{kV}, \tag{6.16}$$

where

$$F_1 = \frac{a + \cosh x^2}{1 + a \cosh x} - \frac{1}{1 - \tilde{\eta}^2}. \tag{6.17}$$

From Equation 6.16, a simple expression for χ_b in the paraelectric phase is obtained:

$$\chi_b = \frac{f_b}{2T} \frac{1}{a - v_1/4kT}. \tag{6.18}$$

The free energy per molecule in the cluster approximation is given by

$$f = -\frac{w}{4} - kT \left\{ \ln(\cosh x + a) + \tfrac{1}{2}\ln(1 - \tilde{\eta}^2) \right\} + \frac{v_1 \tilde{\eta}^2}{8}. \tag{6.19}$$

Calculations using Equation 6.19 yield the entropy and the specific heat of the crystals.

Analysis of the thermodynamic properties of CDP-type crystals (on the basis of the obtained equations) shows that the temperature of the second-order phase transition is defined here by

$$e^{-w/kT_c} = \frac{v_1}{4kT_c}. \tag{6.20}$$

The dynamic susceptibility is given by

$$\chi_b(\omega) = \chi_b - i\omega \int_0^\infty \Phi_b(t)e^{-\omega t}\, dt + \chi_b(\infty). \tag{6.21}$$

Here $\Phi_b(t)$ is the relaxational function (Equation 2.8) written as

$$\Phi_b(t) = \lim_{E \to 0} \frac{\mu}{V} \frac{d\eta_t}{dE} = \chi_b e^{-t/\tau}, \tag{6.22}$$

The contribution of electronic polarization and phonons is $\chi_b(\infty)$.

The real and imaginary parts of the complex dielectric permittivity are found by substituting Equation 6.22 into Equation 6.21:

$$\varepsilon_b'(v) = \varepsilon_b(\infty) + \frac{\chi_b}{1 + (\omega\tau)^2}, \tag{6.23}$$

$$\varepsilon_b''(v) = \frac{\chi_b \omega\tau}{1 + (\omega\tau)^2}, \tag{6.24}$$

where $\omega = 2\pi\nu$. The static dielectric permittivity follows from Equation 6.23:

$$\varepsilon_b(0) = \varepsilon_b(\infty) + \chi_b. \qquad (6.25)$$

The experimental data of CDP-type crystals will be discussed on the basis of the cluster theory where the direct tunneling effects are not taken into account because of the negative evidence of proton-tunneling (Ichikawa and Motida 1988). Proton-tunneling effects on the thermodynamic properties of quasi-one-dimensional ferroelectrics have been discussed by Zinenko (1985). However, Zinenko's paper contains some inaccuracies and the thermodynamic properties are less sensitive to the changes of the theory parameters than dynamical properties. Besides, the tunneling in quasi-one-dimensional ferroelectrics is strongly suppressed by the short-range interactions. The theoretical parameters, w and v_1, and variable parameters, f_b and τ_0, are determined by comparing the calculated data for T_c, $P_s(T)$, $\varepsilon_b'(\nu, T)$, $\varepsilon_b''(\nu, T)$, and $\varepsilon_b(0)$ with the experimental values (Deguchi et al. 1982; Kozlov et al. 1984; Kriukova 1984; Grigas et al. 1985; Levitsky et al. 1986; Grigas et al. 1988c) so that the best fit may be obtained between the theoretical and experimental values in a wide temperature range. For crystals with various degrees of deuteration the averaged parameter

$$w(x) = (1 - x)w_H + xw_D \qquad (6.26)$$

and corresponding parameters $v_1(x)$, $f_b(x)$ and $\tau_0(x)$ have been employed. Alternatively, $w(x)$ and $v_1(x)$ may be determined by Equations 6.26 and 6.20. The parameters of the theory obtained for the various x values are presented in Table 6.1, where parameters R_1 and R_2 enter the constant τ_0 (Equation 6.3) in the following way:

$$\tau_0 = [R_1 + R_2(\Delta T - 1)] \times 0^{-16}\,\text{s}, \qquad \Delta T = T - T_c, \qquad (6.27)$$

Using Equations 6.23 and 6.24 to calculate $\varepsilon_b'(\nu, T)$ and $\varepsilon_b''(\nu, T)$ with the corresponding parameters from Table 6.1, shows that the cluster approximation gives a good quantitative description of the experimental data for both slightly (Levitsky et al. 1986) and highly (Grigas et al. 1988c) deuterated CPD and other H-bonded ferroelectrics (Mizeris et al. 1990).

Figure 6.3 shows the results for only some of the frequencies of the CsD_2PO_4 crystal ($x = 0.94$). Figure 6.4 shows a complete dielectric spectrum of the CsD_2PO_4 crystal in the region of fundamental ferroelec-

Table 6.1. The theory parameters which describe the soft mode dynamics and the static properties of $Cs(H_{1-x}D_x)_2PO_4$ crystals

No	x	T_c (K)	w/k (K)	v_1/k (K)	$T < T_c$ f_b	R_1 (s)	R_2 (s/K)	$T > T_c$ f_b	R_1 (s)	R_2 (s/K)
1	0.98	268.3	1080	19.16	583	622	0	1030	950	−4.13
2	0.95	266	1069	18.84	577	606	0	1018	925	−4.03
3	0.94	263.5	1065	18.51	575	551	23	1014	917	−3.99
4	0.82	249.8	1018	16.97	550	538	0	966	819	−3.58
5	0.14	172	753	8.63	–	–	–	695	260	0
6	0.03	159.3	711	7.27	390	100	3.3	500	121	0.4
7	0	156	700	7.02	384	136	0	640	167	0

tric dispersion in the paraelectric phase. The ferroelectric dispersion shows a monodispersive nature, and the soft relaxational mode dynamics over a wide frequency range are successfully described by the cluster theory.

Figure 6.5 shows the results calculated from Equation 6.14 of the inverse soft mode relaxation time for different CDP-type crystals. On approaching the Curie temperature, the relaxation time undergoes a more intensive critical slowing down than in KDP-type crystals and the soft mode relaxation time is about 10 times higher than that inKDP-type crystals. The anisotropy of the interactions and the resulting short-range order along the chains may be responsible for the slow soft mode dynamics in CDP-type crystals (Blinc and Shmidt 1984).

Also, the cluster approximation effectively describes the static properties of CDP-type crystals (Levitsky et al. 1986; Grigas et al. 1988c). Note that in the vicinity of the phase transition temperature, there is a weak dependence of the characteristics of these crystals on the parameter w (see Equation 6.1). This dependence becomes considerable only at $T \gg T_c$. One should note that conclusions about the possibility of the theoretical description of the experimental results could be made only by comparing a number of the experimental data with theoretical results in a wide temperature region. For this the dynamic dielectric properties should be investigated in the frequency range of 10^6 to 10^{11} Hz.

Figure 6.6 shows the inverse soft mode relaxation time for CsH_2PO_4 calculated by Equation 6.14 using the parameters given in

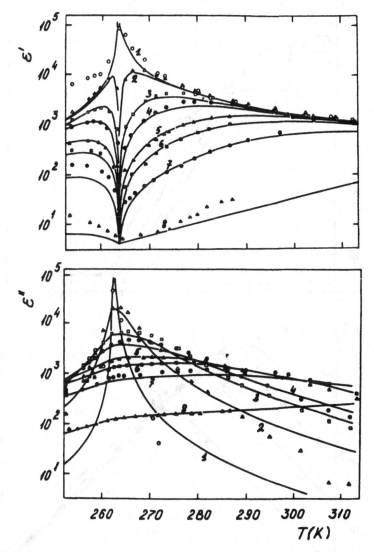

Figure 6.3. Temperature dependence of ε_b' and ε_b'' of CsD_2PO_4 crystal at frequencies (GHz): 1) 0.0051; 2) 0.0724; 3) 0.251; 4) 0.413; 5) 0.73; 6) 1.044; 7) 2.0; and 8) 11.5. Dots are experimental values. Lines are theoretical values. From Grigas et al. (1988c).

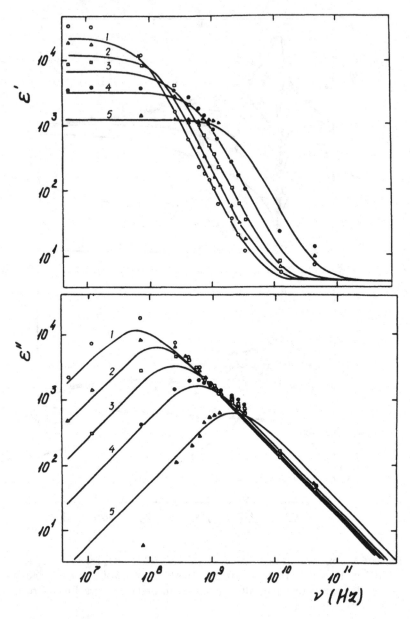

Figure 6.4. Frequency dependence of ε_b' and ε_b'' of CsD_2PO_4 at temperatures $T - T_c$ (K): 1) 3; 2) 5.6; 3) 10; 4) 19.4; 5) 45. Dots are experimental values. Lines are theoretical values. From Grigas et al. (1988c).

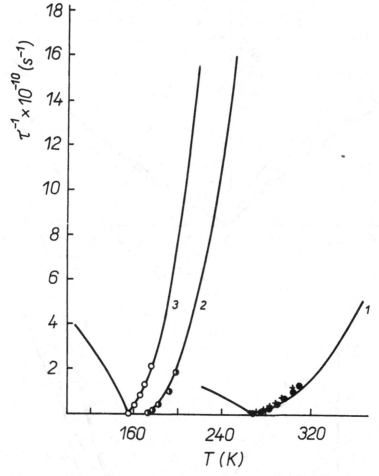

Figure 6.5. Temperature dependence of the inverse soft mode relaxation time of $Cs(H_{1-x}D_x)_2PO_4$. Dots are experimental values for $x = 0.98$ (1), 0.14 (2), and 0 (3). Lines are theoretical values. From Levitsky et al. (1986).

Table 6.2. Curves 2 and 4 are calculated using the theory parameters given by Kanda et al. (1982). Curve 4 at $T < T_c$ is calculated by Equation 12 of Kanda et al. (1982), and curve 3 is calculated using the parameters given by Deguchi et al. (1982). Finally, curve 5 is obtained from submillimeter range data by Kriukova (1984). The curve is obtained by least squares fitting.

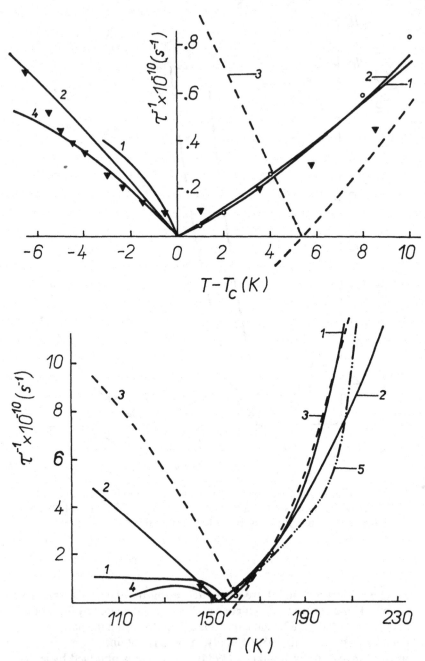

Figure 6.6. Comparison of the temperature dependence of inverse relaxation time for CDP with the different parameters taken from Table 6.2. The number of each curve corresponds to the number for the parameter set in Table 6.2. From Levitsky et al. (1986).

Table 6.2. Comparison of the theory parameters of CsH_2PO_4 for relaxation time description (Figure 6.6)

No	w/k (K)	v_1/k (K)	$\tau_0(s)$ $T < T_c$	$T > T_c$
1	700	7.02	1.36×10^{-14}	1.67×10^{-14}
2	468	27.6		1.9×10^{-13}
3	531	24.0		6.7×10^{-14}
4	468	27.6		1.9×10^{-13}
5	The curve is obtained by least square method			

The expressions for χ_b and τ described by Žumer (1980) are transformed to

$$\chi_b = \frac{1 + a - a^2}{a(1 + a - a^2)\dfrac{v_1}{4kT}} \cdot \frac{A}{T}, \quad A = 2\pi f_b, \qquad (6.28)$$

$$\tau = \tau_0 \frac{1 + a - a^2}{2a\left[a - (1 + a - a^2)\dfrac{v_1}{4kT}\right]}, \quad v_1 = 4J_\perp. \qquad (6.29)$$

Equations 6.28 and 6.29 have no singularities at the Curie point, which is the result of an approximate decoupling of the chain of equations for the time-dependent distribution functions. Only when

$$a \ll 1, \quad (1 + a - a^2 \simeq 1)$$

are the expressions (6.14) and (6.18) obtained from Equations 6.28 and 6.29.

Finally, the direct accounting by Watarai and Matsubara (1984) of tunneling effects in CDP crystals does not improve the agreement of theory with the experimental data for $\varepsilon_b^{-1}(T)$ and $\tau(T)$ in the vicinity of T_c compared to the cluster theory.

Thus, the cluster theory—treating the phase transition as the result of instability in a deuteron (proton) subsystem—gives a quantitative description of the relaxational soft mode dynamics and the static properties for deuterated and even undeuterated crystals without taking into account the tunneling effects.

6.3. MICROWAVE DIELECTRIC DISPERSION IN PbHPO$_4$

Lead hydrogen phosphate PbHPO$_4$ (LHP) and its deuterated isomorph LDP are monoclinic hydrogen-bonded ferroelectrics which, at the Curie temperatures $T_c^H = 310$ K and $T_c^D = 452$ K, undergo the second-order proper ferroelectric phase transition $P2/c \rightarrow Pc$ of order-disorder type (Negran et al. 1974), similar to that of the CDP and DCDP ferroelectrics. The crystal structure consists of two formula units in the primitive cell. The PO$_4$ groups are linked into isolated one-dimensional chains along the c-axis by short O—H(D)$\cdots$O bonds, which are coupled to each other by long-range dipolar forces only. The isomorphous compound is PbHAsO$_4$. The large isotope effect on T_c, neutron (Nelmes 1980), and NMR (Ermak et al. 1989) studies show that the ordering of protons (deuterons) in the H-bonds, as in the case of CDP/DCDP, triggers the ferroelectric phase transition.

Despite the simple structure of LHP, some basic aspects of the transition have not been discussed until now. The initially proposed centrosymmetric structure for paraelectric phase has not always been confirmed by subsequent structural studies. In addition to previously mentioned studies, LHP has been studied by Raman (Ohno and Lockwood 1989), hyper-Raman (Shin et al. 1990), IR (Kock and Happ 1980), second harmonic generation (Keens and Happ 1988), and microwave techniques. According to some authors (Lockwood et al. 1987; Ohno and Lockwood 1989), the space group of LHP remains Pc (C_s^2) above T_c. However, the existence of Pc symmetry above T_c is due to the presence of an internal biasing field produced by oriented defects with non-zero charge and dipole moment (Ermak et al. 1989). This field breaks the symmetry of the paraelectric phase.

The second problem is the type of the soft ferroelectric mode related to the phase transition of LHP. There is no soft mode in the IR region that could account for the static permittivity. Kroupa et al. (1987) examined the dielectric spectrum in the 6 to 20 cm^{-1} (180 to 600 GHz) range and found that dielectric dispersion is caused by a relaxational soft mode. From the microwave measurements, Kock and Happ (1980) and Happ et al. (1985) concluded that the soft mode is a heavily damped mode with a frequency-dependent damping constant and the proton dynamics cannot be explained by a simple relaxational mode. Dielectric dispersion studies by Deguchi and Nakamura (1988) in the range 1 kHz to 1 GHz show the relaxation of the Debye type with a single relaxation time, which is slightly higher than the results obtained by Shin et al. (1990). However, the results below 1 GHz give information about a relaxational soft mode only in the vicinity of T_c, while

Table 6.3. The parameters of LHP and LDP

Crystal	T_c (K)	w/k (K)	v_1/k (K)	f
LDP	452	1450	73.11	50
LHP	310	1160	29.4	40
	$\tau_0 = [123 + 0.08(\Delta T - 1)] \times 10^{-16}$ s			$(T > T_c)$
	$\tau_0 = [12 + 0.06(\Delta T + 1)] \times 10^{-16}$ s			$(T < T_c)$

submillimeter spectra give information only about a tail of dielectric dispersion.

Microscopic models of the phase transition in LHP and LDP have been discussed by Carvalko and Salinas (1978), and others.

Relaxational dynamics and the phase transition in LHP and LDP can be treated on the basis of the Hamiltonian (6.1) in the two-particle cluster approximation. The parameters of the cluster theory (discussed in detail in Section 6.2) are presented (Mizeris et al. 1990) in Table 6.3.

Figure 6.7 shows the experimental temperature dependence of the real and imaginary parts of the complex dielectric permittivity of one of the LHP crystals along the a-axis at several frequencies. The maximum value of ε' in the MHz frequency region depends on the internal bias electric field and varies in the range $(1 \div 10) \times 10^3$. The permittivity follows the Curie–Weiss law, and the ratio of the Curie–Weiss constants $C_p/C_f = 2$ confirms the second order of the phase transition. Figure 6.8 shows a complete dielectric spectrum of LHP in the paraelectric phase. As in other H-bonded ferroelectrics, the region of a main relaxational dielectric dispersion is between 10^8 and 10^{11} Hz. An internal bias electric field causes a distribution of relaxation times. The average relaxation frequency, $1/2\pi\tau$, on approaching $T_0 = 308.6$ K softens up to 2.2 GHz and accounts for all static permittivity of LHP and its temperature dependence.

The distribution of the relaxation times decreases the maximum value of the static permittivity near T_c, widens the dispersion region, and eliminates a minimum of ε' at T_c caused by the critical slowing down of the polarization fluctuations in the dispersion region of the second-order phase transition. Such a minimum of ε' follows from Equation 2.11. In the samples with a small internal bias field in the vicinity of T_c there is a minimum of ε' in the region of 15 to 600 GHz.

The relaxational soft mode in the paraelectric phase is evidently associated with the flipping motion of H-atoms between two potential

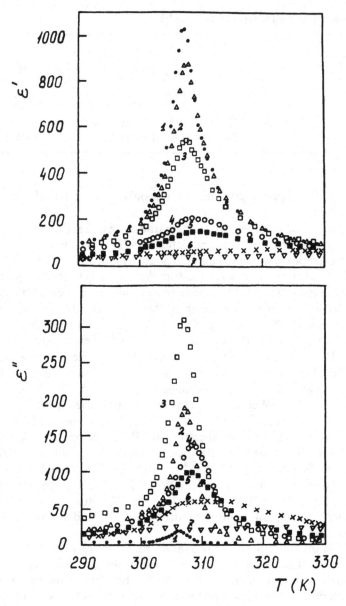

Figure 6.7. Temperature dependence of ε' and ε'' of LHP crystal at frequencies (GHz): 1) 0.05; 2) 0.62; 3) 2.1; 4) 8.4; 5) 15; 6) 54; and 7) 77. From Mizeris et al. (1992).

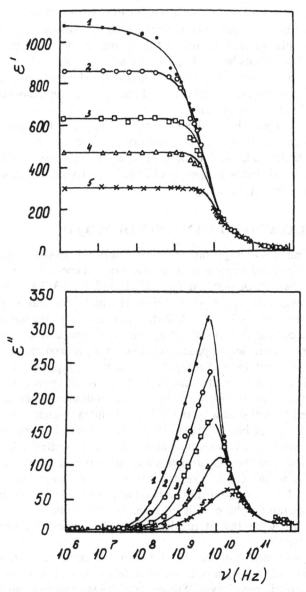

Figure 6.8. Frequency dependence of ε' and ε" of LHP crystals at temperatures $T - T_c$ (K): 1) 0.16; 2) 1.66; 3) 2.66; 4) 4.16; and 5) 7.16. From Mizeris et al. (1992).

minima in the O–H...O bonds linking PO_4 groups. The ferroelectric transition in LHP and LDP is primarily order-disorder in character. Proton/deuteron ordering drives the transition. However, the heavy-atom displacements below T_c, produced as a secondary effect, result in some features (Ohno and Lockwood 1989), which are displacive in nature. Nevertheless, the relaxational mode makes the main contribution to the static permittivity.

Two-particle cluster theory, treating the phase transition as the result of instability in the proton subsystem, describes the relaxational soft mode dynamics and static properties of these crystals. Dynamic acoustic and dielectric properties of $PbHPO_4$ have been discussed also by Mizaras et al. (1994a).

6.4. RELAXATIONAL DYNAMICS OF RbD_2PO_4

Monoclinic RbD_2PO_4 is one of the less studied quasi-one-dimensional hydrogen-bonded ferroelectrics. It undergoes successive phase transitions $P2_1/m \rightarrow P2_1/c \rightarrow P2_1$ at $T_{c1} = 377$ K and $T_{c2} = 317$ K as temperature decreases (Hagiwara et al. 1984). The high temperature phase is isomorphous to CsD_2PO_4 while the intermediate phase has a super-structure comparable to CsD_2PO_4 with four molecular units per unit cell. The nonferroelectric phase transition at T_{c1} is accompanied by the unit cell constant doubled along the c-axis, while the phase transition at T_{c2} is related to b-directed atomic displacements and the occurrence of a superlattice structure with the unit cell constant doubled along the a-axis.

The PO_4 tetrahedra, as in CDP, are bonded to each other by two types of hydrogen bonds. The shorter O–D–O bonds form zigzag quasi-one-dimensional chains running along the b-axis, while the longer O–D–O bonds link the PO_4 groups along the c-axis (see Figure 6.1). The deuterons in the chains are ordered at $T < T_{c2}$ and disordered at $T > T_{c2}$, while those along the c-axis are ordered at all temperatures. The low temperature phase with eight molecular units per unit cell is ferroelectric, while the intermediate and high temperature phases are paraelectric.

Experimental data on the static and dynamic properties of the monoclinic RbD_2PO_4 crystal—such as temperature dependence of the static dielectric permittivity (Sumita et al. 1984), spontaneous polarization (Osaka et al. 1983), dynamic dielectric properties (Mizeris et al. 1987), and NMR studies (Topič et al. 1984)—reveal that the phase transitions are of second order. At T_{c2}, there is an order-disorder ferroelectric transition and the Debye-type dielectric dispersion is caused by a single relaxational soft mode.

The ferroelectric phase transition and the static dielectric permittivity of the RbD_2PO_4 crystal have been theoretically discussed by Komukae and Makita (1985) using the quasi-one- dimensional Ising model and assuming two sublattices. In this model, the short-range intrachain interactions are exactly accounted for, and the long-range interchain interactions between deuterons are approximated by a mean-field. Ising model parameters have been found that give a good description of the temperature dependence of the static dielectric permittivity in the intermediate phase.

The dynamic dielectric properties of the monoclinic RbD_2PO_4 crystals will be considered below in the range of fundamental ferroelectric dispersion, 10^6 to 10^{11} Hz, using a dynamical model in the two-particle cluster approximation (Levitsky et al. 1990). As in the case of CsD_2PO_4 crystals, the deuteron system in RbD_2PO_4, moving along O—D...O bonds, which form zigzag chains along the b-axis of the crystal will be considered. The Bravais lattice of the low temperature phase ($T < T_{c2}$) is assumed to have a unit cell doubled along the a- and c-axes. The unit cell consists of four chains ($\alpha = 1, \ldots, 4$). Every chain consists of two neighboring PO_4 tetrahedra with two short hydrogen bonds belonging to one of them (tetrahedra of A-type). Hydrogen bonds of the other tetrahedron (B-type) belong to the two nearest structure elements which surround it.

In an external electric field the Hamiltonian of the deuteron system of RbD_2PO_4, taking into account short- and long-range interactions, has the form

$$
\hat{H} = -\frac{w}{2} \sum_{q_i q_j} \sum_\alpha \sigma_1^{z\alpha}(R_{q_j}) \sigma_2^{z\alpha}(R_{q_j})(\delta_{R_{q_i} R_{q_j}} + \delta_{R_{q_i} + r_\alpha, R_{q_j}})
$$

$$
-\frac{1}{8} \sum_{\substack{q_i q_j \\ f_i f_j}} \sum_\alpha J_{f_i f_j}^{\alpha\alpha}(R_{q_i} R_{q_j}) \sigma_{f_i}^{z\alpha}(R_{q_i}) \sigma_{f_j}^{z\alpha}(R_{q_j})
$$

$$
-\frac{1}{8} \sum_{\substack{q_i q_j \\ f_i f_j}} \sum_{\substack{\alpha\alpha' \\ (\alpha \neq \alpha')}} K_{f_i f_j}^{\alpha\alpha'}(R_{q_i} R_{q_j}) \sigma_{f_i}^{z\alpha}(R_{q_i}) \sigma_{f_j}^{z\alpha'}(R_{q_j}) \tag{6.30}
$$

$$
-\frac{1}{2} \sum_{q_i f_j} \sum_\alpha (\mu_\alpha E) \sigma_{f_i}^{z\alpha}(R_{q_i}).
$$

The first term describes the short-range configuration interactions of deuterons in the chains near the A-type tetrahedra (first Kronecker symbol) and those of the B-type (second Kronecker symbol). The second and third terms describe the effective long-range interaction of deuterons in the same and different chains, respectively, including, indirectly, the interactions through the lattice vibrations. μ_α is the effective dipole moment of the αth chain in the unit cell along the b-axis. Other designations are the same as in Equation 6.1. Chains 1 and 2, which are bonded by the long hydrogen $O-D-\ldots O$ bonds and in which deuterons are ordered at all temperatures, form the a-type sublattice. Chains 3 and 4 form the b-type sublattice. The existence of the two sublattices follows from the experimental data and the crystal symmetry analysis. A cluster consists of two ordering hydrogen bonds near the A- and B-type tetrahedra. Also, w is a parameter of the cluster. First, the main results of the thermodynamic properties of monoclinic RbD_2PO_4 crystals, as obtained from Equation 6.30, are presented.

The spontaneous polarization is given by

$$P_s(T) = \frac{1}{V} \sum_{f_j} \sum_\alpha \mu_\alpha \tfrac{1}{2} \langle \sigma_{f_i}^{z\alpha} \rangle = \frac{2}{V} [\mu_a \tilde{\eta}_a - \mu_b \tilde{\eta}_b], \qquad (6.31)$$

where V is the volume of the unit cell and the order parameters $\tilde{\eta}_a$ and $\tilde{\eta}_b$ satisfy Equation 6.32:

$$\tilde{\eta}_a = \frac{\sinh x_a}{a + \cosh x_a}, \quad \tilde{\eta}_b = -\frac{\sinh x_b}{a + \cosh x_b}, \quad a = e^{-w/kT}, \qquad (6.32)$$

where

$$x_a = \frac{1}{2} \ln \frac{1 + \tilde{\eta}_a}{1 - \tilde{\eta}_a} + \frac{v_a}{4kT} \tilde{\eta}_a - \frac{v_{ab}}{4kT} \tilde{\eta}_b + \frac{(\mu_a E)}{2kT},$$

$$x_b = -\frac{1}{2} \ln \frac{1 + \tilde{\eta}_b}{1 - \tilde{\eta}_b} + \frac{v_{ab}}{4kT} \tilde{\eta}_a - \frac{v_b}{4kT} \tilde{\eta}_b + \frac{(\mu_b E)}{2kT}, \qquad (6.33)$$

$$v_a = v_{11} + v_{12}, \quad v_b = v_{11} + v_{34}, \quad v_{ab} = v_{13} + v_{14}.$$

Note that the following relations are used to obtain Equation 6.31:

$$v_{11} = \sum_{q_j f_j} J_{f_i f_j}(R_{q_i} R_{q_j}),$$

$$v_{12} = \sum_{q_j f_j} K_{f_i f_j}^{12}(R_{q_i} R_{q_j}) = \sum_{q_j f_j} K_{f_i f_j}^{21}(R_{q_i} R_{q_j})$$

$$= \sum_{q_j f_j} K_{f_i f_j}^{34}(R_{q_i} R_{q_j}) = \sum_{q_j f_j} K_{f_i f_j}^{43}(R_{q_i} R_{q_j}),$$

$$v_{13} = \sum_{q_j f_j} K_{f_i f_j}^{13}(R_{q_i} R_{q_j}) = \sum_{q_j f_j} K_{f_i f_j}^{31}(R_{q_i} R_{q_j})$$

$$= \sum_{q_j f_j} K_{f_i f_j}^{24}(R_{q_i} R_{q_j}) = \sum_{q_j f_j} K_{f_i f_j}^{42}(R_{q_i} R_{q_j}),$$

$$v_{14} = \sum_{q_j f_j} K^{14}_{f_i f_j}(R_{q_i} R_{q_j}) = \sum_{q_j f_j} K^{41}_{f_i f_j}(R_{q_i} R_{q_j})$$

$$= \sum_{q_j f_j}^{\infty} K^{23}_{f_i f_j}(R_{q_i} R_{q_j}) = \sum_{q_j f_j} K^{32}_{f_i f_j}(R_{q_i} R_{q_j}),$$

$$v_a \neq v_b \quad (T < T_{c2}); \quad v_a = v_b = \bar{v} \quad (T < T_{c2}), \tag{6.34}$$

and also definitions:

$$\tilde{\eta}_a = \langle \sigma^{z1}_{f_j}(R_{q_j}) \rangle = \langle \sigma^{z2}_{f_j}(R_{q_j}) \rangle,$$

$$\tilde{\eta}_b = -\langle \sigma^{z3}_{f_j}(R_{q_j}) \rangle = -\langle \sigma^{z4}_{f_j}(R_{q_j}) \rangle, \tag{6.35}$$

$$\mu_a = \mu_1 = \mu_2, \quad \mu_b = \mu_3 = \mu_4,$$

obtained from the symmetry analysis of the crystal.

From Equations 6.31 and 6.32 the static susceptibility is given by

$$\chi(0, T) = \lim_{E \to 0} \frac{dP(T)}{dE} = \frac{1}{4T} \frac{f_a\left(F_b - \dfrac{v_b}{4kT}\right) + f_b\left(F_a - \dfrac{v_a}{4kT}\right) + 2(f_a f_b)^{1/2} \dfrac{v_{ab}}{4kT}}{\left(F_a - \dfrac{v_a}{4KT}\right)\left(F_b - \dfrac{v_b}{4kT}\right) - \left(\dfrac{v_{ab}}{4kT}\right)^2}, \tag{6.36}$$

where

$$f_l = \frac{\mu_l^2}{kV}, \quad l = a, b,$$

$$F_l = \frac{(a + \cosh \tilde{x}_{lo})^2}{1 + a \cosh \tilde{x}_{lo}} - \frac{1}{1 - \tilde{\eta}_{lo}^2}, \tag{6.37}$$

$$x_l|_{E=0} = x_{lo}, \quad \tilde{\eta}_l|_{E=0} = \tilde{\eta}_{lo}.$$

Equation 6.36 agrees with the result obtained by Komukae and Makita (1985). The free energy per molecule in the cluster approximation is given by

$$f = \frac{F}{4kN} = 2\frac{w}{k} + \frac{1}{8}\frac{v_a}{k}\tilde{\eta}_{ao}^2 + \frac{1}{8}\frac{v_b}{k}\tilde{\eta}_{bo}^2 - \frac{1}{4}\frac{v_{ab}}{k}\tilde{\eta}_{ao}\tilde{\eta}_{bo}$$

$$- T\{\ln(a + \cosh x_{ao}) + \ln(a + \cosh x_{bo}) \tag{6.38}$$

$$- \ln[1/(1 - \tilde{\eta}_{ao}^2)^{1/2}] - \ln[1/(1 - \tilde{\eta}_{bo}^2)^{1/2}]\}.$$

The analysis of Equation 6.38 shows that the ferroelectric phase transition in RbD_2PO_4 takes place at a temperature that is defined by the relation

$$a^2 - a\frac{v_a(1+y)}{4kT_{c2}} + \frac{v_a^2 y - v_{ab}}{(4kT_{c2})^2} = 0, \quad y = \frac{v_b}{v_a}. \tag{6.39}$$

Equation 6.38 enables the calculation of the entropy and the specific heat of the crystal.

Note that the model describes only the phase transition in monoclinic RbD_2PO_4 from the intermediate to the low temperature phase (which is related to the ferroelectric ordering of the deuterons in the a- and b-sublattices). Obviously, the character of the ordering of deuterons at the phase transition temperature T_{c1} does not change.

The treatment of the dynamic properties of monoclinic RbD_2PO_4 crystals in the frame of a two-particle cluster approximation (based on the Glauber (1963) model) will be limited to the case of small deviations from equilibrium. It will represent the distribution η_a and η_b, the effective field, and x_a and x_b (Equation 6.33) as the sum of the equilibrium terms, their time-dependent fluctuating terms η_{at}, η_{bt}, and E_t. A set of equations for fluctuating terms of the distribution functions in the two-particle cluster approximation is given by

$$\frac{d}{dt}\eta_{at} = N_{1a}\eta_{at} + N_{2a}N_{bt} - N_a\frac{(\boldsymbol{\mu}_a\boldsymbol{E}_t)}{2kT},$$

$$\frac{d}{dt}\eta_{bt} = N_{2a}\eta_{at} + N_{1a}\eta_{bt} - N_b\frac{(\boldsymbol{\mu}_b\boldsymbol{E}_t)}{2kT}, \tag{6.40}$$

where

$$N_{1l} = \frac{1}{\tau_0}\left(1 - P_{l0}\frac{2\phi_l}{2\phi_l - R_l} - \frac{v_l}{4kT}\frac{R_l\phi_l}{2\phi_l - R_l}\right),$$

$$N_{2l} = -\frac{1}{\tau_0}\frac{v_{ab}}{4kT}\frac{R_l\phi_l}{2\phi_b - R_b}, \quad N_l = \frac{1}{\tau_0}\frac{R_l\phi_l}{2\phi_b - R_b}. \tag{6.41}$$

The temperature-dependent constant is τ_0 (see Equation 6.27), effectively defining the interaction of the deuterons with the heat bath, and

$$P_{l0} = \frac{1 - a^2}{1 + a^2 + 2a\cosh\tilde{x}_{l0}}, \quad P_{l1} = -\frac{4a(1 - a^2)\sinh\tilde{x}_{l0}}{(1 + a^2 + 2a\cosh\tilde{x}_{l0})^2},$$

$$R_l = P_{l1}\tilde{\eta}_{l0} + L_{l1}, \quad L_{l1} = \frac{4a[2a + (1 + a^2)\cosh\tilde{x}_{l0}]}{(1 + a^2 + 2a\cosh\tilde{x}_{l0})^2}, \tag{6.42}$$

$$\phi_l = 1 - \tilde{\eta}_{l0}^2.$$

Solving the set of Equations 6.40 gives

$$\eta_{at} = C_1 e^{-t/\tau_1} + C_2 e^{-t/\tau_2} + \frac{E_t}{2kT} \frac{D_a}{D},$$

$$\eta_{bt} = \tilde{C}_1 e^{-t/\tau_1} + \tilde{C}_2 e^{-t/\tau_2} + \frac{E_t}{2kT} \frac{D_b}{D},$$

(6.43)

where $C_1, C_2, \tilde{C}_1$ and $\tilde{C}_2$ are constants, τ_1 and τ_2 are the relaxation times of polarization fluctuations:

$$\tau_{1,2}^{-1} = \tfrac{1}{2}\{(N_{1a} + N_{1b}) \pm [(N_{1a} + N_{1b})^2 - 4(N_{1a}N_{1b} - N_{2a}N_{2b}\cdots)]^{1/2}\},$$

(6.44)

and

$$D_a = i\omega N_a \mu_a + N_a N_{1b}\mu_a - N_b N_{2a}\mu_b,$$

$$D_b = i\omega N_b \mu_b + N_b N_{1a}\mu_b - N_a N_{2b}\mu_a,$$

$$D = (i\omega)^2 + (N_{1a} + N_{1b})i\omega + (N_{1a}N_{1b} - N_{2a}N_{2b}).$$

(6.45)

The dynamic susceptibility of the crystal is given by

$$\chi = \frac{2}{V}\left[\mu_a \lim_{E \to 0} \frac{d\eta_{at}}{dE} + \mu_b \lim_{E \to 0} \frac{d\eta_{bt}}{dE}\right] = \frac{\chi_1}{1 + i\omega\tau_1} + \frac{\chi_2}{1 + i\omega\tau_2},$$

(6.46)

where the dielectric contributions are defined as

$$\chi_1 = \frac{1}{2T}\frac{\tau_1\tau_2}{\tau_2 - \tau_1}\{f_a N_a + f_b N_b - \tau_1[f_a N_a N_{1b} + f_b N_b N_{1a}$$

$$- (f_a f_b)^{1/2}(N_b N_{2a} + N_a N_{2b})]\},$$

$$\chi_2 = \frac{1}{2T}\frac{\tau_1\tau_2}{\tau_2 - \tau_1}\{f_a N_a + f_b N_b - \tau_2[f_a N_a N_{1b} + f_b N_b N_{1a}$$

$$- (f_a f_b)^{1/2}(N_b N_{2a} + N_a N_{2b})]\}.$$

(6.47)

The expressions for the real and imaginary parts of the complex dielectric permittivity of the RbD_2PO_4 crystal are given by

$$\varepsilon'(v, T) = \varepsilon_\infty + \frac{\chi_1}{1 + (\omega\tau_1)^2} + \frac{\chi_2}{1 + (\omega\tau_2)^2},$$

$$\varepsilon''(v, T) = \frac{\chi_1\omega\tau_1}{1 + (\omega\tau_1)^2} + \frac{\chi_2\omega\tau_2}{1 + (\omega\tau_2)^2}.$$

(6.48)

Here $\omega = 2\pi v$, ε_∞ (as in Equation 6.23) is the dielectric contribution of the high frequency phonons and electronic polarization. The static permittivity follows from Equation 6.48:

$$\varepsilon(0, T) = \varepsilon_\infty + \chi(0, T). \tag{6.49}$$

Thus, the expressions in the two-particle cluster approximation for the thermodynamic and dynamic properties of monoclinic RbD_2PO_4 crystals have been obtained. The experimental data (Mizeris et al. 1987) will be discussed on the basis of the cluster theory. The parameters of the theory, w, v_a, v_b, and v_{ab}, and variable parameters f_a, f_b, and τ_0 can be determined by comparing the calculated data for T_{c2}, $\varepsilon(0, T)$, $\varepsilon'(v, T)$ and $\varepsilon''(v, T)$ with experiment. However, evaluation of the parameters of the long-range interactions v_{12} and v_{34}, and the temperature dependence of the spontaneous polarization of the sublattices (Baranov et al. 1987) allow simplification of the RbD_2PO_4 crystal model, taking $v_a \simeq v_b = \bar{v}$ and $\gamma = \mu_b/\mu_a = 0.99$. Then, from Equation 6.32, when $E = 0$ the order parameter is given by

$$\tilde{\eta}_0 = \tilde{\eta}_{a0} = \tilde{\eta}_{b0} = \frac{\sinh \tilde{x}_0}{a + \cosh \tilde{x}_0}, \tag{6.50}$$

where

$$\tilde{x}_0 = \frac{1}{2}\ln\frac{1 + \tilde{\eta}_0}{1 - \tilde{\eta}_0} + \frac{\bar{v} - v_{ab}}{4kT}\tilde{\eta}_0. \tag{6.51}$$

From Equations 6.39 and 6.50, when $\tilde{\eta}_0 \to 0$ the equation for the phase transition temperature T_{c2} in the simplified model is defined similarly to Equation 6.20 by

$$e^{-w/kT_{c2}} = \frac{\bar{v} - v_{ab}}{4kT_{c2}}. \tag{6.52}$$

The expression of the state dielectric susceptibility can also be simplified. From Equation 6.36 it becomes

$$\chi(0, T) = \frac{f_a}{2T}\frac{1}{F - \dfrac{\bar{v} + v_{ab}}{4kT}}(1 + \gamma)^2 - \frac{f_a}{2T}\frac{1}{F - \dfrac{\bar{v} - v_{ab}}{4kT}}(1 - \gamma)^2. \tag{6.53}$$

The main contribution to $\chi(0, T)$ gives the first term because $(1 - \gamma)^2 \ll 1$. Only in the close vicinity of T_{c2} (when $T - T_{c2} < 10^{-5}$ K) does the second term of Equation 6.53 exceed the first and its value go to infinity as $T \to T_{c2}$. This result coincides with the results obtained by Komukae and Makita (1985) using the Ising model.

Equations 6.44, 6.47 and 6.48 can also be simplified. The reciprocal relaxation time is expressed by

$$\tau_{1,2}^{-1} = \frac{1}{\tau_0} \frac{R\phi}{2\phi - R} \left\{ \frac{2\phi - R - 2\phi P_0}{R\phi} - \frac{\bar{v} \pm v_{ab}}{4KT} \right\}, \qquad (6.54)$$

and the dielectric contributions $\chi_{1,2}$ in Equation 6.47 become

$$\chi_{1,2} = \frac{f_a}{2T} \frac{(1 \pm \gamma)^2}{\dfrac{2\phi - R - 2\phi P_0}{R\phi} - \dfrac{\bar{v} \pm v_{ab}}{4kT}}. \qquad (6.55)$$

When $v_a = v_b = \bar{v}$, one can obtain simplified equations for P_0, R and ϕ from Equation 6.42.

The parameters of the theory are determined by comparing calculated and experimental values of T_{c2}, $\varepsilon(0, T)$, $\varepsilon'(v, T)$ and $\varepsilon''(v, T)$ in a wide temperature region and are presented in Table 6.4. The parameters R_1 and R_2 enter the constant τ_0 in the following way:

$$\tau_0 = \begin{cases} R_1 + (\Delta T_1 - 1)R_2 \\ R_1 + (\Delta T_2 - 1)R_2 \\ R_1 + (-\Delta T_1 - 1)R_2 \end{cases} \times 10^{-16}(s), \quad \begin{matrix} \Delta T_1 = T - T_{c1}, T > T_{c1}, \\ \Delta T_2 = T - T_{c2}, T_{c2} < T < T_{c1} \\ T < T_{c2}. \end{matrix}$$

Figure 6.9 shows the temperature dependence of the real and imaginary parts of the complex dielectric permittivity $\varepsilon^* = \varepsilon' - i\varepsilon''$ for the RbD_2PO_4 crystal along the b-axis for some of the frequencies. The peak value of ε' at 1 MHz is about 8×10^3. This value is regarded as the static dielectric permittivity $\varepsilon(0)$ and approximates the value shown at 50 MHz. The dynamic dielectric properties for RbD_2PO_4 crystals—obtained by Komukae and Makita (1985) in the 1 MHz to 1 GHz frequency range and the Ising model parameters are slightly different. According to designations in this chapter they are: $w/k = 2400\,K$, $(v_a - v_{ab})/k = 0.786\,K$, and $(v_a + v_{ab})/k = 0.755\,K$.

Figure 6.10 shows a complete dielectric spectrum of a RbD_2PO_4 crystal in the region of fundamental ferroelectric dispersion. Due to the small value of $(1 - \gamma)^2$ in Equation 6.53, the dielectric contribution, χ_2, of the relaxation mode with time constant, τ_2, is negligible in comparison with the dielectric contribution of the relaxation mode with time constant τ_1. As a result, the ferroelectric dispersion has a monodispersive nature and the soft relaxational mode dynamics over a wide frequency range are successfully described by the cluster theory.

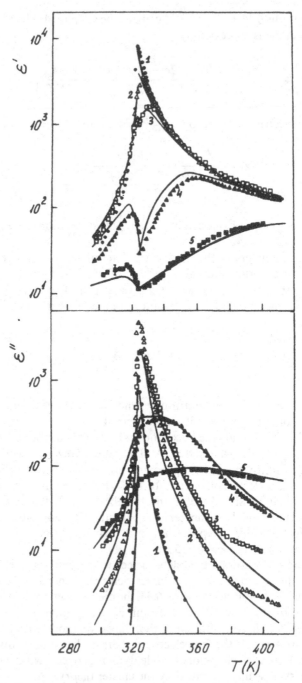

Figure 6.9. Temperature dependence of ε' and ε'' of RbD_2PO_4 at frequencies (GHz): 1) 0.05; 2) 0.5; 3) 1.25; 4) 10; and 5) 52. Dots are experimental values. Lines are theoretical values. From Levitsky et al. (1990).

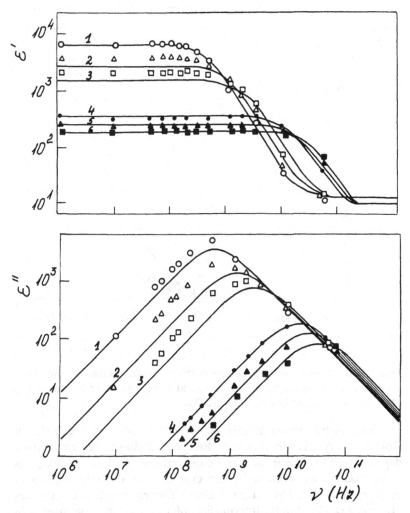

Figure 6.10. Frequency dependence of ε' and ε'' of RbD_2PO_4 at temperatures (K): 1) 325; 2) 329; 3) 334; 4) 364; 5) 377; and 6) 394. Dots are experimental values. Lines are theoretical values. From Levitsky et al. (1990).

Figure 6.11 shows the temperature dependence of the reciprocal relaxation time, τ_1^{-1}, calculated from Equation 6.54, as well as the experimental values of the relaxation time determined by means of Cole–Cole plots. The relaxation time on approaching T_{c2} undergoes a more intense critical slowing down (as in CsD_2PO_4-type crystals), and the soft relaxational mode is of much lower frequency than in KDP crystals.

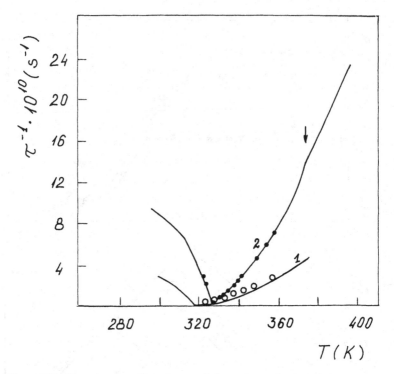

Figure 6.11. Temperature dependence of the reciprocal relaxation time. Dots are experimental values. Lines are theoretical values. 1) data of Komukae and Makita (1985); 2) from Levitsky et al. (1990).

The cluster approximation also describes the static properties of RbD_2PO_4 crystals. To illustrate, Figure 6.12 shows the temperature dependence of the reciprocal static dielectric permittivity $\varepsilon^{-1}(0, T)$ as a function of temperature for the crystals with different degrees of deuteration obtained by various authors. The temperature dependences of $\varepsilon^{-1}(0, T)$ and τ^{-1} in the intermediate phase are clearly nonlinear as in DCDP-type ferroelectrics. Smaller parameters, w and $\bar{v} + v_{ab}$, describe the behavior of $\varepsilon(0, T)$ in the high temperature phase ($T > T_{c1}$).

However, within the frame of the model, the value of the spontaneous polarization (with the parameters from Table 6.4) is equal to $P_{sa} = 7.5\,\mu C/m^2$ at $T = 303$ K. This value is nearly one half the experimental value but larger than the theoretical Ising model result (Komukae and Makita 1985).

Finally, the two-particle cluster approximation also gives a quantitative description of the ferroelectric relaxation dynamics in RbD_2PO_4 crystals.

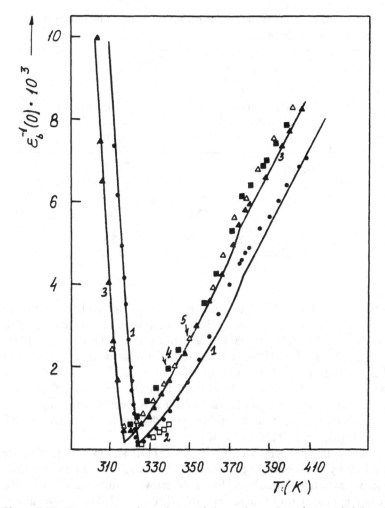

Figure 6.12. Temperature dependence of the reciprocal static permittivity: 1) data from Mizeris et al. (1987); 2) data from Pykacz et al. (1984); 3) data from Komukae and Makita (1985); 4) data from Sumita et al. (1984), 5) data from Sumita et al. (1981). Lines are theoretical values. The kinks in Figures 6.11 and 6.12 at 377 K are caused by the nonferroelectric phase transition.

6.5. MICROWAVE DIELECTRIC PROPERTIES OF TlH$_2$PO$_4$

Thallium dihydrogen phosphate TlH$_2$PO$_4$ (TDP) is the heaviest representative of the KDP-type hydrogen-bonded crystals. The crystal structure of TDP is similar to the structure of CsH$_2$PO$_4$. Yet TDP shows

Table 6.4. The theory parameters which describe the dynamic and static properties of RbD_2PO_4 crystals

$\bar{v} - v_{ab}$	$\bar{v} + v_{ab}$			$T < T_{c2}$		
w/k (K)	k (K)	k (K)	f_a	R_1 (s)	R_2 (s/K)	T_{c2} (K)
1900	3.68	3.53	3.9	0.90	0.022	324

$T_{c2} < T < T_{c1}$				$T > T_{c1}$				
f_a	R_1 (s)	R_2 (s/K)	T_{c1} (K)	w/k (K)	$(\bar{v}/v_{ab})/k$ — f_a		R_1 (s)	R_2 (s/K)
14.5	2.8	0.013	377	1800	3.03	22.6	6.9	0.052

substantial differences in structure and in properties. At room temperature, TDP belongs to the monoclinic system with the space group C_{2h} (Nelmes and Choudhary 1981). Phase transitions in TDP at $T_c = 230$ K and DTDP at $T_c = 350$ K are first-order antiferroelectric transitions (Yasuda et al. 1980). They are caused by the ordering of H-bonds along the b-axis, which link already ordered $H-PO_4$ chains and form two-dimensional sheets perpendicular to the a-axis. Such an ordering causes the antiferroelectric state. Yoshida et al. (1984) have found a ferroelastic phase transition in TDP at 357 K.

Dielectric properties of TDP and DTDP at low frequencies are masked by sufficiently high dielectric loss above room temperature (Narasaiah and Choudhary 1987). Microwave investigations of the dielectric properties of these crystals (Mizeris et al. 1989) have shown that the components of the dielectric permittivity, ε'_a and ε'_c, are small and depend slightly on temperature. Figure 6.13 shows the temperature dependence of ε' along the monoclinic b-axis in the frequency range of 1 MHz to 78 GHz. At $T_{c1} = 230$ K, the permittivity ε' exhibits a discontinuous increase. It also increases with increasing temperature at $T > T_{c1}$. In the vicinity of T_{c1}, ε' does not depend on frequency as one can expect in the case of an antiferroelectric phase transition. At $T_{c2} = 357$ K, the anomaly of ε' is caused by a ferroelastic phase transition. The results suggest a second-order phase transition.

The behavior of dielectric permittivity at the structural phase transitions phenomenologically can be explained by the Landau theory.

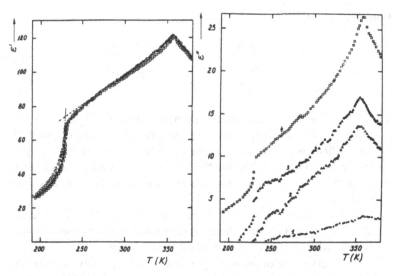

Figure 6.13. Temperature dependence of ε_b' and ε_b'' of TDP at frequencies (GHz): 1) 8.7; 2) 40; 3) 48.7; and 4) 78. From Mizeris et al. (1989).

The free energy (Gibbs function) can be expressed as

$$G = G_0 + \tfrac{1}{2}\alpha(T - T_c)\eta^2 + \tfrac{1}{4}\beta\eta^4 - \tfrac{1}{2}a\eta^2 E_b^2 - \tfrac{1}{2}\chi_{22}^0 E_b^2, \qquad (6.56)$$

which takes into consideration the squared electric field invariant, E_b^2. Only the term of the squared interaction of E and η causes the shape of dielectric anomalies. The interaction of higher orders are insignificant near T_c. According to Equation 6.56, polarization along the b-axis:

$$P = (a\eta_0^2 + \chi_{22}^0)E_b, \qquad (6.57)$$

appears only if there is an electric field applied in this direction. The phase transition causes the appearance of the order parameter:

$$\eta_0^2 = \frac{\alpha(T - T_c) + aE_b^2}{\beta}, \qquad (6.58)$$

which excites a homogeneous sheer deformation along the b-axis. The temperature dependence of dielectric permittivity follows from Equations 6.57 and 6.58:

$$\varepsilon_b = \begin{cases} \varepsilon_b^0 & , \quad T > T_c \\ \varepsilon_b^0 - \dfrac{a\alpha(T - T_c)}{\beta}, & T < T_c \end{cases} \qquad (6.59)$$

where $\varepsilon_b^0 = 1 + \chi_{22}^0$ is the dielectric permittivity at $T > T_{c1}$.

When $\alpha < 0$, the temperature dependence of the dielectric permittivity for the first-order structural phase transition causes an upward jump to the permittivity ε_b, while for the second-order phase transition the increase of temperature gives rise to a break according to the data shown in Figure 6.13.

With the increase of electric field strength near $T < T_{c1}$, the temperature T_{c1} decreases, and the high field induces the first-order phase transition, which causes the double hysteresis loops.

The fluctuations of the order parameter near T_{c1} cause the temperature dependence of ε_b' to deviate from linearity according to $\Delta \varepsilon_b \sim \langle \eta^2 \rangle$, where $\Delta \varepsilon_b = \varepsilon_b^0 - \varepsilon_b$. The dependence $\varepsilon_b^0(T)$ is shown by the dashed line on Figure 6.13. The values of ε_b' and ε_b'' of DTDP crystals and their temperature dependences at microwaves are similar to those for TDP.

A high value of the dielectric permittivity and considerable dielectric loss in the millimeter wave region as well as the temperature dependence show that the ferroelastic phase transition at $T_{c2} = 357$ K is also associated with the softening of one of the phonon modes. The parameters of the mode (see Equation 5.4) calculated from the experimental results at $T = 360$ K are: $v_s = 367$ GHz, $\gamma = 381$ GHz, $\Delta \varepsilon = 116$, and $\varepsilon_\infty = 4.5$. As long as the phase transition at T_{c2} is not the ferroelectric one, the mode softens slightly when approaching T_{c2}. However, this softening probably causes the anomaly of ε_b' at T_{c2} and the linear temperature dependence of ε_b^0 in the ferroelastic phase.

7 OTHER ORDER-DISORDER
FERROELECTRICS

Dynamics of disordered materials is a wide branch of solid state physics. It is concerned with slow (in comparison with phonon frequencies) order-disorder processes in ferroelectrics, orientational glasses, superionics, and others, which manifest themselves at microwaves.

In past decades, a number of microwave studies of the dielectric relaxation in ferroelectrics have been reported. Only a few of them have proven to be reliable for the unambiguous determination of the relaxational mode parameters and can distinguish between various theoretical models. This is related on a large scale with the experimental difficulties in covering the wide frequency range where ferroelectric dispersion takes place, and in calculating the dielectric parameters from the measured data when the loss is high. As a result, the mechanism of the phase transitions—even in the first ferroelectric Rochelle salt—is still under discussion despite the excellent results by Horioka and Abe (1979), which clearly show that the dielectric dispersion of the Debye-type, with a single relaxation time over the complete range of temperatures, reflects an overdamped motion of protons in an asymmetric double-well potential. However, submillimeter spectra (Petzelt et al. 1987) reveal that the relaxational soft mode has an unusual softening law from 255 K down to 180 K. Below 180 K, it changes into a well-underdamped oscillator with a very low damping at low temperatures.

Nevertheless, recent studies have shown that in many order-disorder ferroelectrics, the dipolar relaxation process is of a simple Debye-type with a single relaxation time (Grigas et al. 1991)—as in the classical $NaNO_2$ (Hatta 1968)—and that the molecular field approximation holds. However, there are crystals in which more than one subsystem is involved in the ferroelectric dynamics.

This Chapter describes ferroelectric dispersion at microwaves of both types of new or less studied order-disorder ferroelectrics. The selected examples are based on the preferences of the author and are by no means complete.

7.1. DIGLYCINE NITRATE

Diglycine nitrate, $(NH_2CH_2COOH)_2HNO_3$, (DGN) is a less studied ferroelectric crystal of the TGS family. The first attempt at measuring the ferroelectric dispersion of TGS in the microwave region was done by Lurio and Stern (1960). Later, Hill and Ichiki (1963) published a more detailed study, which for a long time was the most cited work for the dispersion of TGS. Nevertheless, many further publications on the ferroelectric dispersion of TGS appeared later. These did not confirm their broad Gaussian distribution of relaxation times and were incomplete. Luther (1973) and Takayama et al. (1984) have shown that Debye-type ferroelectric dispersion in the close vicinity of T_c begins in the MHz region and ranges to millimeter waves far from T_c. They have also showed that the single soft relaxational mode, related to the flipping motion of dipoles, explains the complete dynamical origin of the order-disorder phase transition in TGS. However, according to Petzelt et al. (1987), two soft relaxational modes are involved in the ferroelectric dynamics of TGS.

DGN is a monoclinic crystal with two molecular units per unit cell. At about $T_c = 206$ K, it undergoes the second-order proper ferroelectric phase transition $P2_1/a \leftrightarrow Pa$ of order-disorder type (Hoshino and Sato 1963)—similar to that of TGS. The structure of DGN is less complex than that of TGS. Although there is no unique polar axis for the structure of DGN, the maximum value of the spontaneous polarization has been found along the [101] direction.

X-ray and neutron diffraction (Sato 1968) as well as NMR studies (Blinc et al. 1970) reveal the presence of two kinds of strongly hydrogen bonded glycine groups: glycinium ions and zwitter-ions, and NH_3 groups. All of these may contribute to the spontaneous polarization and static dielectric permittivity $\varepsilon(0)$. These glycine and NH_3 groups are linked by a complicated network of hydrogen bonds. Recent crystal structure analysis has shown that paraelectric phase of DGN consists of small polar regions compensating each other. A considerable disorder of NO_3 groups has been observed (Lukaszewicz 1994). The flipping of the

H atom from one equilibrium position in the short hydrogen bond (close to the oxygen atom of glycine A) to another (close to that of glycine B) changes the nature of these two glycine ions (i.e, glycine A becomes the zwitter-ion, while glycine B becomes the glycinium ion with associated rotation of the NO_3 groups). This ferroelectric "flipping" mode of glycine units seems to be responsible for the crystal instability at T_c and relaxational dynamics of DGN crystals.

The experimental results on the temperature dependence of the thermodynamical properties of DGN crystals (i.e., low-frequency dielectric permittivity (Igoshin et al. 1984), elastic coefficients (Gavrilova et al. 1982), spontaneous polarization, electrocaloric effect and specific heat (Strukov et al. 1986)) are in good agreement with the Landau theory. As for the dynamic dielectric properties, Kolodziej et al. (1978) have observed relaxation of the Debye-type in the vicinity of T_c below 10 GHz with a single relaxation time, while Leschenko et al. (1989) have observed relaxation up to 30 GHz with the distribution of relaxation times.

Reexamination of the dielectric dispersion of DGN in a wider temperature and frequency region (Sobiestianskas et al. 1989; Czapla et al. 1990) has shown that the ferroelectric dispersion caused by the relaxational soft mode begins in the MHz frequency region and ranges into millimeter waves (Figure 7.1).

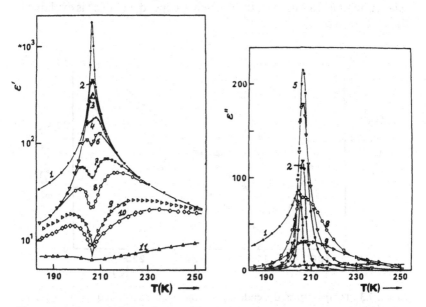

Figure 7.1. Temperature dependence of ε' and ε'' along [101] of DGN crystal at frequencies (GHz): 1) 0.001; 2) 0.14; 3) 0.36; 4) 0.76; 5) 0.94; 6) 1.15; 7) 2.5; 8) 4.0; 9) 9.0; 10) 12; and 11) 42.0. From Sobiestianskas et al. (1989).

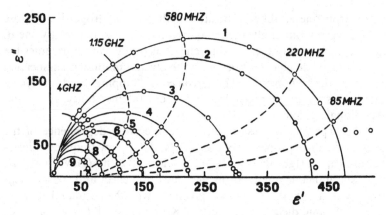

Figure 7.2. Cole–Cole plot of ε^* in the paraelectric phase at temperatures $T - T_c$(K): 1) 0.4; 2) 1; 3) 2; 4) 3; 5) 4; 6) 5; 7) 7; 8) 10; and 9) 14. From Sobiestianskas et al. (1989).

The Cole–Cole plots of observed dielectric relaxation (Figure 7.2) can be represented quite well by semicircles with their centers approximately on the real axis. They indicate that the dielectric dispersion (both close to T_c and away from T_c,) is characterized by a single soft mode relaxation time. In this case, the frequency dependence of ε^* is explained by the Debye equation (2.11).

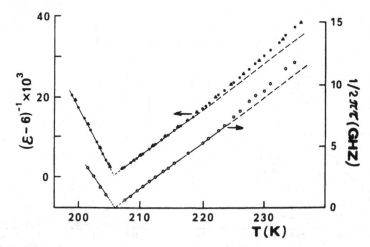

Figure 7.3. Temperature dependence of the soft mode frequency (○) and adiabatic clamped permittivity at 100 MHz (◄). From Sobiestianskas et al. (1989).

The value of $\varepsilon(0)$ in the vicinity of T_c, obtained by extrapolation of the Cole–Cole arc plot, is less than the experimental value obtained at low frequencies. As in the case of TGS crystals, this may be caused by the contribution of the domain wall motion.

Figure 7.3 shows the parameters, $1/\Delta\varepsilon$ and $1/(2\pi\tau)$, in the Debye equation, calculated by the least square method using results presented in Figure 7.1. The soft mode relaxation frequency on approaching T_c varies according to the law $v_s = 0.34(T - T_c) + 0.002(T - T_c)^2$ GHz for $T > T_c$. The dielectric contribution follows the Curie–Weiss law. Linear dependence is valid for $T - T_c < 15$ K. However, one should note that nonlinear dependence on temperature of $1/\Delta\varepsilon$ and $1/(2\pi\tau)$ might be caused by the temperature dependence of ε_∞ ($\varepsilon_\infty = 6$ for $T - T_c < 15$ K), the contribution of the hard phonon modes, and electronic polarization. If, in the crystal, a temperature-dependent phonon mode of higher frequency exists, then even with a small dielectric contribution it causes a nonlinear dependence of $1/\Delta\varepsilon$ and v_s on the temperature. The Curie–Weiss law for adiabatic clamped dielectric permittivity shows that the dielectric contribution of the relaxational mode accounts for all values of the clamped permittivity and its temperature dependence.

A complete dielectric spectrum of DGN (Figure 7.4) shows that, over wide temperature and frequency ranges (10^7 to 10^{11} Hz), the

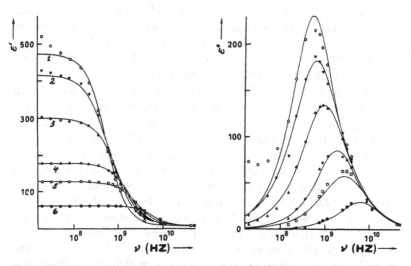

Figure 7.4. Frequency dependence of ε' and ε'' of DGN at temperatures $T - T_c$ (K): 1) 0.4; 2) 1; 3) 2; 4) 4; 5) 6; and 6) 14. The continuous curves are the relaxator fits with the parameters given in Figure 7.3. The points are experimental values. From Sobiestianskas et al. (1989).

ferroelectric dispersion in DGN shows a monodispersive relaxational nature and can be described by the Debye dispersion relation. The ferroelectric mode is mainly associated with the flipping of the hydrogen atom between two potential minima in the short O—$H \cdots O$ hydrogen bonds linking two glycine ions. This "flipping" mode makes the main contribution to the static permittivity. The magnitude of the activation energy is $\Delta U = 0.029 \, eV = 1.6 \, kT_c$. It is similar to the value of ΔU for TGS (Takayama et al. 1984) and is reasonable for the flipping motion of H atoms for the hydrogen bonded order-disorder type ferroelectrics.

Finally, the results confirm that the phase transition in DGN is a pure case of order-disorder type and may be treated as the result of the crystal instability in a proton subsystem.

7.2. TELLURIC ACID AMMONIUM PHOSPHATE

Telluric acid ammonium phosphate, $Te(OH)_6 \cdot 2(NH_4H_4PO_4)(NH_4)_2$-$HPO_4$, (TAAP) is another monoclinic H-bonded crystal, which at the Curie temperature $T_c = 321 \, K$ (Guillot Gauthier et al. 1984) undergoes a proper ferroelectric phase transition $P2/n \leftrightarrow Pn$. X-ray studies (Averbuch-Pouchot and Durif 1984) reveal a complicated network of hydrogen bonds linking TeO_6, PO_4, and NH_4 groups. Four types of hydrogen bonds of different length (linking PO_4 tetrahedra) are present in the structure of the TAAP crystal (Stadnicka 1994). All the hydrogen bonds are asymmetric and the protons are ordered in the ferroelectric phase. In the paraelectric phase, one hydrogen bond is symmetrical with a double potential well for protons and two other hydrogen bonds become symmetrically equivalent (with the same length).

NMR measurements of ^{31}P nuclei (Gaillard et al. 1984) have shown that the phase transition is of the second order and is connected with the ordering of protons in hydrogen bonds. Raman spectra (Shashikala et al. 1989) have confirmed that the groups TeO_6, PO_4, and NH_4 do not undergo changes that could trigger the phase transition. The absence of a soft mode in the Raman spectra suggests the order-disorder nature of the phase transition involving hydrogen bonds. The spontaneous polarization $P_s = 2.0 \, \mu C/cm^2$ (Cach and Czapla 1990) in the ferroelectric phase appears along the $[101]^*$ direction, which is close to the $[101]$-axis. The low-frequency dielectric permittivity is strongly anisotropic. The approximately equal values of ε_a and ε_c diverge at T_c according to the Curie–Weiss law, whereas ε_b exhibits only a small anomaly. The static ferroelectric properties of TAAP are similar to those for TGS.

The ferroelectric dispersion (Sobiestianskas et al. 1992a) in the vicinity of T_c begins in the megahertz frequency region and ranges up to

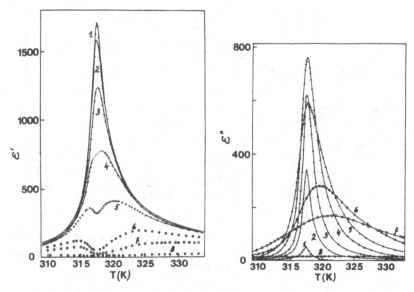

Figure 7.5. Temperature dependence of ε' and ε'' of TAAP crystal for various frequencies (GHz): 1) 0.05; 2) 0.092; 3) 0.86; 4) 2; 5) 4; 6) 8.4; 7) 15; and 8) 77. From Sobiestianskas et al. (1992a).

short millimeter waves as in other H-bonded ferroelectrics. Figure 7.5 shows the temperature dependences of the complex dielectric permittivity along the [101] direction within the ferroelectric dispersion region. The Curie temperature in this crystal is $T_c = 317.3$ K. The flat minimum of ε' at T_c indicates the influence of impurities and surface inhomogeneities on the critical slowing-down process. The ferroelectric dispersion near T_c is almost over at 77 GHz. However, in the far paraelectric phase, it ranges up to the short millimeter wave region.

Despite a complicated network of hydrogen bonds, the dielectric relaxation in the whole paraelectric phase of TAAP is characterized by a single relaxation time (Figure 7.6). Surface layers and spatial inhomogeneities of the crystal cause a distribution of relaxation times only at $T - T_c < 0.5$ K.

On approaching the Curie temperature, the correlated proton relaxation time reaches the value $\tau = 9 \times 10^{-11}$ s. The relaxation frequency $\nu_s = 1/(2\pi\tau)$ varies according to the law $\nu_s = 0.94(T - T_0)$ GHz in the paraelectric phase, where $T_0 = 315.7$ K (Figure 7.7). The critical slowing-down of the relaxation frequency is in agreement with the Curie–Weiss law.

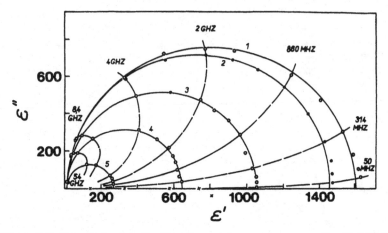

Figure 7.6. Cole–Cole plots for TAAP crystal at temperatures $T - T_c$ (K): 1) 0.35; 2) 0.6; 3) 1.6; 4) 3.6; and 5) 10.6. Crosses on the ε' axis mark the centers of the semicircles. From Sobiestianskas et al. (1992a).

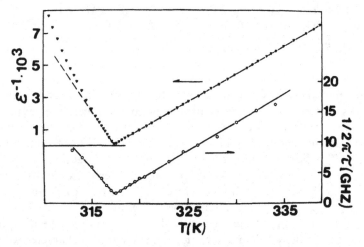

Figure 7.7. Temperature dependence of the relaxational mode frequency (o) and permittivity (▼) at 1 MHz of TAAP. From Sobiestianskas et al. (1992a).

A complete dielectric spectrum of TAAP in the region of ferroelectric dispersion is shown in Figure 7.8. The relaxational soft mode dynamics reveal themselves in the frequency region below 10^{11} Hz, which is characteristic for H-bonded ferroelectrics.

Thus, TAAP is an H-bonded order-disorder ferroelectric in which the phase transition is assumed to be triggered by proton ordering in

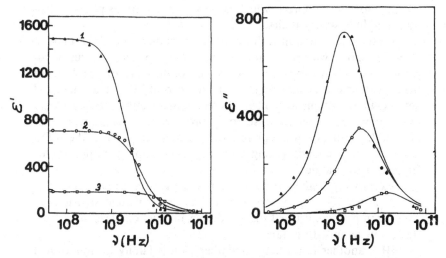

Figure 7.8. Frequency dependence of ε' and ε'' of TAAP at temperatures $T - T_c$ (K): 1) 0.6; 2) 3; and 3) 16.5. The continuous curves are monodispersive relaxator fits; dots are experimental results. From Sobiestianskas et al. (1992a).

a double-well potential of a hydrogen bond. The frequency of the relaxational flipping proton mode makes the main contribution to the high static permittivity, which follows the Curie–Weiss law.

7.3. BETAINE COMPOUNDS

The betaine addition compounds belong to a class of molecular crystals which show a fascinating variety of ordered low temperature states (Albers 1988). The zwitter-ion (dipole ion), betaine $(CH_3)_3N^+CH_2COO^-$, is the third amino acid (following glycine and sarcosine) in the family of compounds displaying a number of interesting phase transitions of different appearance with dielectric anomalies. Interesting properties of these compounds were discovered and have been studied by the ferroelectrics group in Saarbrücken, Germany. Betaine phosphate (BP), betaine arsenate (BA), betaine phosphite (BPI), and betaine calcium chloride dihydrate (BCCD) are the most interesting of them from the point of view of ferroelectric or antiferroelectric properties and incommensurate structures. A review of the experimental results of phase transitions in betaine compounds has been given by Schaack (1990). In general, weak intermolecular as well as intramolecular (between betaine and inorganic groups) bonds make the structure of betaine crystals very sensitive to external forces, which is the origin of various phase transitions and high susceptibility.

All betaine compounds show similar crystallochemical features. In all adducts, the zwitterion betaine forms a quasi-planar molecule comprising all nonhydrogen atoms except two of the CH_3 groups. Most of the structures can be interpreted as being formed by parallel chains of betaine-anion-betaine sequences. Therefore, they exhibit excellent cleavage planes and strongly anisotropic properties. Most known compounds have orthorombic or monoclinic symmetry. As in the KDP family, hydrogen bonds linking the phosphate groups are important for ferroelectric dynamics.

Extensively studied antiferroelectric BP is isomorphous with ferroelectric BA. They form also a complete series of solid solutions in which competing electrical dipole–dipole interactions lead to a frustrated low temperature state.

BPI, another ferroelectric member of this family of amino acid compounds, shows close structural affinity to BP and BA (Fehst et al. 1993). In contrast to BP, in BPI (Figure 7.9), the hydrogen atom H12 forms hydrogen bonds from the oxygen O2 in the organic part to O3 located in the inorganic component. The oxygen atom O5, which is attached to the P atom in BP, is replaced by H14 in BPI. Thus, only one hydrogen bridge connects the amino acid to the inorganic part of the molecule. BPI also forms mixed crystals with BP of a dipolar glass state at intermediate concentrations.

Ferroelectric dynamics of these quasi-linear hydrogen-phosphate (arsenate) chains have not been completely studied.

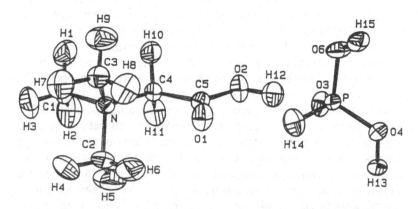

Figure 7.9. Atomic positions in BPI. From Fehst et al. (1993).

At $T_{c1} = 365$ K, betaine phosphate exhibits an improper ferroelastic order-disorder type phase transition (PT) with doubling of the unit cell in the c-direction. At $T_{c2} = 86$ K, BP undergoes a second-order isostructural PT with a large anomaly of $\varepsilon(0)$. Submillimeter spectra reveal a relaxational excitation that looks like a ferroelectric soft mode (Goncharov et al. 1988). However, in the submillimeter spectra, only a tail of the dielectric dispersion has been observed. Neither the dielectric contribution nor its temperature dependence explain the high value of $\varepsilon(0)$ nor its temperature dependence. It was supposed that an additional soft mode should exist in the millimeter wave region. However, microwave studies by Brückner et al. 1988 have shown that one relaxator describes the dielectric relaxation over the frequencies 10 MHz to 400 GHz.

At $T_{c1} = 411$ K, betaine arsenate exhibits a pseudoproper ferroelastic PT, while at $T_{c2} = 119$ K it undergoes a proper ferroelectric PT with the Curie–Weiss type anomaly of $\varepsilon(0)$ (the peak value of 10^6). Phenomenological theory of this PT was worked out by Maeda (1988). Freitag et al. (1985) have found a monodispersive relaxation at microwaves. In submillimeter spectra (Goncharov et al. 1988), a second relaxational soft mode with $\Delta\varepsilon = 100$ has been found. Two soft relaxational modes are also revealed in DBA crystals. Deuteration decreases the frequencies of the two modes polarized along the c-axis. Microwave studies of DBA have been performed by Brückner (1989).

BCC and BCCD exhibit the richest sequence of various commensurate and IC phases between 46 and 164 K. The dielectric response function has not been well studied in detail, but its complexity shows the submillimeter spectra (Goncharov et al. 1988): an intensive absorption line at 450 GHz in the high temperature phase on cooling decreases in frequency and intensity. Below 125 K the situation reverses. Below 116 K the lines are split into a quantity of lines lying above 450 GHz. In general, some features of Raman, IR, and submillimeter spectra indicate that the ferroelectric dynamics of all these compounds involve two temperature dependent relaxational excitations and, hence, two subsystems of the ordered particles. Theoretical aspects of phase transitions in betaine compounds are reviewed by Dvořak (1990).

The less studied betaine phosphite, like BA and BP, exhibits a high temperature paraelastic phase. Below a second-order ferroelastic PT at $T_{c1} = 365$ K, the betaine molecules arrange in an antiferrodistortive pattern. The ferroelectric PT of the second order occurs at $T_{c2} = 220$ K (Fehst et al. 1993). These authors have found two types of identical structures: with $T_{c2} = 220$ K or 196 K. Nevertheless, the permittivity in the paraelectric phase of the different samples could be described in the framework of the anisotropic Ising model with almost identical

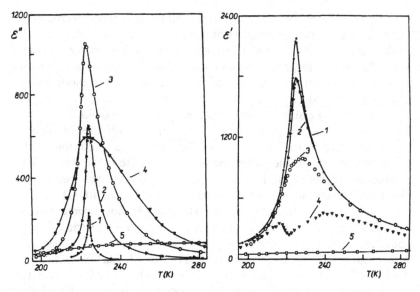

Figure 7.10. Temperature dependence of ε' and ε'' of BPI at frequencies (GHz): 1) 0.1; 2) 1.0; 3) 3.8; 4) 8.8; and 5) 77. From Sobiestianskas et al. (1993).

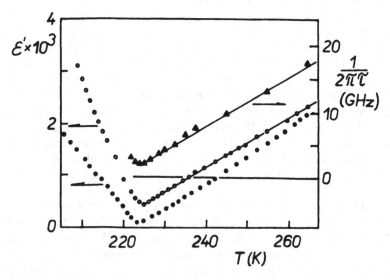

Figure 7.11. Temperature dependence of reciprocal static permittivity ε^{-1}(●), permittivity at 100 MHz (○) and relaxation frequency (◄). From Sobiestianskas et al. (1993).

parameters. Dacko et al. (1992) have found in the same sample additional dielectric and pyroelectric anomalies at 196 K.

Microwave investigations (Sobiestianskas et al. 1993) have shown that the main ferroelectric dispersion occurs at microwaves and exhibits (Figure 7.10) monodispersive Debye-type relaxation. The dielectric contribution of the relaxational mode (Figure 7.11) completely accounts for the value of the static permittivity and its temperature dependence. The results confirm that the ferroelectric PT in BPI is of the order-disorder type.

7.4. DICALCIUM STRONTIUM PROPIONATE

Dicalcium strontium propionate, $Ca_2Sr(C_2H_5CO_2)_6$, (DSP) and its deuterated analog DDSP are examples of crystals that show the long time required to determine the mechanism of a phase transition when it is not triggered by the ordering of hydrogen bonds. At temperatures $T_c^H = 281.6$ K and $T_c^D = 279.5$ K, these crystals undergo the ferroelectric phase transition $P4_12_12 \rightarrow P4_1$. The phase transition mechanism has attracted the interest of many researchers since ferroelectricity was first discovered by Mathias and Remeika in 1957. The phase transition of DSP has been well described phenomenologically by Dvořak and Ishibashi (1976) as an example of improper ferroelectrics. A disordered or rotational motion of CH_3 radicals was reported by NMR and X-ray studies (see Yagi 1990 and references therein). A relaxational Debye-type dielectric dispersion occurs in DSP and DDSP below 1 GHz with a single relaxation time (Nakamura et al. 1990; Banys et al. 1992c; Nakamura and Kashida 1993). This critical slowing-down explains the temperature dependence of $\varepsilon(0)$ and shows an order-disorder type PT. A long relaxation time associated with heavy dipoles (responsible for the ferroelectric dynamics), suggested at first that the PT and relaxational dielectric dispersion were caused by a flipping motion of the propionate molecule. However, the phase transition temperature of 4 to 6 K is obtained by the mean-field approximation assuming that the propionate groups are interacting dipoles and evaluating their volume and the dipole moment from P_s at low temperatures. Consequently, either the phase transition is driven by other forces or the macroscopic spontaneous polarization P_s is merely the difference of polarization of two sublattices.

In order to obtain direct evidence for the role of the molecular motion in the phase transition mechanism, completely deuterated crystals have been studied by Brillouin and Raman scattering, microwave, optical, neutron diffraction, and other measurements (see Yagi 1990). As a result, the role of the molecular motion in the phase transition mechanism is concluded to be a secondary one. According to Nakamura and Deguchi's two component polarization model (1994) the ferroelectric phase transition of DSP is accompanied by the order-disorder behavior of the methyl groups and the atomic displacements of the ionic atoms, which are coupled to each other. The spontaneous polarization is mainly caused by the atomic displacements, which relates only indirectly to the ferroelectric phase transition. Ultrasonic study of ferroelectric PT in DDSP (Valevičius et al. 1994) supports this idea. The polarization fluctuations relaxation time follows the Curie–Weiss law and is similar to one obtained from the dielectric measurements. The similar mechanism of the ferroelectric phase transition is in DDSP and DLP ($T_c = 330$ K).

Another peculiarity of DSP and DDSP crystals is an internal bias field. In this case, an external electric field compensate the internal one, and as a result, the dielectric permittivity depends on the magnitude of the external electric field. A maximum of ε occurs at $E \neq 0$ (Banys et al. 1992c). The internal field points to a macroscopic noncentrosymmetry of the paraelectric phase. Phenomenologically, it can be accounted for by introducing an odd term in the order parameter expansion of free energy. This gives good agreement with experimental data. The origin of the biasing field in the paraelectric phase is still an open question. It looks like it is caused more by the peculiar dipole network than by defects.

7.5. RUBIDIUM HYDROGEN SULFATE AND SELENATE

Now considered are some characteristic cases of ferroelectric dispersion that are not related to the proton dynamics in the hydrogen bonds. Rubidium hydrogen sulfate, $RbHSO_4$, and its deuterated isomorph, $RbDSO_4$, belong to such crystals. At the Curie temperature $T_c^H = 265$ K and $T_c^D = 251$ K, these crystals undergo the second-order ferroelectric phase transition accompanied by a change in symmetry $P2_1/c \rightarrow Pc$ (Aschmore and Petch 1975).

The structure consists of two crystallographically independent molecules in the asymmetric unit (Figure 7.12). Each of the independent sulfate tetrahedra of S_1 and S_2 type is linked to a translationally equivalent sulfate by acentrically ordered hydrogen bonds in the paraelectric phase forming chains parallel to the b-axis.

Ferroelectricity in $RbHSO_4$ is caused by the ordering of the sulfate groups of S_1-type which move in the asymmetrical two-minimum potential wells. Therefore, the phase transition is of the order-disorder type and is accounted for by both the phenomenological theory (Nakamura and Kajikawa 1978) and by statistical models (Blat and

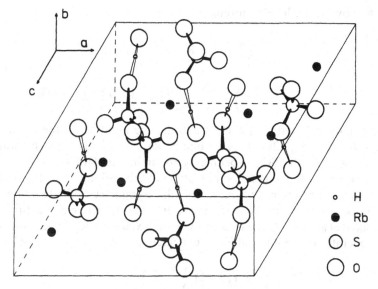

Figure 7.12. Atomic positions in RbHSO$_4$.

Zinenko 1981) of the Mitsui type for Rochelle salt, which explain both the thermodynamic (Ichikawa 1979; Nakamura et al. 1977) and dynamic characteristics in the molecular field (Ozaki 1980) and clusters (Levitsky et al. 1980) approximation.

Ferroelectric dispersion in RbHSO$_4$ has been investigated by a number of authors. In the frequency range of 10 kHz to 9.5 GHz, Ozaki (1980) discovered a monodispersive dielectric relaxation of the Debye type, which continues at higher frequencies as well. According to Krasikov et al. (1981), at $v > 5$ GHz, another nonrelaxational excitation appears at microwaves. Raman spectra (Toupry et al. 1981) do not show any high-frequency soft excitations of the crystal lattice. Ambrazevičienė et al. (1983) measured the dielectric spectra of RbHSO$_4$ and RbDSO$_4$ in the 118 to 510 GHz range and found that the ferroelectric dynamics are also described by a relaxation mechanism with a relaxation time close to that obtained by Ozaki (1980).

Investigations in the main region of ferroelectric dispersion—10 to 100 GHz (Paprotny et al. 1984a)—in conjunction with the data gathered over the lower frequency range where the dispersion begins and the submillimeter range where it ends, have produced the complete dielectric spectrum of RbHSO$_4$ determined by ferroelectric dynamics.

The dynamics of a crystal characterized by an asymmetric one-particle two-minimum potential can be expressed (Paprotny et al.

1984a) by the model Hamiltonian

$$\hat{H} = -\frac{1}{2}\sum_{i,j}\sum_{f,f'=1}^{2} R_{ij}(ff')S_i^z(f)S_j^z(f')$$

$$-\Delta\sum_i [S_i^z(1) - S_i^z(2)] - (\mu E)\sum_i \sum_{f=1}^{2} S_i^z(f). \tag{7.1}$$

Here, indices 1 and 2 at $S_i^z(1,2)$ indicate the numbers of the sublattices: $R_{ij}(11) = R_{ij}(22) = J_{ij}$ is the interaction potential of the structural ordering elements (SOEs) belonging to similar sublattices; $R_{ij}(12) = R_{ij}(21) = K_{ij}$ is the interaction potential of SOEs belonging to different sublattices. The quantity Δ represents the asymmetry of the two-minimum potential where the SOEs move; μ is the effective electric dipole moment of a single SOE interacting with the external field E; and $S_i^z(f)$ is the operator of the internal freedom degrees that corresponds to SOE:

$$S_i^z(f) = \frac{1}{2}\begin{pmatrix} 1 & 0 \\ 0 & -1 \end{pmatrix}.$$

The approximation of the molecular field (Žekš et al. 1972) will be discussed. Dynamic characteristics can be obtained from a set of equations for time-dependent single functions of distribution of the SOEs of the crystal. These equations can be presented (Levitsky et al. 1980) as

$$-\alpha\frac{d}{dt}\zeta = \zeta - \frac{1}{2}[L_1 + L_2], \quad 2\zeta = \eta(1) + \eta(2),$$

$$\eta(f) = 2\langle S_i^z(f)\rangle, \tag{7.2}$$

$$-\alpha\frac{d}{dt}\sigma = \sigma - \frac{1}{2}[L_1 - L_2], \quad 2\sigma = \eta(1) - \eta(2).$$

Here, α is the temperature dependent constant with the dimension of time that is effective in defining the time scale of dynamic processes, while

$$L_{1,2} = \tanh\left\{\pm\frac{\gamma}{t^0}\pm\frac{a\sigma}{t^0}+\frac{\zeta}{t^0}+\frac{e}{t^0}\right\}, \tag{7.3}$$

where

$$a = \frac{\tilde{K} - \tilde{J}}{\tilde{K} + \tilde{J}}, \quad \gamma = \frac{2\Delta}{\tilde{K} + \tilde{J}}, \quad \tilde{K} = \frac{K}{k}, \quad \tilde{J} = \frac{J}{k}, \quad \tilde{\Delta} = \frac{\Delta}{k},$$

$$t^0 = \frac{4T}{\tilde{K} + \tilde{J}}, \quad e = \frac{2\tilde{\mu}E}{\tilde{K} + \tilde{J}}, \quad \tilde{\mu} = \frac{\mu}{k}, \quad K = \sum_j K_{ij}, \quad J = \sum_j J_{ij}.$$

The precise solution of Equation 7.2 is a complicated task. Therefore, the discussion will be limited to the case of small declinations from the equilibrium state:

$$\zeta = \tilde{\zeta} + \zeta(t), \quad \sigma = \tilde{\sigma} + \sigma(t). \tag{7.4}$$

By substituting Equation 7.4 into Equation 7.2 in the linear approximation for $\zeta(t)$ and $\sigma(t)$, two systems of equations for the equilibrium functions $\tilde{\zeta}$ and $\tilde{\sigma}$ and their fluctuating parts are obtained. The solution of the first set of equations is

$$\tilde{\zeta} = \frac{1}{2}\left[\tanh\left(\frac{\gamma}{t^0} - \frac{a\tilde{\sigma}}{t^0} + \frac{\tilde{\zeta}}{t^0} + \frac{e}{t^0} \right) + \tanh\left(-\frac{\gamma}{t^0} + \frac{a\tilde{\sigma}}{t^0} + \frac{\tilde{\zeta}}{t^0} + \frac{e}{t^0} \right) \right],$$

$$\tilde{\sigma} = \frac{1}{2}\left[\tanh\left(\frac{\gamma}{t^0} - \frac{a\tilde{\sigma}}{t^0} + \frac{\tilde{\zeta}}{t^0} + \frac{e}{t^0} \right) - \tanh\left(-\frac{\gamma}{t^0} + \frac{a\tilde{\sigma}}{t^0} + \frac{\tilde{\zeta}}{t^0} + \frac{e}{t^0} \right) \right]. \tag{7.5}$$

The spontaneous polarization is defined as

$$P_S = 2N\mu\tilde{\zeta}. \tag{7.6}$$

Accordingly, the static susceptibility:

$$\chi_0 = \lim_{E \to 0} \frac{dP}{dE} = \frac{\mu^2}{kV} \times \frac{4\chi^*(0)}{\tilde{K} + \tilde{J}}, \quad \chi^*(0, T) = \lim_{E \to 0} \frac{d\tilde{\zeta}}{de},$$

(the volume of the unit cell is V as in Equation 6.31) is expressed by

$$\chi^*(0) = \frac{(1 - \tilde{\zeta}^2 - \tilde{\sigma}^2)(t^0)^{-1} + a(1 - \tilde{\zeta}^2 - \tilde{\sigma}^2)(t^0)^{-2} - 4a\tilde{\zeta}^2\tilde{\sigma}^2(t^0)^{-2}}{1 + (a-1)(1 - \tilde{\zeta}^2 - \tilde{\sigma}^2)(t^0)^{-1} - a(1 - \tilde{\zeta} - \tilde{\sigma}^2)(t^0)^{-2} + 4a\tilde{\zeta}\tilde{\sigma}^2(t^0)^{-2}}. \tag{7.7}$$

For the disordered phase ($\tilde{\zeta} = 0$), $\chi^*(0)$ has a simpler expression and coincides with the known results (Žekš et al. 1972).

By solving the second set of equations for $\zeta(t)$ and $\sigma(t)$, and by using the following relations

$$\chi(\omega) = \chi(0) - i\omega \int_0^\infty \Phi(t)e^{i\omega t}\, dt, \tag{7.8}$$

$$\Phi(t) = f\frac{4}{\tilde{K} + \tilde{J}} \lim_{E \to 0} \frac{d\zeta(t)}{dE}, \quad f = \frac{\mu^2}{Vk}, \tag{7.9}$$

the real and imaginary parts of the complex dielectric permittivity are obtained (see Equation 2.11):

$$\varepsilon'(\omega) = \varepsilon_\infty + \frac{w_1}{1 + (\omega\tau_1)^2} + \frac{w_2}{1 + (\omega\tau_2)^2},$$

$$\varepsilon''(\omega) = \frac{w_1\omega\tau_1}{1 + (\omega\tau_1)^2} + \frac{w_2\omega\tau_2}{1 + (\omega\tau_2)^2}, \tag{7.10}$$

where the relaxation time $\tau_{1,2}$ is expressed by

$$\frac{\alpha}{\tau_{1,2}} = \frac{1}{2}\{M_{11} + M_{22} \pm [(M_{11} - M_{22})^2 + 4M_{12}M_{21}]^{1/2}\}, \tag{7.11}$$

and

$$w_1 = f\frac{4}{\tilde{K} + \tilde{J}}w_1^*, \quad w_2 = f\frac{4}{\tilde{K} + \tilde{J}}w_2^*. \tag{7.12}$$

The notations used here are

$$w_1^* = \left\{\left(M_{11} - \frac{\alpha}{\tau_2}\right)\chi^*(0) + M_{12}\chi^*\right\}\left(\frac{\alpha}{\tau_1} - \frac{\alpha}{\tau_2}\right)^{-1},$$

$$w_2^* = \left\{\left(M_{11} - \frac{\alpha}{\tau_1}\right)\chi^*(0) + M_{12}\chi^*\right\}\left(\frac{\alpha}{\tau_1} - \frac{\alpha}{\tau_2}\right)^{-1}, \tag{7.13}$$

where

$$M_{11} = 1 - \frac{1}{t^0}(1 - \tilde{\zeta}^2 - \tilde{\sigma}^2), \quad M_{12} = -\frac{2a}{t^0}\tilde{\zeta}\tilde{\sigma},$$

$$M_{22} = 1 - \frac{1}{t^0}(1 - \tilde{\zeta}^2 - \tilde{\sigma}^2), \quad M_{21} = \frac{2a}{t^0}\tilde{\zeta}\tilde{\sigma}, \tag{7.14}$$

$$\chi^* = \frac{d\sigma}{de}\bigg|_{e\to 0} = -\frac{2\dfrac{\tilde{\zeta}\tilde{\sigma}}{t^0}}{1 + \dfrac{a-1}{t^0}(1 - \tilde{\zeta}^2 - \tilde{\sigma}^2) - \dfrac{a}{t^{02}}(1 - \tilde{\zeta}^2 - \tilde{\sigma}^2)^2 + 4\dfrac{a\tilde{\zeta}\tilde{\sigma}}{t^{02}}}. \tag{7.15}$$

It is necessary to note that the relation

$$w_1 + w_2 = \chi(0) \tag{7.16}$$

is satisfied.

At the paraelectric phase $w_1 = \chi(0)$, and at the low temperature phase $w_1 \gg w_2$. Thus, at the paraelectric phase, only the first relaxation mode makes a contribution to dielectric dispersion. The same mode also makes the main contribution at the low temperature phase.

The temperature dependence of the spontaneous polarization P_s, the relaxation time τ, the static susceptibility $\chi(0)$, and the complex dynamic dielectric permittivity $\varepsilon^*(\nu)$ is determined by the temperature dependence of the parameters $\tilde{\zeta}$ and $\tilde{\sigma}$ in Equation 7.5 in the case where $e \to 0$ (which, in turn, are related to the power parameters $\tilde{K}$, $\tilde{J}$ and $\tilde{\Delta}$). For RbHSO$_4$, the parameters $\tilde{K} = 616$ K, $\tilde{J} = 777.2$ K and $\tilde{\Delta} = 244$ K. For RbH$_{0.3}$D$_{0.7}$SO$_4$, the parameters are $\tilde{K} = 614$ K, $\tilde{J} = 720$ K and $\tilde{\Delta} = 244$ K. These provide a satisfactory description of the thermodynamic properties. Besides, Equations 7.7, 7.10 and 7.11 include the variable parameters α and f, which have been defined by the experimental results obtained by Ozaki (1980) and Paprotny et al. 1984a (and which are equal for RbHSO$_4$):

$$f = 4.1 \quad (T < T_c),$$

$$\alpha = 9.4\tau_0^*, \quad \tau_0^* = 10^{-14}\,\text{s},$$

$$f = 6.15 \quad (T > T_c).$$

And for RbH$_{0.3}$D$_{0.7}$SO$_4$, $f = 4.33$ and $\alpha = 9\tau_0$.

Figures 7.13 and 7.14 show the experimental and theoretical results of the temperature dependence of the complex dielectric permittivity over the frequency range of ferroelectric dispersion in RbHSO$_4$ and RbH$_{0.3}$D$_{0.7}$SO$_4$, respectively.

The temperature dependence of the reciprocal static susceptibility and the relaxation time, calculated from Equations 7.7 and 7.11, are presented in Figure 7.15 (where the dots also show the results obtained by assuming the Debye equation (2.11) on the basis of the experimental results). The polarization relaxation process undergoes a critical slowing-down and, near T_c, the critical index approaches 1. This result means that the critical dynamic phenomena in RbHSO$_4$ can be explained by both the classical phenomenological theory and the mean-field approximation of the Mitsui model. At temperatures remote from T_c, which is defined from equations

$$\frac{1}{t_c} = \cos h\left(\frac{\gamma - a\tilde{\sigma}}{t_c}\right), \quad t_c = \frac{4T_c}{\tilde{K} + \tilde{J}},$$

the temperature dependence of $\chi^{-1}(0)$ and τ^{-1} deviates from the linearity predicted by the Landau theory.

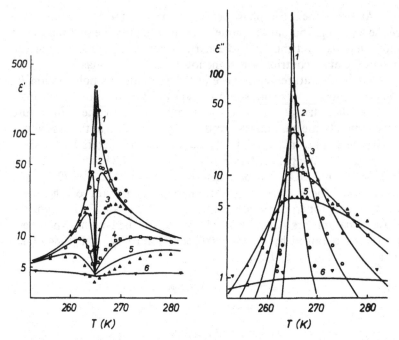

Figure 7.13. Temperature dependence of the complex dielectric permittivity of $RbHSO_4$ at frequencies (GHz): 1) 0.455; 2) 3.27; 3) 9.5; 4) 16.05; 5) 42; and 6) 253. Dots are experimental values. Lines represent theoretical values calculated by Equation 7.10. From Paprotny et al. (1984a).

Figure 7.16 shows a complete dielectric spectrum of the $RbHSO_4$ crystal in the region of the ferroelectric dispersion. Deuteration changes a dielectric spectrum of the ferroelectric dispersion in $RbDSO_4$ very slightly (Grigas et al. 1984b). Over a wide frequency range, the ferroelectric dispersion shows a monodispersive relaxational nature and is successfully described by the Mitsui model. It attests to the fact that the Mitsui model describes the character of the micromechanism of polarization in $RbHSO_4$.

Thus, the microwave investigations confirm that the phase transition in $RbHSO_4$ and $RbDSO_4$ is a pure case of the order-disorder type, and its ferroelectric dynamics are described by monodispersive relaxational dispersion within the framework of the model with an asymmetric one-particle two-minimum potential.

Rubidium hydrogen selenate, $RbHSeO_4$, at $T_c^H = 446\,K$ and $T_c^D = 375\,K$ undergoes the successive phase transitions $C6 \rightarrow C6 \rightarrow C$ in order of descending temperature. The phase at $T > T_c^H$ is a superionic

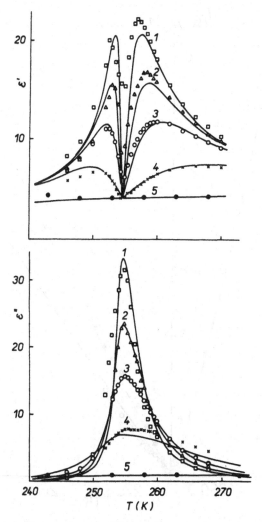

Figure 7.14. Temperature dependence of the complex dielectric permittivity of $RbH_{0.3}D_{0.7}SO_4$ at frequencies (GHz): 1) 8.72; 2) 12.50; 3) 18.75; 4) 41.70; and 5) 330. Dots are experimental. Lines are theoretical. From Grigas et al. (1984b).

one. The phase at $T < T_c^D$ is ferroelectric-ferroelastic, while the phase between T_c^H and T_c^D is the ferroelastic phase. The ferroelastic phase transition at T_c^D is supposed to be improper one and related to the ordering of one type of hydrogen bonds and such a transformation of SeO_4 tetrahedra, which makes the distances Se—O nonequivalent; the hydrogen bonds of another type are ordered at all temperatures. Dynamical

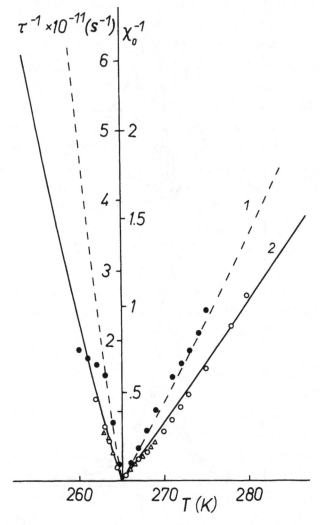

Figure 7.15. Temperature dependence of the reciprocal static susceptibility (dotted line) and the relaxation time (solid line) of $RbHSO_4$. Points show values calculated by Debye equation from experimental results. From Paprotny et al. (1984a).

dielectric properties of $RbHSeO_4$ are different than those of $RbHSO_4$. The dielectric dispersion of the Debye type along the ferroelectric b-axis in the 1 to 100 GHz frequency range is caused by a single relaxational mode; its frequency in the paraelectric phase is about 12 GHz and does not depend critically on temperature (Sobiestianskas et al. 1994).

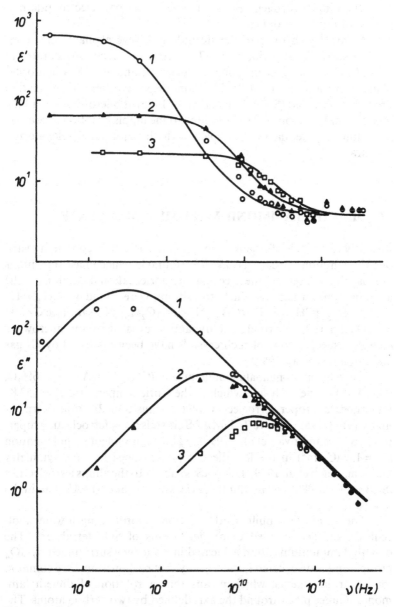

Figure 7.16. The complete dielectric spectrum of RbHSO$_4$ in the region of ferroelectric dispersion at temperatures (K): 1) 266; 2) 268; and 3) 263. Dots represent experimental values; lines are theoretical. From Paprotny et al. (1984a).

Rubidium hydrogen selenate is one of the best electrooptic materials for nonlinear optics.

Ammonium hydrogen selenate undergoes several phase transitions: at $T_s \simeq 416\,K$ a superionic, at $T_i \simeq 260\,K$ an incommensurate, at $T_c \simeq 251\,K$ a first order improper ferroelectric and at $97\,K$ a first order phase transition to glass state. The monodispersive dielectric relaxation process is observed (Sobiestianskas et al. 1995) at microwaves in all the phases, which is supposed to be related to the flipping motion of protons in a double potential well coupled with the motion of selenate tetrahedra.

7.6. DIMETHYLAMMONIUMALUMINUM SULFATE

Recently, a number of organic-inorganic ferroelectrics containing various alkylammonium cations have been found in which phase transitions are usually related to the freezing of reorientational motion of the alkylammonium cations. Such transitions were found in $CH_3NH_3^+$, $(CH_3)_2NH_2^+$, $(CH_3)_3NH^+$, $(CH_3)_4N^+$ and $(C_2H_5)_4N^+$ salts (see Czapla and Dacko 1993). Ferroelectric dynamics revealed by the microwave dielectric spectroscopy of such crystals have been discussed by Grigas and Sobiestianskas (1994).

Dimethylammoniumaluminum sulfate, $(CH_3)_2NH_2Al(SO_4)_2 \cdot 6H_2O$, (DMAAS) is one such salt, which at the Curie temperature $T_c = 152\,K$, undergoes a proper ferroelectric phase transition $2/m \leftrightarrow m$ (Kirpichnikova et al. 1989). Above T_c, DMAAS crystals show ferroelastic properties (Kirpichnikova et al. 1990). The spontaneous polarization ($P_s = 1.6\,\mu C/cm^2$) in the ferroelectric phase appears in the symmetry plane, m, (Cach et al. 1989). DMAAS is similar to the alums ferroelectrics (Sekine et al. 1988) except that there is less water in DMAAS than in the alums.

The structure is built of Al^{3+} cations coordinating 6 water molecules, hydrogen bonded to oxygen atoms of SO_4 tetrahedra. The dimethylammonium cation is located in a vacancy surrounded by SO_4 groups, and is disordered in such a way that carbon atoms are common for all arrangements, which means that a rotation of dimethylammonium takes place around the axis defined by two carbon atoms. The rotation is hampered by weak hydrogen bonds N—H···O resulting in four arrangements related in pairs to each other by a symmetry center, with occupancy parameters of about 0.40 (N1) and 0.10 (N2) (Pietraszko et al. 1993).

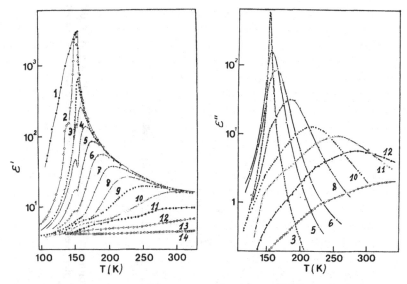

Figure 7.17. Temperature dependence of ε' and ε'' of DMAAS crystal for various frequencies. (kHz): 1) 0.12 and 2) 4; (MHz): 3) 1; 4) 4; 5) 11; 6) 28; 7) 70; 8) 173; and 9) 428; (GHz): 10) 1.2; 11) 3.5; 12) 9; 13) 26; and 14) 56. From Sobiestianskas et al. (1991).

The absence of a soft mode in the Raman spectra (Torgashev et al. 1990) suggests, as in the case of alums, an order-disorder nature of the phase transition that is caused by changes in the orientational mobility of the sulfate and dimethylammonium groups (Kiosse et al. 1990a, 1990b).

Figure 7.17 shows the temperature dependence of the complex dielectric permittivity along the ferroelectric x-axis at various frequencies. The value of the permittivity is highest along this axis. In the lower frequency region the dielectric permittivity is affected by the domain wall motion. A small minimum of ε' at T_c in the dispersion region indicates that the phase transition is far from being of the second order.

The ferroelectric dispersion begins in the kilohertz region near T_c and ranges up to the millimeter wave region at high temperatures (Figure 7.18). The region of ferroelectric dispersion is nearly the same as in the best known representative of ferroelectric alum, MASD (Makita and Sumita 1971).

As-grown DMAAS crystals usually contain an internal bias field, which causes the macroscopic noncentrosymmetry of the paraelectric

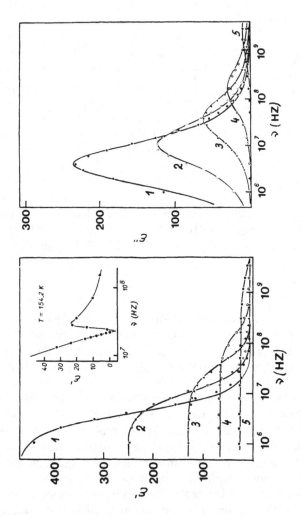

Figure 7.18. Frequency dependence of ε' and ε'' of DMAAS at various temperatures $T - T_c$ (K): 1) 3.8; 2) 8.8; 3) 20; 4) 40; and 5) 104. Dots are experimental; lines are monodispersive relaxator fits. The insert shows piezoelectric dispersion in the paraelectric phase. From Sobiestianskas et al. (1991).

phase, and, as a result, a piezoelectric resonance reveals itself in the whole paraelectric phase.

Over the 10^6 to 10^{10} Hz frequency range, the ferroelectric dispersion shows a monodispersive character. The distribution of the relaxation times appears near T_c only (Sobiestianskas et al. 1991). The relaxation time reaches the value 1.6×10^{-7} s at T_c. The critical slowing-down process accounts for the static dielectric permittivity which follows the Curie–Weiss law with the Curie constants $C^p = 2810$ K and $C^f = 695$ K in the paraelectric and ferroelectric phases, respectively. The ratio of the Curie constants $C^p/C^f = 4.05$ shows that the phase transition is nearly a critical one. Analysis of the temperature dependence of the spontaneous polarization (Cach et al. 1990) also shows that it follows the typical law for a critical phase transition

$$P_s \sim (T_c - T)^{1/4}.$$

Ultrasonic investigations (Sobiestianskas et al. 1991) reveal a large influence of the term $P_s^2 u^2$ in the free energy of the DMAAS crystal (u is the deformation of the ultrasonic wave). In this case, the increase of the velocity in the ferroelectric phase can be described as

$$\Delta v = BP_s^2.$$

The behavior $(\Delta v)^2 \sim (T_c - T)$ (Figure 7.19) confirms a critical phase transition in the DMAAS crystal. Because of the extremely low relaxation time, the condition $\omega\tau > 1$ is fulfilled at 10 MHz (Blinc and Žekš 1979). Therefore, there is no critical slowing-down when the ultrasonic wave propagates perpendicularly to the P_s-axis.

When the dimethylammonium groups are disordered in the paraelectric phase (Kiosse et al. 1990a and 1990b), it can be assumed that the phase transition is related to the ordering of these big groups in which the motion is responsible for the Debye-type dielectric dispersion. The value of the activation energy of the dipoles in the paraelectric phase is equal to $\Delta E = 0.112$ eV $= 8.5 \, kT_c$. The temperature dependence of the relaxation time calculated with this activation energy—using the dynamical Ising model (Blinc and Žekš 1979)—is in good agreement with the experimental values obtained from the results presented in Figure 7.17.

Finally, from the results above it follows that the DMAAS crystal undergoes a nearly critical ferroelectric phase transition of the order-disorder type. The ferroelectric dynamics are described to a good approximation in terms of the Debye-type monodispersive relaxation

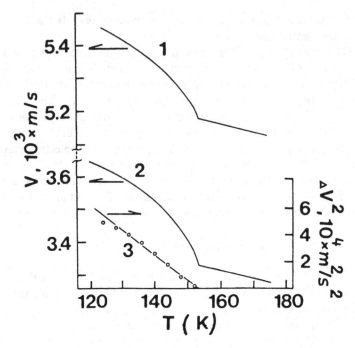

Figure 7.19. Temperature dependence of the longitudinal ultrasonic velocity perpendicular to the P_s-axis (1), along the P_s-axis (2), and the square of the velocity along the P_s-axis (3) at 10 MHz. From Sobiestianskas et al. (1991).

(which is supposed to be associated with the motion of the dimethylammonium groups). The heavy dipoles cause an extremely long relaxation time. The relaxational mode makes the main contribution to the high static permittivity of these crystals.

7.7. DIMETHYLAMMONIUMGALLIUM SULFATE

By isomorphous replacement of Ga for Al in DMAAS, the new compound dimethylammoniumgallium sulfate, $(CH_3)_2NH_2Ga(SO_4)_2 \cdot 6H_2O$, was obtained (Andreyev et al. 1991) having similar properties and structure (Pietraszko et al. 1993). This isomorphous analogue of DMAAS undergoes two phase transitions at $T_{c1} = 136$ K and $T_{c2} \approx 113$ K. The phase between T_{c1} and T_{c2} is ferroelectric, while the paraelectric phase at $T > T_{c2}$ is ferroelastic. The dielectric permittivity, ε'_x, at $T > T_{c1}$ and $T < T_{c1}$ follows the Curie–Weiss law with the Curie constants $C^p = 2930$ K and $C^f = 340$ K (Sobiestianskas et al. 1992b). The value of

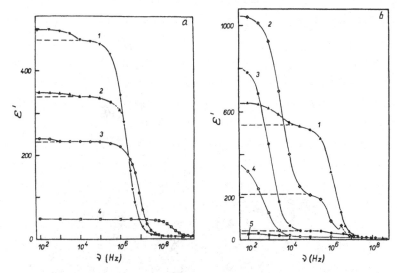

Figure 7.20. Frequency dependence of ε'_x of DMAGaS: a) in the paraelectric phase at: 1) 137 K; 2) 145 K; and 3) 200 K; b) in the ferroelectric phase at: 1) 135 K; 2) 131 K; 3) 125 K; 4) 115 K; and 5) 105 K. The dashed lines show a contribution of the polar phase clusters in paraelectric phase and domain walls in ferroelectric phase. From Sobiestianskas et al. (1992b).

the spontaneous polarization is $P_s = 1.94\,\mu\mathrm{C/cm^2}$. Upon cooling the crystal at the temperature 106 K, P_s vanishes suddenly. Upon heating, P_s appears at 118 K. The first-order phase transition at T_{c2} is supposed to be related to the freezing of reorientational motions of the dimethylammonium cations.

As in DMAAS, the ferroelectric dispersion occurs at rather low frequencies (Figure 7.20), and is well described by the Debye equation (2.11) with the single relaxation time. The contribution of hard modes and electronic polarization is $\varepsilon_\infty = 6$. The relaxation time of the dipole system reaches the value 1×10^{-7} s at T_{c1}. The low value of τ suggests that the motion of the heavy dimethylammonium groups is responsible for the ferroelectric dispersion, and that the phase transition is triggered by the ordering of these groups. The activation energy of the dipoles in the paraelectric phase is $0.09\,\mathrm{eV} = 7.6\,kT_{c1}$.

This ferroelectric, DMAGaS, is the characteristic example of crystals in which the ferroelectric dynamics overlap with the domain-wall dynamics (in the ferroelectric phase) and polar-phase cluster dynamics (above T_{c1}), which considerably increase the low-frequency permittivity

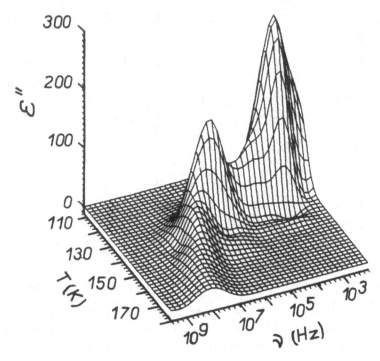

Figure 7.21. A computer graphic representation of the dielectric spectrum of DMAGaS crystal. From Sobiestianskas et al. (1992b).

upon approaching T_{c1}. The relaxation time of the latter process at 120 K reaches the value of 5×10^{-4} s. Figure 7.21 shows three-dimensional computer graphics of the dielectric spectrum of the DMAGaS crystal. On cooling, the relaxational soft mode comes from microwaves and, in the vicinity of T_{c1}, causes the maximum dielectric loss. The second, larger maximum represents the loss of the domain-wall dynamics. These two processes overlap in the spectrum of the real part of dielectric permittivity.

Finally, the DMAGaS crystal is a proper order-disorder type ferroelectric. The ferroelectric dispersion of the Debye type (associated with the motion of the dimethylammonium groups) in the vicinity of T_{c1} and the ferroelectric phase overlap with the domain-wall dynamics and cause the complex dielectric spectrum.

In organic-inorganic compounds the ferroelectric dynamics can be more complicated than those discussed above. For instance, two relaxations are observed, one at 2.7 GHz, the other at 48 GHz at room temperature in $CH_3NH_3HgCl_3$ (Ech-Chaoui et al. 1993). The high-frequency relaxation is linked to the motion of the dipoles from which the ferroelectricity originates.

7.8. ANTIMONY AND BISMUTH ORGANIC-INORGANIC CRYSTALS

Antimony and bismuth form a number of organic-inorganic halogenide ferroelectrics that show similar structural features to those discussed in Chapter 4. They belong to the family with the general formula $[(CH_3)_nNH_{4-n}]_3X_2Y_m$ ($X = Sb$ or Bi; $Y = Cl$, Br or I) and are also low-dimensional (see e.g., Miniewicz et al. 1990). For instance, the monoclinic $(P2_1/a)$ dimethylammonium nanochlorodiantimonate $[NH_2(CH_3)_2]_3Sb_2Cl_9$ (DMACA)—which, at $T_c = 242$ K, exhibits fer-roelectric PT—and dimethylammonium nanobromodiantimonate (DMABA)—which exhibits ferroelectric phase transition at $T_{c1} = 164$ K and structural phase transition at $T_{c2} = 228$ K—are formed of the infinitive $(Sb_2Cl_9^{3-})$ or $(Sb_2Br_9^{3-})$ layers, respectively. The methylam-monium cations occupy the cavities formed in the inorganic sublattice and are linked by weak hydrogen bonds to halogen atoms. The cations possess a considerable amount of freedom for a reorientational motion, which can excite phase transitions.

Orthorombic isomorphic bismuth ferroelectric compounds $(CH_3NH_3)_5Bi_2Br_{11}$ (MAPBB) and $(CH_3NH_3)_5Bi_2Cl_{11}$ (MAPCB) show a similar sequence of phase transitions:

$$Pcab \xrightarrow{T_{c1}} Pca2_1 \xrightarrow{T_{c2}} ?.$$

In MAPBB, $T_{c1} = 311.5$ K, $T_{c2} = 77$ K; in MAPCB $T_{c2} = 308$ K, $T_{c2} = 170$ K.

Near T_{c1} temperature, the static dielectric permittivity of both crystals shows a large anomaly in the c-direction. In MAPBB, the Curie–Weiss like behavior is also observed at T_{c2}. In MAPCB, only a stepwise change of the permittivity occurs.

Structural investigations show that the structure is built of bi-octahedra $Bi_2X_{11}^{5-}$. The $CH_3NH_3^+$ cations are spanned between these anion. The ordering of methylammonium cations plays a substantial role for the phase transitions: one pair of cations C3–N3 is disordered in the paraelectric phase and become ordered at T_{c1}; another pair, C12–N12, is disordered in both phases.

Dielectric investigations (Pawlaczyk et al. 1992) in the frequency range 100 Hz to 90 GHz have revealed three relaxators, which determine the dielectric behavior of these crystals: relaxator 1 shows dielectric dispersion in the frequency range of 1 to 100 MHz, and relaxator 2 shows it in the range of 1 to 10 GHz. The relaxator contribution, $\Delta\varepsilon_1$, prevails near T_{c1} and obeys the Curie–Weiss law.

At $T < 250$ K, the dielectric properties of both crystals are determined by the contribution $\Delta\varepsilon_2$. Near T_{c2}, the contribution of the third relaxator becomes noticeable.

Thus, the mechanism of the phase transitions is rather complicated. The C3–N3 cations, disordered above T_{c1}, are ordered below this temperature and cause the order-disorder phase transition at T_{c1}. Cooperative interactions between C3–N3 cations are the basis of the critically temperature dependent contribution $\Delta\varepsilon_1$. Cooperative ordering interactions of the cations C12–N12 are involved in the dielectric behavior far below T_{c1} and in the mechanism of the PT at T_{c2}. The dielectric contribution, $\Delta\varepsilon_3$, in MAPBB corresponds to the dielectric response of the interacting dipoles C12–N12. In MAPCB a freezing of the motion of C12–N12 cations causes a stepwise vanishing of the contribution of $\Delta\varepsilon_2$ near T_{c2}. The freezing suppresses the cooperative ordering interactions between cations C12–N12 yielding the contribution $\Delta\varepsilon_3$ of extremely small strength.

Trimethylammonium nanochlorodiantimonate [$(CH_3)_3NH]_3Sb_2Cl_9$ (TMACA) exhibits two phase transitions around 200 and 365 K, and ferroelectric properties along the ferroelectric c-axis below 365 K down to at least 125 K. A quasi-elastic neutron scattering study (Urban et al. 1992) has shown that the phase transitions are connected with releasing reorientational degrees of freedom of the $(CH_3)NH_2^+$ cations and uniaxial 120 jumps of both the cation and its methyl groups. The temperature dependence of ε^* on frequency is shown in Figure 7.22. The relaxational dielectric dispersion (Sobiestianskas et al. 1992c) consists of two parts: the first relaxation occurs below 3 GHz, and the second one above this frequency. The dielectric contribution, $\Delta\varepsilon_1$, for the first relaxator and the relaxation frequency, $v_1 = (2\pi\tau_1)^{-1}$, behave critically (Figure 7.23) with temperature (i.e., $v_1 = 32\,(T - 360.5)$ MHz) and cause the Curie–Weiss behavior of $\Delta\varepsilon_1$. The dielectric contribution of the second relaxator, $\Delta\varepsilon_2 = 60$, depends slightly on temperature.

The results show that the ferroelectric PT in this crystal is also of the order-disorder type. The relaxational processes are related to the reorientational motion of the trimethylammonium cations and its methyl groups. The relaxation time near T_c reaches the value 2×10^{-9} s and is characteristic of such heavy dipoles. The cooperative interactions

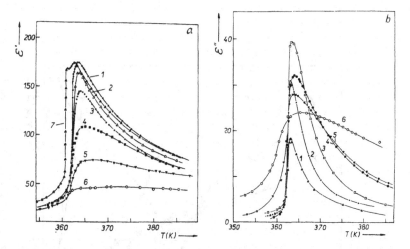

Figure 7.22. Temperature dependence of ε' and ε'' along the c-axis of TMACA crystal upon heating at frequencies: a) (MHz): 1) 1; 2) 9; 3) 30; 4) 130; and 5) 600; 6) 4 GHz; and 7) on cooling at 1 MHz; b) (MHz): 1) 9; 2) 30; 3) 97; 4) 320; and 5) 600; and 6) 4 GHz. From Sobiestianskas et al. (1992c).

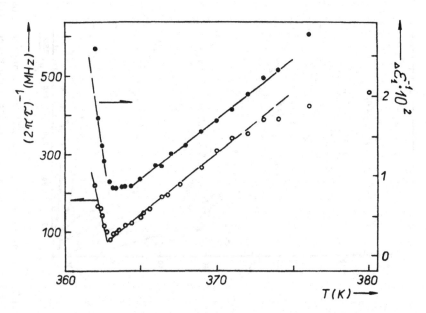

Figure 7.23. Temperature dependence of the relaxation frequency (o) and dielectric contribution (●) for the first relaxator in TMACA. From Sobiestianskas et al. (1992c).

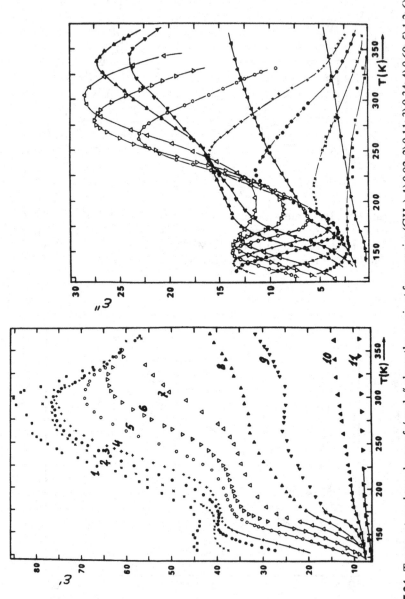

Figure 7.24. Temperature dependence of ε' and ε'' along the c-axis at frequencies (GHz): 1) 0.02; 2) 0.11; 3) 0.34; 4) 0.60; 5) 1.2; 6) 2.0; 7) 3.7; 8) 8.8; 9) 11; 10) 42; and 11) 77. From Sobiestianskas et al. (1990).

between the cations is the origin of the critical temperature dependence of $\Delta\varepsilon_1$. In a sense, the situation seems to be similar to MAPBB and MAPCB.

7.9. TOWARDS ORIENTATIONAL GLASS

A solid consisting of a regular lattice, some of whose sites are occupied by constituents containing a dipole or quadrupole moment, is called orientational glass. These moments have orientational degrees of freedom, they interact with one another and, below some freezing temperature, their motion slows and they freeze into a configuration devoid of long-range order (Höchli et al. 1990). The slowing of reorientations at the freezing temperature is observed in dielectric measurements.

This examination of crystals containing ammonium cations and undergoing various phase transitions concludes with two examples which show the orientational glass features. One of them is the $[(CH_3)_2NH_2]_5Cd_3Cl_{11}$ crystal. Primary investigations have shown that the crystal is of *Cmcm* symmetry with four formula units in the unit cell. Dielectric permittivity shows anomalies at the temperatures 260, 180 and 127 K. One of the dimethylammonium cations is disordered, and the 180 degree flip-flop motions are possible.

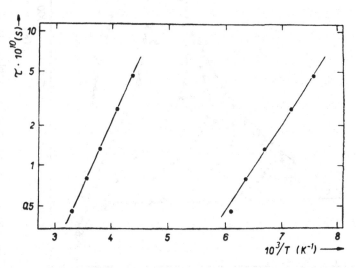

Figure 7.25. Relaxation times vs. temperature. From Sobiestianskas et al. (1990).

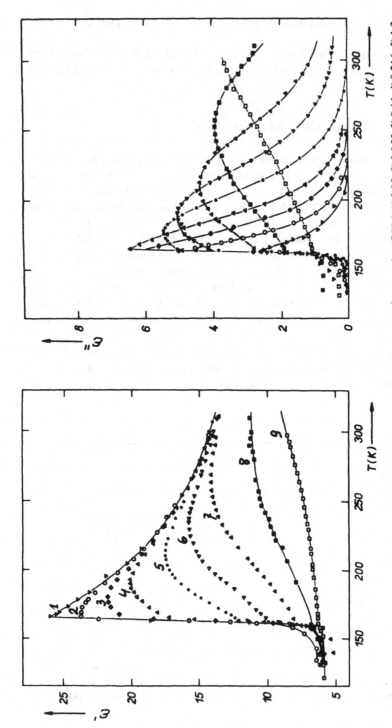

Figure 7.26. Temperature dependence of complex permittivity of N_2H_5Al-alum at frequencies (GHz): 1) 0.02; 2) 0.055; 3) 0.11; 4) 0.216; 5) 1.2; 6) 2; 7) 4.2; 8) 8.9; and 9) 42. From Grigas et al. (1990b).

Dielectric dispersion (Figure 7.24) in this crystal is caused by two different relaxators (Sobiestianskas et al. 1990). The relaxation responsible for the dielectric loss at high temperatures overlaps with the relaxation responsible for the loss at low temperatures. In the low temperature range, ε' remains nearly constant. As frequency increases, maxima of the $\varepsilon'(T)$ and $\varepsilon''(T)$ curves shift to the higher temperature region, similar to the orientational glass behavior. The complex dielectric permittivity can be described by Equation 2.14 with $\beta = 1$ above 300 K, and $\beta = 0.89$ at 150 K. Two different relaxation times (Figure 7.25) slow down and cause complicated dielectric spectrum.

N_2H_5Al-alum also shows orientational glass behavior (Figure 7.26). The dielectric dispersion can be described by Equation 2.13 with the parameter, α, increasing from the value $\alpha = 0.1$ at $T = 320$ K to $\alpha = 0.35$ at $T = 160$ K. The relaxation frequency depends nonlinearly on temperature (Grigas et al. 1990b; Czapla et al. 1990) and causes distorted Cole–Cole plots.

8 PROTONIC CONDUCTORS

8.1. INTRODUCTION

Ionic conductors over a wide frequency range exhibit frequency-dependent dielectric permittivity, electrical conductivity, and dielectric loss, each of which provide information about the orientational adjustment of dipoles and the translational changes of mobile carriers. Protonic conduction in dielectrics is a particular case of ionic conduction in solid electrolytes. It is caused by two kinds of ionic motions: namely, oscillatory motions of the H-bond and diffusive motions from site to site (i.e., transfer of the H-bond). The information about the short characteristic times of the first type of motion can be obtained by vibrational spectroscopy, while information about long characteristic times of the second type of motion can be obtained from tracer diffusion or low frequency electrical conductivity measurements. Microscale dynamics of protons are obtained by NMR and neutron scattering measurements.

Frequency and temperature dependencies of dielectric permittivity and electrical conductivity allow the gap between local and long range dynamics to be filled. The results can be analyzed in the light of diffusion and hopping models.

The experimental techniques described in Chapter 3 allow the investigation of the dynamic properties of the complex dielectric permittivity or conductivity of solid electrolytes or superionics.

As conductivity of dielectrics in an alternating electric field is complex, the density of the electric current is given by

$$I = \sigma^*(\omega)E = i\omega\varepsilon_0\varepsilon^*(\omega)E. \tag{8.1}$$

In the case of the linear dielectric, the real part of permittivity, $\varepsilon'(\omega)$, and the imaginary part of conductivity give the components of displacement current. The imaginary part of permittivity, $\varepsilon''(\omega)$, and the real part of the conductivity, $\sigma(\omega)$, cause the dielectric loss

$$\varepsilon''(\omega) = \frac{\sigma(\omega)}{\varepsilon_0\omega}, \qquad (8.2)$$

which linearly decreases with the increase of frequency as long as the conductivity is frequency independent.

The density of current in the order-disorder type dielectrics (see Chapter 2) with the Debye-type polarization is given by

$$I = i\omega\varepsilon_0\left[\varepsilon_\infty + \frac{\varepsilon(0) - \varepsilon_\infty}{1 + \omega^2\tau^2} - i\frac{\omega\tau(\varepsilon(0) - \varepsilon_\infty)}{1 + \omega^2\tau^2}\right]E, \qquad (8.3)$$

where the first two terms in the square brackets are the real part, while the last term is the imaginary part of dielectric permittivity. The conductivity, σ^p, caused by the dipole polarization processes, is given by

$$\sigma^P(\omega) = \varepsilon_0\omega^2\tau\frac{\varepsilon(0) - \varepsilon_\infty}{1 + \omega^2\tau^2}. \qquad (8.4)$$

In the frequency range of $\omega\tau \ll 1$, $\varepsilon'' \sim \omega$, and $\sigma^P(\omega) \sim \omega^2$. At frequencies of $\omega\tau \gg 1$, $\varepsilon'' \sim 1/\omega$, and $\sigma^P(\omega) = const$.

When in the Debye-type materials both the frequency independent d.c. conductivity, σ_0, and the frequency dependent conductivity (caused by dipole polarization and given by Equation 8.4) coexist, the total conductivity is

$$\sigma(\omega) = \sigma_0 + \sigma^P(\omega). \qquad (8.5)$$

At frequencies $\omega\tau \ll 1$, the conductivity $\sigma_0 \gg \sigma^P(\omega)$ and the total conductivity $\sigma(\omega) = const$. At $\omega\tau \gg 1$, $\sigma^P(\omega) \gg \sigma_0$ and again $\sigma(\omega) = const$.

The dipole absorption, showing Cole–Cole arcs, was found in the pellets of some ionic conductors by impedance spectroscopy (Colomban 1992). However, neglect of the conditions of the quasi-stationary electric field distribution (Equation 3.10) in the sample yields doubtful results. The dipole absorption in the crystals of proton conductors, in general, is of non-Debye-type. In that case at lower frequencies the translational motion of ions contribute to the loss according to Equation 8.2. At higher frequencies the total conductivity is of the form

$$\sigma(\omega) = \sigma_0 + A\omega^n, \qquad (8.6)$$

where $n < 1$ and A is a constant.

When $\sigma_0 \gg \sigma^P(\omega)$, $\varepsilon'' \sim 1/\omega$ and $\sigma(\omega) = const.$ At $\sigma^P(\omega) > \sigma_0$, $\varepsilon'' \sim \omega^{n-1}$ and $\sigma(\omega) \sim \omega^n$. The "universal dynamic response" in disordered solids, which predominates in the conductivity spectra of ionic conductors, has been discussed in detail by Jonscher (1983).

Impedance spectroscopy of solid electrolyte materials used for solid oxide fuel cells (SOFC) and chemical sensors has been discussed by Macdonald (1987). Today the SOFC technique has proven its technical and ecological feasibility. New processing techniques in combination with modified and new materials are under development to improve performance and to meet future economic targets. Considerable contributions to the development of SOFC materials such as perovskites for electrodes, ZrO_2 and CeO_2 as electrolytes is made by Prof. L.J. Gauckler at ETH Zürich (see e.g., Gauckler and Sasaki 1994; Gauckler et al. 1994; Gödickemeier et al. 1994).

Section 8.2 describes the simplest model of hopping conductivity in the protonic conductors. The following sections include: 1) a brief review of the existing approaches to the dynamic conductivity of ionic conductors (Section 8.3); 2) the experimental results of the temperature dependence of complex dielectric permittivity (Section 8.4); 3) electrical conductivity (Section 8.5); and 4) the frequency dependence of the permittivity and conductivity of some representatives of protonic conductors (Section 8.6), namely, $CsDSO_4$ and $CsDSeO_4$, $Me_3H(SeO_4)_2$ and $Cs_5H_3(SO_4)_4 \times 2H_2O$. A model of the protonic conductivity in $Rb_3H(SeO_4)_2$ is presented in Section 8.7.

8.2. LATTICE GAS MODEL OF HOPPING CONDUCTIVITY

The microscopic description of ionic conductivity can be based on several microscopic models, which can be divided in two main groups: discrete hopping and continuous flow models. The first is valid when the transfer time of the ion, τ_t, is much shorter than the mean residence time, τ_r ($\tau_t \ll \tau_r$). Otherwise, when $\tau_t \lesssim \tau_r$, the second group of models is more appropriate. The ratio τ_t/τ_r could be extracted from the ion-density maps, if available. In general, the ratio is larger for materials with a very small activation energy for conductivity. For a number of the protonic conductors the proton density maps cannot be obtained. However, the measured activation energy in the superionic phase $E_a \leqslant 0.20\,eV$

(Section 8.5) points to the discrete-hopping nature of the conductivity. The simplest model of hopping conductivity in the protonic conductors is a lattice gas model.

The lattice gas model describes a system of interacting particles, which are located at fixed sites on the regular lattice. For the protonic conductors the particles are represented by protons while the sites are represented by hydrogen bonds (actual and virtual). The proton can break the H-bond due to thermal vibrations and form another bond in the neighboring site. This process is referred to as the "jump" of the proton between two sites. All the movements of heavy atoms, which possibly assist the jump, are believed to influence only the height of the effective potential barrier between the sites and are not considered explicitly further. Actually, the H-bond is known to be characterized by the double-well potential minimum. This feature leads to the concept of interacting electric dipoles, giving rise to the ferroelectric behavior in H-bonded crystals (Chapters 6 and 7). However, the proton conductivity, not the arrangement of dipoles, is of primary interest. Thus, the particular structure of the H-bond is neglected and treated as a single potential well. This approach seems plausible since the ferroelectric phenomena take place at much lower temperatures than the superionic transition. This suggests that the dipolar interaction is much weaker and plays a minor role in the superionic disordering. Under the above mentioned assumptions, the model Hamiltonian takes the form

$$H = \sum_{i>j} V_{ij} n_i n_j - \sum_j E_i n_i, \qquad (8.7)$$

where $n_i = 1$ if the site i is occupied by a proton (otherwise $n_i = 0$), V_{ij} are the pair interaction potentials between the protons (these may be screened coulomb interactions or phonon-mediated elastic coupling and are usually short-ranged), and E_i is the binding energy of the H-bond. The sum is over all configurations of the protons among the available sites.

The thermodynamic properties of the model given by Equation 8.7 are obtained by averaging of the canonical ensemble with the total amount of particles fixed or the grand canonical ensemble with the chemical potential fixed.

The time evolution of the microscopic state actually determines the thermodynamic equilibrium as well as a gradual change of the thermodynamic state under the change of the environment. This process is governed by the corresponding kinetic equation. For the lattice gas model, as well as for the Ising model, the kinetics can be described in terms of the master equation (Glauber 1963)

$$\frac{d}{dt} p(\{n\}_i, t) = -\sum_j w_{ij} p(\{n\}_i, t) + \sum_j w_{ji} p(\{n\}_j, t), \qquad (8.8)$$

where $p(\{n\}_i, t)$ is the probability of finding the system in the i-th microscopic state $\{n\}_i$ (a certain configuration of empty and occupied sites), and w_{ij} is the jumping rate between microscopic states $\{n\}_i$ and $\{n\}_j$. The sums run over states $\{n\}_j$, which differ from the state $\{n\}_i$ by the occupancy of a single site or by the interchange of the occupancies of the single pair (jump of a single atom). The master equation approach (Equation 8.8) is based on two assumptions (Beyeler et al. 1979): 1) the immobile part of the system relaxes during the residence time of a mobile ion; and 2) the ion thermalizes after a jump and cannot make several jumps during an elementary process. Under these assumptions, the jump probability, w_{ij}, of the ion depends only on the current microscopic state and does not depend on the preceding jump. These assumptions are not applicable in some cases for high frequencies and low E_a (e.g., microwave conductivity of AgJ). However, for the protonic conductors these assumption seem to be quite valid at frequencies up to 10^{11} Hz due to relatively high E_a and long τ_r.

The expression of the jump rate, w_{ij}, can have various forms. In the thermodynamic equilibrium, the following relation must hold:

$$w_{ij} p(\{n\}_i) = w_{ji} p(\{n\}_j). \qquad (8.9)$$

In the thermodynamic equilibrium, the probability of a microscopic state is proportional to the Boltzman factor $\exp(-E_i/kT)$, where E_i is the energy of microscopic state $\{n\}_i$. Thus, the jump rates should satisfy the relation

$$\frac{w_{ij}}{w_{ji}} = \exp[(E_i - E_j)/kT]. \qquad (8.10)$$

This is the only restriction on w_{ij}. The detailed expression must correspond to the particular nature of the elementary jump process. For the ionic conductivity in the lattice-gas model, a thermally activated jump over a saddle-point of the potential barrier is proposed as an elementary process (Kikuchi 1966). In this case, the jump rate can be written as

$$w_{ij} = v_0 \exp[(E_t - U)/kT], \qquad (8.11)$$

where v_0 is an attempt frequency ($\sim 10^{11}$ to 10^{13} Hz), U is the height of the potential barriers in the absence of interionic coupling, and E_t is the total interaction energy of the current ion with its neighbors (a linear combination of V_{ij} from the Hamiltonian given by Equation 8.7). In Equation 8.11, it is supposed that the interaction does not affect the saddle-point energy but alters the energy of the potential minimum. However, Equation 8.11 is a simplification of the elementary jump, and it is commonly used in the analytic calculations of the transport phenomena as well as in the Monte–Carlo simulation (Murch 1982). Some alternative expressions were also proposed by Kutner et al. (1982) and Singer and Peshel (1980). It is apparent that Equation 8.9 satisfies the detailed balance requirement for Equation 8.10.

No dispersion of the hopping ionic conductivity is expected in the one-component system of noninteracting ions distributed on the equivalent lattice sites (Murch and Thorn 1977) since no temporal evolution of the ion distribution can occur under the electric field.

8.3. DISPERSION OF IONIC CONDUCTIVITY

The one-component noninteracting lattice-gas system with two types of sites (having different potential depths) or two types of barriers (with different heights) exhibits the Debye-type dispersion with a single relaxation time (Ishii 1990), followed by the conductivity (Equation 8.5) determined mainly by the higher barrier. The conductivity, $\sigma(\omega)$, in this case (see Section 8.1) has flat regions in the low and high frequency limits, $\sigma(0)$ and $\sigma(\infty)$ respectively. The latter represents the local movement of the ions between higher potential barriers. The exact expression of $\sigma(\omega)$ for the one-dimensional system of noninteracting ions as proposed by Ishii (1990) is

$$\sigma(\omega) = \frac{n(ea)^2}{kT} \left[\frac{1}{\langle \Gamma^{-1} \rangle} + \left(\frac{\Delta \Gamma}{4 \langle \Gamma \rangle} \right)^2 - \frac{i4\omega \langle \Gamma \rangle}{4 \langle \Gamma \rangle - i\omega} \right], \qquad (8.12)$$

where $\langle \Gamma^{-1} \rangle = (\Gamma_0^{-1} + \Gamma_1^{-1})/2$, $\langle \Gamma \rangle = (\Gamma_0 + \Gamma_1)/2$, $\Delta \Gamma = \Gamma_1 - \Gamma_0$, and Γ_0 and Γ_1 are the jump rates over higher and lower barriers, respectively. Thus, the limiting frequencies of dispersion ω_1 and ω_2 are equal to

$$\omega_2 \approx 2\Gamma_1 \qquad \omega_1 \approx 4(\Gamma_0 \Gamma_1)^{1/2} \qquad (8.13)$$

in the limit $\Gamma_1 \gg \Gamma_0$.

Usually, the high frequency plateau is not clearly seen experimentally (Section 8.6), since ω_2 lies well in the region of phonon frequencies while the slope at the dispersion region $\omega_1 < \omega < \omega_2$ can be described by the exponent $n \leq 1$ (the "universal dynamic response" after Jonscher (1983), see also Section 8.1).

There are two basic approaches to the problem of the "universal dynamic response" in one-component systems. The first supposes the existence of a number of various potential barriers and considers the distribution of relaxation times, while each particular relaxation time gives the Debye response. Obviously, one could always accept the distribution $p(T)$, which fits the observed experimental data. For example, a flat distribution of activation energies yields the frequency dependence $\sigma \sim \omega^n$ with $n = 1$ (Giuntini et al. 1988; Elliot 1987). This approach seems to be quite acceptable for systems with a certain amount of static structural disorder of the lattice (glasses, amorphous compounds, and others). However, it is somewhat difficult to adopt such an approach for dynamically disordered protonic conductors, where a fast protonic transport in high-temperature phases is caused by dynamic disorder of the H-bond network and rotation of $HS(Se)O_4^-$ groups (Section 8.7).

The second approach considers collective phenomena due to ionic interactions. These interactions are certainly of major importance in protonic conductors since they are essential for ion ordering during the transition from superionic to ordered phase. Due to collective phenomena, the system rearranges after each individual jump. This rearrangement should give rise to the frequency dependence of ionic conductivity even in the systems with all sites and potential barriers equivalent. Hill and Jonscher (1983) proposed a model of quantum-tunneling rearrangement for dielectric response, which can also be transferred to ionic conductivity. This model can describe the non-Debye response in a very general way. However, it requires some assumptions that are beyond the lattice-gas model and cannot be definitely substantiated for the particular structure. The "jump-relaxation" model (Funke and Hoppe 1990) considers a lattice-gas with long-range coulomb interactions of ions. The interaction is treated in terms of Debye–Hückel atmosphere. The ion jumps between potential minima of the lattice, and the additional cage potential due to interaction with surrounding ions is imposed. At high frequencies, the conductivity depends on the local movement of the ion within the cage, while the low frequency transport of the ion requires a simultaneous movement of the cage potential and, therefore, a continuous rearrangement of the surrounding ions. The conductivity is obtained from the linear response theory via the correlated backward hopping rate, which corresponds to the velocity auto-correlation function. The cross-terms of the correlation function are neglected. The model gives a remarkable description of the low frequency part of $\sigma(\omega)$ for $RbAg_4I_5$, though no structural details are involved in the model. However, the assumptions, made in the formalism of the model, are valid only for dilute systems when the number of available sites is much larger than the number of mobile ions. A detailed analysis of the model also shows that $\sigma(\omega)$ depends heavily on the value of the somewhat indefinite parameter $\rho = R_0(N/V)^{1/3}$, where V/N is the volume per ion, and R_0 is the distance between the ion and the boundary of the surrounding Debye–Hückel atmosphere (see Funke and Hoppe 1990). The choice of this parameter to a rather limited extent enables the theoretical curves to be fitted to various experimental data. However, the model is quite valuable for a semiquantitative understanding of the "universal" behavior of $\sigma(\omega)$ and can also describe a corresponding $\varepsilon'(\omega)$ behavior due to rearrangement of charged ions in the applied field.

The first extensive study of the ionic conductivity of the concentrated lattice gas with interionic coupling was made by Kikuchi and Sato (1971). They used their Path Probability Method (PPM), based on the master equation with the jump rates (Equation 8.11), for the description of the ionic conductivity of the Na-β'' and Na-β alumina solid electrolytes. They found that the ionic conductivity can be expressed as follows:

$$\sigma(0) = v_0 e^{-U/kT} \frac{e^2}{3^{1/2}kTc} nVWf_I, \tag{8.14}$$

where n is the mobile ion concentration, V is the vacancy availability factor, W is the bond breaking factor representing the effect of interionic coupling, and c is the distance between adjacent two-dimensional layers of the lattice. The last factor, f_I, was later shown to be the only frequency-dependent factor (see e.g., Murch 1982), so that

$$f_I = \frac{\sigma(0)}{\sigma(\infty)}, \tag{8.15}$$

if only the effects included in the master equation are considered. It was also shown that the factor, f_I, is not a real correlation factor but rather a percolation efficiency of the ionic current. The percolation efficiency, f_I, was found to be unity for the Na-β'' alumina in the disordered state when all lattice sites are equivalent. For Na-β alumina with two types of lattice sites having different binding energies, f_I is not unity and has a minimum near $n = 1/2$ with temperature-dependent depth. It was later generalized (Ishii et al. 1986) that for the nontrivial f_I in the framework of PPM (and the frequency dependent conductivity), the existence of several nonequivalent sublattices (or different mobile species) is essential. However, the results of Monte-Carlo simulation of ionic conductivity in Na-β'' alumina (Murch and Thorn 1977) also revealed the nontrivial f_I value. Thus, the original version of PPM is inadequate in treating the frequency-dependent ionic conductivity in a disordered one-component system with equivalent sites. Later, some improvements of the PPM were proposed (Sato and Kikuchi 1983; Sato 1990), which offered a better description of the ionic conductivity and the mixed alkali effect in binary systems. However, no such improvements applied to the one-component lattice gas are known.

The details of $\sigma(\omega)$ in the binary systems were also investigated by PPM (Ozeki and Ishikawa 1986; Ishii et al. 1986). The Debye-type response was obtained in all cases, giving rise to the power law $\sigma(\omega) \sim \omega^2$ in the middle of dispersion region. The index $n = 2$ is far above the widely observed "universal response" $n \leq 1$.

This was a brief review of the existing approaches to the dynamic conductivity of ionic conductors.

8.4. TEMPERATURE DEPENDENCE OF COMPLEX DIELECTRIC PERMITTIVITY

Consider the general features of the experimental behavior of dielectric permittivity and loss in the vicinity of the first-order superionic phase transitions in the anhydrous monoclinic protonic conductors.

Early studies of the dielectric properties of $CsHSO_4$ crystals at low frequencies by Komukae et al. (1981) showed that both real and imaginary parts of the complex dielectric permittivity[1] ε_a^*, ε_b^*, and ε_c^* increase discontinuously at the transition temperature to the superionic phase T_c and, at 10 kHz, show magnitudes on the order of 10^3 for $T > T_c$. According to Wolak and Czapla (1981), the dielectric permittivity ε_c^* of $CsHSeO_4$ at T_c shows an abrupt jump to about 350 at 1 MHz and remains almost independent of temperature both at $T < T_c$ and $T > T_c$. According to Yokota et al. (1982), both ε_b' and ε_b'' of $CsHSeO_4$ at 1 MHz exhibit sharp peaks at T_c. Above T_c, ε' is on the order of 10^2 and increases with the increase of temperature.

Microwave studies of these crystals and their deuterated analogues have been performed by Mizeris et al. (1988). Figure 8.1 shows the temperature dependence of ε' and ε'' along the a-axis of the $CsDSO_4$ crystal measured during heating at different frequencies. At T_c, ε' exhibits a peak and is weakly dependent on temperature in the superionic phase.

---

[1] Monoclinic axes are used for crystals investigated in this chapter.

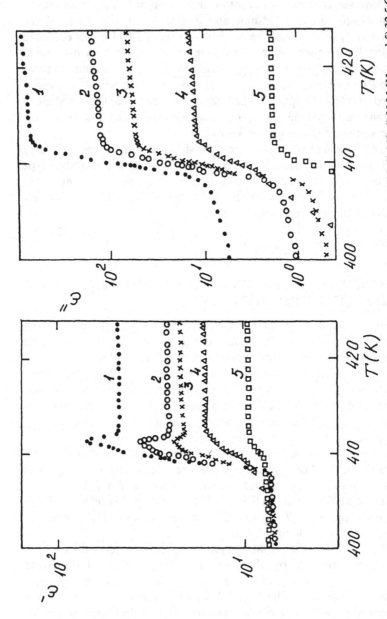

Figure 8.1. Temperature dependences of ε_a' and ε_a'' of CsDSO$_4$ crystal at frequencies: 1) 15; 2) 95; 3) 240; 4) 985 MHz; and 5) 32.5 GHz. From Mizeris et al. (1988).

The imaginary part of permittivity at all frequencies and the real part at frequencies $v > 240$ MHz exhibit an upward jump at the transition to the superionic phase. Similar values and behavior are exhibited by $CsHSO_4$, $CsHSeO_4$, and $CsDSeO_4$ crystals. Contrary to Colomban (1992), no Curie–Weiss law in the high temperature phases was observed, as superionic conductivity is not associated with ferroelectricity. Also, the dielectric relaxation in the protonic conductors has nothing to do with ferroelectric relaxation.

Microwave studies of $Rb_3H(SeO_4)_2$ crystals by Mizeris et al. (1991) up to 10^{11} Hz have shown (Figures 8.2 and 8.3) that the behavior of permittivity and loss in the vicinity of the phase transition temperature and in the superionic phase qualitatively is similar to that shown in Figure 8.1. Figure 8.4 shows the dielectric parameters of $Cs_5H_3(SO_4)_4 \times 2H_2O$ crystal in the superionic plane. Microwave and ultrasonic studies show that this crystal below T_c exibits a dipole glass state (Mizaras et al. 1994b).

Different behavior is observed among the high frequency and microwave permittivities and dielectric loss of these protonic conductors, although all crystals show a large increase of the permittivity and loss in the superionic phase. In the several megahertz region, the dielectric permittivity and loss are affected to a certain degree by electrode effects. At or above 100 MHz the intrinsic permittivity and loss are predominant. The loss is negligibly small in the ordered dielectric phase and high ($\tan\delta > 1$) in the superionic phase, where the loss depends on frequency according to Equations 8.2 and 8.6. The real part of the intrinsic dielectric permittivity of all these crystals in the ordered phase is small. An abrupt increase of the dielectric permittivity at T_c can be understood on the basis of a modified q-component Potts model within the framework of the molecular field approximation. The proton (deuteron) transfer process consists of two steps: SO_4 (SeO_4) group rotation inducing cross-linking between the chains, and the translation of protons between the two equilibrium states in the new O—H(D)—O bond. These two steps contribute to the dynamic dielectric permittivity and its temperature dependence, which can be quantitatively understood by referring to the model by Urbonavičius et al. (1982b). When this model is applied for protons, it can be assumed that if a proton is in the site, then its average dipole momentum is equal to zero. When leaving the site the proton acquires the dipole moment p. For the first-order phase transition, the temperature dependence of the permittivity is given by

$$\varepsilon(0) = \varepsilon_\infty + \frac{eN}{kT_c q - KN}, \tag{8.16}$$

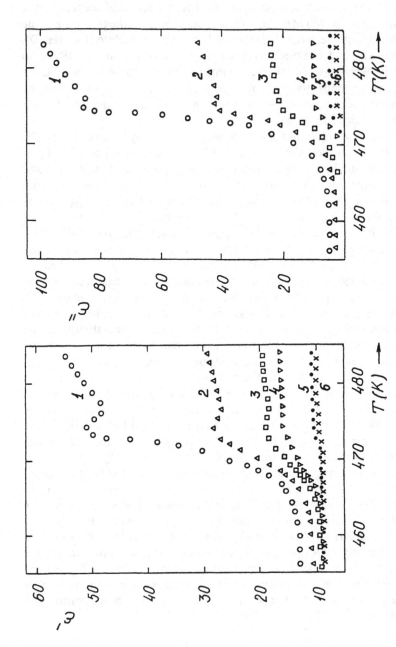

Figure 8.2. Temperature dependences of ε' and ε'' of $Rb_3H(SeO_4)_2$ crystals along the a-axis at frequencies: 1) 50; 2) 110; 3) 250; 4) 550 MHz; 5) 47; and 6) 77 GHz. From Mizeris et al. (1991).

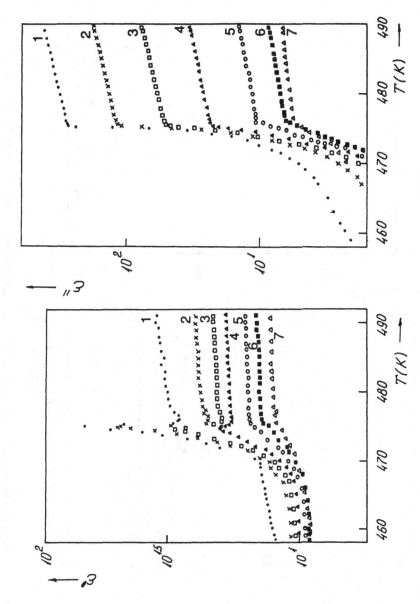

Figure 8.3. Temperature dependences of ε' and ε'' of $Rb_3H(SeO_4)_2$ crystals along the c-axis at frequencies: 1) 20; 2) 47; 3) 109; 4) 253; 5) 620 MHz; 6) 1.2; and 7) 4 GHz.

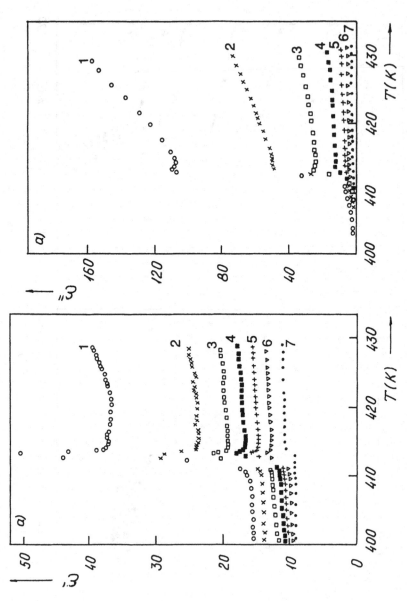

Figure 8.4. Temperature dependences of ε' and ε'' for $Cs_5H_3(SO_4)_4 \times 2H_2O$ at frequencies: 1) 20; 2) 39; 3) 75; 4) 146; 5) 283 MHz; 6) 1.2; and 7) 70.15 GHz.

where k is the Boltzman constant, N is the density of the disordered protons, $e = 4\pi p^2/v$, v is the volume occupied by one proton, q is the number of the states for disordered protons, $K = I_{11}^{(2)} - I_{11}^{(1)}$, $I_{11}^{(1)}$ and $I_{11}^{(2)}$ are interaction potentials of parallel and nonparallel orientated dipoles respectively.

It follows from Equation 8.16 that, when $kT_c q > KN$, the jump of N (i.e., the jump of conductivity at T_c) causes an upward jump of dielectric permittivity. The larger the K, the larger is the magnitude of the $\Delta\varepsilon = \varepsilon(0) - \varepsilon_\infty$. At microwave frequencies the magnitude of the jump is $\Delta\varepsilon = 3$ for AgI (Roemer and Luther 1981), $\Delta\varepsilon = 1.5$ for Ag_2HgI_4, and $\Delta\varepsilon = 1.2$ for $CuHgI_4$ (Wong et al. 1981). Due to a relaxational dispersion, the jump of decreases with the increase of frequency. Proton-relaxation dynamics in the superionic phase cause the decrease of permittivity with the increase of temperature. Meanwhile, the rotation of SO_4 (SeO_4) groups increases static permittivity with temperature. The contributions of these two processes are different at different frequencies. When the latter process prevails at lower frequencies, the permittivity increases with the increase of temperature. When both processes compensate each other, the dielectric permittivity depends weakly on temperature. An upward jump of permittivity at the first-order structural phase transition is also predicted by the phenomenological Landau theory.

Close to T_c, all crystals show an extra peak of ε' that does not originate from jumps between the two states of the first-order transition scenario. The peaks are reminiscent of a long-range order and arise from finite-lifetime collective excitations of protons at that particular frequency.

Triammonium hydrogen bisulfate and biselenate crystals differ from their analogues as they undergo a sequence of successive phase transitions, including ferroelectric, improper ferroelastic and superionic ones (Lukaszewicz and Pietraszko 1992). Figure 8.5 shows the temperature dependence of ε' and ε'' for the stoichiometric $(NH_4)_3H(SeO_4)_2$ crystal along the b-axis. The influence of nonstoichiometric acid growing solutions on the phase transitions and physical properties of the crystals have been discussed by Augustyniak and Hoffmann (1992). There are two high temperature trigonal superionic phases. At $T > T_{c1}$, all protons (including those of ammonium ions) undergo diffusion motion. According to NMR studies at $T < T_{c1}$, only protons of the hydrogen bonds participate in fast diffusion (Moskvich et al. 1988). At T_{c2}, dielectric permittivity and loss show an upward jump as in other proton conductors. At the second order phase transition, T_{c1}, ε' shows an upward break for lower frequencies in accordance with the phenomenological Landau theory.

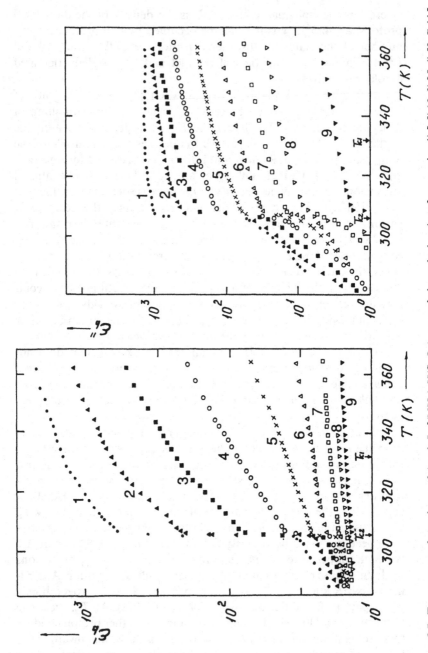

Figure 8.5. Temperature dependence of ε' and ε'' of $(NH_4)_3H(SeO_4)_2$ crystal along the b-axis at frequencies: 1) 5; 2) 10; 3) 25; 4) 50; 5) 115; 6) 290; and 7) 550 MHz; 8) 1.2; and 9) 2 GHz. From Augustyniak and Grigas (1994).

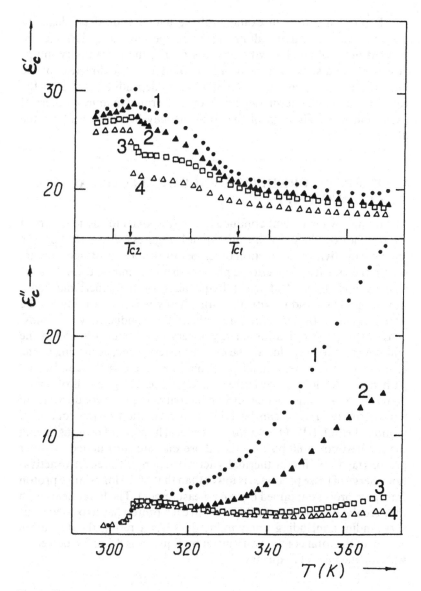

Figure 8.6. Temperature dependence of ε' and ε'' of $(NH_4)_3H(SeO_4)_2$ crystal along the c-axis at frequencies: 1) 25; 2) 50; 3) 250; and 4) 550 MHz. From Augustyniak and Grigas (1994).

Figure 8.6 shows an example of the temperature dependence of the complex permittivity along the non-superionic axis. Despite the upward jump of conductivity and loss at T_{c2}, the permittivity jumps downward and shows a downward break at T_{c1}. The direction of the jump in the phenomenological Landau theory depends on the sign of the constant, a, in the term $(a\eta^2 E^2)/2$ of a thermodynamical potential (Equation 6.56). The sign of the constant may be different along the different axis.

8.5. TEMPERATURE DEPENDENCE OF ELECTRICAL CONDUCTIVITY

The frequency dependent conductivity corresponds to the real part of the complex conductivity in Equation 8.1. Figures 8.7 to 8.10 show the ionic conductivity as a function of temperature of the protonic conductors in the vicinity of superionic phase transition temperatures. Drastic changes are observed at lower frequencies for the superionic phase transition. The conductivity in both phases varies according to the Arrhenius law. In the superionic phase the conductivity is almost constant (i.e., the activation energy is very low, about 0.20 eV) in the millimeter wave range for all the crystals investigated, according to the Drude model of a very small potential barrier along the conducting pathway. In the low temperature ordered phase, a large gap is observed between low and high frequency conductivity. The high conductivity in the microwave region can be related to relaxation of the rotational motion of HSO_4 ($HSeO_4$), and/or to the local hopping of protons having the smallest potential barriers (i.e., those that are surrounded by other protons favorable for the thermally activated jump). The effective activation energy of these processes is lower than that of the long-range proton transport process obtained from σ_0 measurement. The latter deals with the percolation efficiency problem when each proton has to pass various surroundings, including unfavorable ones. As a result, the activation energy of the total conductivity given by Equations 8.5 or 8.6 decreases with the increase of frequency.

8.6. FREQUENCY DEPENDENCE OF PERMITTIVITY AND CONDUCTIVITY

In the wide frequency range below phonon frequencies, the dielectric permittivity of protonic conductors (Figure 8.11) shows a pronounced relaxational non-Debye-type dispersion. In the entire frequency range

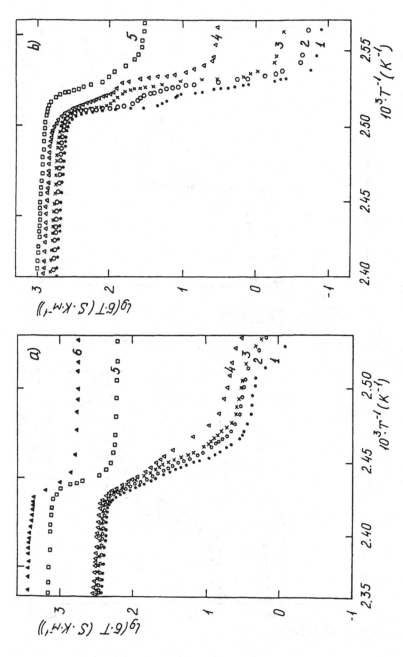

Figure 8.7. Ionic conductivity as a function of $1/T$ of: a) CsDSO$_4$ at frequencies: 1) 15; 2) 95; 3) 240; 4) 985 MHz; 5) 32; 6) 52 GHz; b) CsDSeO$_4$ crystal at frequencies: 1) 15; 2) 95; 3) 240; 4) 980 MHz; 5) 3.9 GHz. From Mizeris et al. (1988).

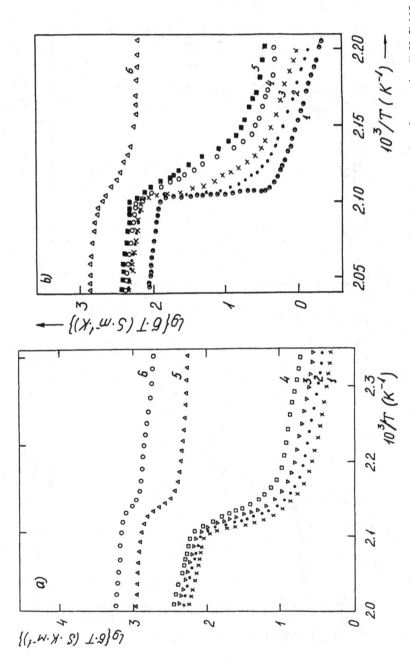

Figure 8.8. Temperature dependence of electrical conductivity of $Rb_3H(SeO_4)_2$ crystal: a) along the a-axis at frequencies: 1) 50; 2) 110; 3) 250; 4) 550 MHz; 5) 47; 6) 77 GHz; b) along the b-axis at: 1) 1; 2) 20; 3) 47; 4) 620 MHz; 5) 1.2; 4 GHz. From Mizeris et al. (1991).

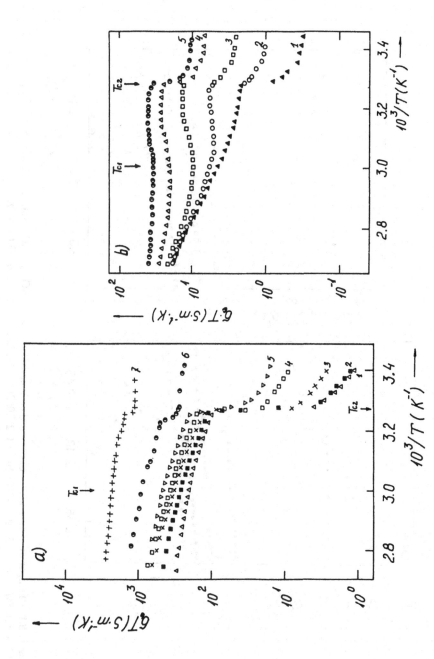

Figure 8.9. Temperature dependence of electrical conductivity of $(NH_4)_3 - H(SeO_4)_2$ crystal: a) along the b-axis at frequencies: 1) 10; 2) 25; 3) 115; 4) 550 MHz; 5) 1.2; 6) 8.7; 7) 77.3 GHz; b) along the c-axis at frequencies: 1) 50; 2) 110; 3) 250; 4) 550 MHz; 5) 1.2 GHz. From

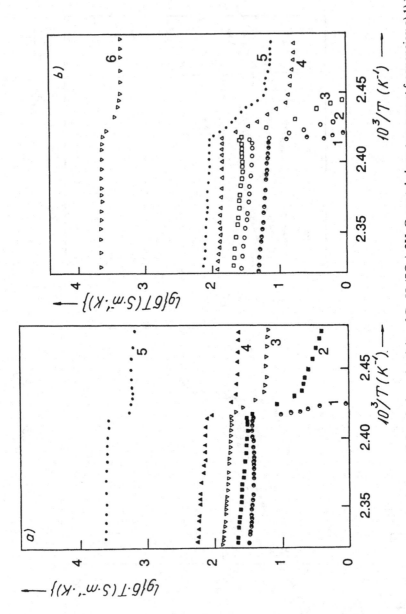

Figure 8.10. Temperature dependence of electrical conductivity of $Cs_5H_3(SO_4)_2 \cdot 2H_2O$ crystal along two axes at frequencies: a) 1) 1; 2) 110; 3) 550 MHz; 4) 1.2; 5) 70.15 GHz; b): 1) 1; 2) 20; 3) 75; 4) 550 MHz; 5) 1.2; 6) 70.15 GHz.

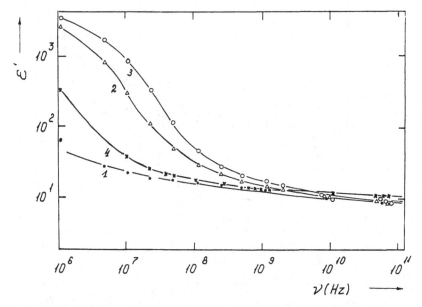

Figure 8.11. Frequency dependence of dielectric permittivity along the b-axis of $(NH_4)_3H(SeO_4)_2$ crystal at temperatures: 1) 300; 2) 315; 3) 345 K; and $Cs_5H_3(SO_4)_42H_2O$ crystal (4).

there are no relaxators which could be assigned to the specific protonic or ionic species with the characteristic relaxation times. Dielectric permittivity caused by hard phonon modes and electronic polarization tends to $\varepsilon_\infty \approx 10$. These results essentially differ from the results obtained by Colomban (1992) by impedance spectroscopy of the powder pellets of many protonic conductors, including the above discussed, where typical Debye-like Cole–Cole plots have been obtained. Electric conductivity (Figure 8.12) depends on frequency according to the "universal dynamic response" given by Equation 8.6. Additional results concerning the dispersion of permittivity and conductivity of these materials can be found in papers by Mizeris et al. (1988, 1991) and Augustyniak and Grigas (1994).

The next section describes the microscopic model of the hopping conductivity in $Rb_3H(SeO_4)_2$ (TRHSe) based on the extended version of PPM. Then some Monte–Carlo results for the disorder phase will be given, and, finally, the qualitative evaluation of $\sigma(\omega)$ in the framework of "jump-relaxation" model will be described for two-dimensional crystals.

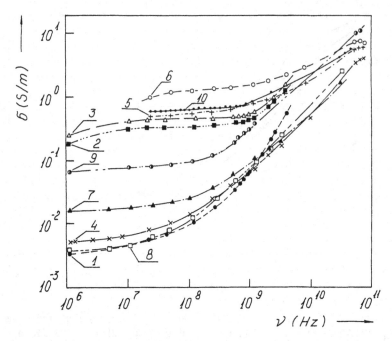

Figure 8.12. Frequency dependence of electrical conductivity of $Rb_3H(SeO_4)_2$ at: 470 (1), 480 (2) and 490 K (3); $(NH_4)_3H(SeO_4)_2$ along the *b*-axis at: 300 (4), 315 (5) and 345 K (6) and along the *c*-axis at 315 (7) and 345 K (8); $Cs_5H_3(SO_4)_4 2H_2O$ at 425 K (9); $CsDSO_4$ at 420 K (10).

8.7. MODEL OF $Rb_3H(SeO_4)_2$

An attempt will be made to develop the microscopic model of protonic conductivity based on the master equation approach for the lattice-gas model. The required data are the lattice structure, which is obtained from crystallographic data (Fortier et al. 1985), and the microscopic parameters included in the jump rate expression (Equation 8.11). These are: 1) the interaction potentials between protons; 2) the net potential barrier U between the adjacent sites due to proton interaction with the rigid lattice only; 3) the attempt frequency ν_0. The two latter parameters refer only to the dynamic properties of the system and, in the case when they are not initially known, they can be chosen to fit the experimental values of activation energy and the absolute value of the conductivity at certain temperature of the disordered phase. However, the interaction potentials are of major importance for the equilibrium properties of the system considered and can be evaluated through the analysis of the thermodynamics of the order-disorder phase transition. The very general microscopic theory of the phase transition in TRHSe has already been proposed (Javadov and Salejda 1990), but a large number of adjustable parameters introduced in this model allows only some rather qualitative predictions to be made. The development of dynamic theory based on this model should be commensurate with the impenetrable complexity.

The model of the proton subsystem of TRHSe in the limits of the lattice-gas model with the Hamiltonian given by Equation 8.7 is considered. The short-range pair interaction potentials are introduced in the first three coordination spheres only. No coupling between adjacent (001) planes is considered, since the interplanar distance is outside the third coordination sphere. This is, of course, a simplification since an ordering between planes in the low temperature phase actually exists (Plakida and Salejda 1988). This ordering can be considered as secondary compared to the in-plane ordering. However, the elastic orders of freedom do not appear explicitly in the Hamiltonian. They can be indirectly included in the effective interaction parameters and the binding energy of Equation 8.7. Thus, a simple two-dimensional lattice-gas model with short-range interprotonic coupling is considered. It is readily applicable for the extended PPM formalism, as well as for the Monte-Carlo simulation of the transport.

The (001) plane of virtual H-bonds is presented in Figure 8.13. Only one third of the sites is occupied by protons in $Me_3H(AO_4)_1$ compounds (A = S, Se). In the low temperature phase, protons form an ordered superstructure of H-bonds (Makarova et al. 1987). In order to introduce a quantitative measure of the order, the lattice is divided into three sublattices a, b, and c, which are nonequivalent with respect to the low temperature order. The occupancies C_a, C_b, and C_c of the sublattices give full account of the ordering. Thus, the low temperature phase (ground state) of TRHSe corresponds to $C_a = 1$ and $C_b = C_c = 0$. In the high temperature disordered phase the occupancies of all sublattices are equal: $C_a = C_b = C_c = 1/3$. The effective interaction potentials V_1, V_2, V_3, or V_4 couple the sites in the three nearest coordination spheres (as shown in Figure 8.13). The potentials V_3 or V_4 refer to the same interaction distance. Nevertheless, they are different since lattice sites interacting by V_3 are connected through an intermediate site, and those interacting by V_4 are not. Therefore, V_3 and V_4 can be formally different. The values of the potentials V_i should be obtained a priori from the first principles calculation or may be adjusted by fitting the theoretical predictions to the experimental data. Some initial restrictions of these values can be obtained from the ground state energy analysis: the set of V_i should provide the lowest energy of the experimentally observed ground state

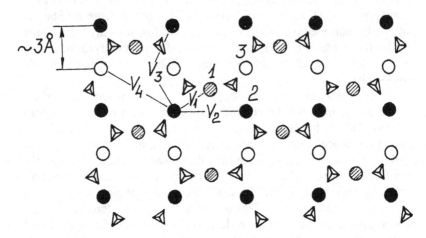

Figure 8.13. Projection of H-bonds system on the (001) plane of $Rb_3H(SeO_4)_2$. $\bigcirc$, $\bullet$, and $\bigcirc$ represent the virtual H-bonds of sublattices a, b, and c, respectively. SeO_4 groups are also shown. Pair interaction potentials in the first three coordination spheres are indicated as V_1, V_2, V_3, and V_4.

among all other possible ground states at a given stoichiometry. The following relations should stabilize the ground state of TRHSe:

$$V_1, V_2, V_3, V_4 > 0, \qquad V_2 < V_3 + V_4, \qquad V_1 > 2V_2. \tag{8.17}$$

It may be noticed that another compound of the class, $Cs_3H(SeO_4)_2$, has another low temperature ordered structure (Phase III), which corresponds to $C_a = C_b = 0$, $C_c = 1$ (Merinov et al. 1988; 1991). For this case the second relation of Equation 8.17 should be changed to $V_2 > V_3 + V_4$. So far this discussion has been limited to TRHSe.

8.7.1. Thermodynamics Properties

Following Equation 8.17, the approximate values of interaction potentials are accepted as $V_1 = 4000$ K, $V_2/V_1 = 0.5$, $V_3/V_1 = 0.25$, $V_4/V_1 = 0.5$. The calculation of equilibrium properties of the model was performed in the twelve-point-cluster approximation of the modified cluster variation method (CVM) (Zubkus et al. 1989). The details of the calculations will not be given here. The above values of V_i yield a second order phase transition between a low temperature ordered phase and a high temperature phase with the disordered system of H-bonds. The

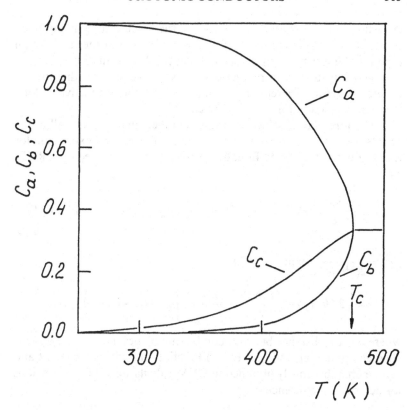

Figure 8.14. Temperature dependence of the sublattice occupancies, calculated by CVM.

transition temperature $T_c = 475$ K is a result of an appropriate choice of the absolute value of V_1. The temperature dependence of sublattice positions C_a, C_b, and C_c are presented in Figure 8.14. By adjusting two relations between V_i in the limits of restrictions in Equation 8.17 one also can obtain a first-order transition (e.g., $V_2/V_1 = 0.3$, $V_3/V_1 = V_4/V_1 = 0.2$). The various multisite correlation functions $\langle n_i n_j \cdots n_k \rangle$ are also obtained during the CVM calculation. They are later used in the calculation of ionic conductivity.

8.7.2. Protonic Conductivity of Rb₃H(SeO₄)₂

The protonic conductivity was calculated in the limits of the master equation approach with the jump rate expression (Equation 8.11). The extended PPM formalism (Kikuchi and Sato 1971) was applied. The extensions of this formalism result in the inclusion of many-body correlations beyond the two-body approximation in order to have an account of interactions in the second and third coordination spheres.

The low temperature phase is anisotropic due to the specific ordered structure. Therefore, two components of the conductivity (along the a- and b-axis of Figure 8.14) are calculated.

The formulae of PPM are rather cumbersome, and not all of the details of this calculation will be presented. The general expression for the conductivity given by Equation 8.14 can be concretized for TRHSe as

$$\sigma_a = \frac{v_0 e^2 3^{1/2}}{4kTc} \exp(-U/kT)[\langle h_1 h_2 \rangle + \langle h_1 h_3 \rangle] \exp[(\mu - E)/kT] f_{I_x},$$

(8.18)

$$\sigma_b = \frac{v_0 e^2}{2(3^{1/2})kTc} \exp(-U/kT)$$

$$\times \left[2\langle h_2 h_3 \rangle + \frac{\langle h_1 h_2 \rangle}{2} + \frac{\langle h_1 h_3 \rangle}{2} \right] \exp[(\mu - E)/kT] f_{I_y},$$

where c is the distance between conducting planes, $h_i = 1 - n_i$, μ is the chemical potential, and $(\mu - E)$ is a function of hydrogen content and can be unambiguously obtained in CVM calculations. The expressions for percolation efficiencies:

$$f_{I_x} \quad \text{and} \quad f_{I_y}$$

contain a large number of different multisite correlation functions $\langle n_i n_j \cdots n_k \rangle$.

The resulting $\sigma(T)$ curves are given in Figure 8.15 for both components and for two limiting cases: low frequency $\sigma(0)$ and high frequency, $\sigma(\infty)$, where the latter corresponds to $f_I = 1$. The low frequency conductivity corresponds to proton transfer over long distance, which is associated with continuous redistribution of neighboring ions. The high frequency conductivity $\sigma(\infty)$ is due to local hopping of the proton and is not related to the changes of its environment.

The interaction potentials V_i are the same as in the previous section. The attempt frequency, $v_0 = 2.2 \times 10^{12}$ Hz, and the net potential barrier, $U = 12,800\,K$ (1.1 eV), are chosen in order to meet the experimental values of the conductivity scale and activation energy of the disordered phase (Section 8.5).

The specific chain-like ordering of protons gives rise to a drastic difference between $\sigma_a(T)$ and $\sigma_b(T)$ curves. While a considerable disper-

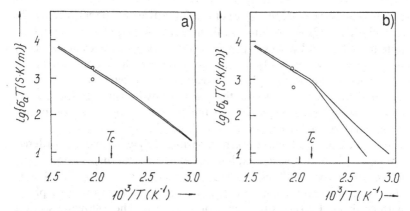

Figure 8.15. Temperature dependence of calculated protonic conductivity: a) along a-axis; b) along b-axis. The upper line is for σ_∞, the lower one is for $\sigma(0)$. Points represent the Monte–Carlo results at $T = 519\,\text{K}$.

sion of conductivity takes place in the ordered phase for σ_b, σ_a shows practically no dispersion in the same phase. This is due to the blocking of the conductivity path in the b direction by protons in the a sublattice, which results in the decrease of $\sigma(0)$. In the a direction, the sublattices b and c form conductivity channels parallel to proton chains of the sublattice a. No obstacles for the long distance transfer occur.

This peculiarity may cause an additional dispersion mechanism in the ordered phase. The possible multidomain structure of the crystal can form a sequence of high and low conductive domains in a certain direction (depending on the orientation of proton chains in each ordered domain). Such a structure may give rise to the low frequency dispersion due to macroscopic space-charge formation at the edges of the high conductive domains (similar to the effects of blocking interfaces). Anyway, the low frequency conductivity of the multidomain crystal should be determined by the σ_b.

The main features of the calculated $\log \sigma_b(1/T)$ are similar to the experimental ones (Figure 8.10). There is a characteristic kink at the phase transition. The activation energy of the ordered phase is larger than that of the disordered phase. At the same time, the activation energy for $\sigma(0)$ is bigger than for $\sigma(\infty)$ at the same temperature. A small deviation from the Arrhenius law takes place at temperature below the phase transition. However, quantitative discrepancies between calculated and experimental results are considerable. First of all, the experimentally observed drop of conductivity $\sigma(0)$ just below the phase transition is much steeper than the calculated one. This discrepancy can be attributed to a relatively weak dependence of the sublattice occupancies below T_c (especially for C_c, see Figure 8.14). A better choice of the interaction potentials V_i could yield steeper temperature dependencies of the occupancies and, consequently, of the conductivity. Such behavior is expected when the critical phase transition is approached. It has been reported (Baranov 1987) that the phase transition in TRHSe can be characterized by the exponents of the critical phase transition rather than by the second order one (in the classical limits).

The second problem concerns the value of $f_I \approx 0.9$ in the disordered phase. It by no means can explain the experimental dispersion of $\sigma(\omega)$ above 10^9 Hz with $\sigma(0)/\sigma(\infty) < 0.2$. As already mentioned, the PPM is somewhat inadequate for the calculation of f_I in the disordered phase with equivalent lattice sites and mobile species. The following section presents the results of Monte–Carlo simulation of the same model in the disordered phase.

8.7.3. Monte–Carlo Simulation of Protonic Conductivity

The calculation is performed on the one-third filled lattice containing 6144 sites. The total amount of protons is fixed, and periodic boundary conditions are used. The averaging was made over 20,000 Monte-Carlo steps. The jump probability is chosen of the form given by Equation 8.11 and the double occupancy of the sites is prohibited. The calculation scheme for $\sigma(0)$ and $\sigma(\infty)$ is the same as in the paper by Ozeki and Ishikawa (1988). From the proton flow under the applied electric field $\sigma(0)$ is obtained. And $\sigma(\infty)$ is calculated from the Kubo formula (Kubo1957) based on the fluctuation-dissipation theorem for the equilibrium state. The obtained values of $\sigma(0)$ and $\sigma(\infty)$ at $T = 519$ K are shown in Figure 8.15. As expected, values of the percolation efficiencies $f_{I_x} = 0.5$ and $f_{I_y} = 0.4$ are much closer to the experimental data, though still insufficient. We hope, that the appropriate choice of the model parameters would also improve the f_i.

The detailed calculation of $\sigma(\omega)$ at finite frequencies also can be based on the fluctuation-dissipation theorem by determining the proton current auto-correlation functions in the thermodynamic equilibrium. However, these calculations need much longer Monte-Carlo runs or larger lattices in order to obtain reliable data. Instead, some qualitative evaluations of $\sigma(\omega)$ for the system of interacting ions using the "jump-relaxation" model (Funke and Hoppe 1990) in the two- dimensional case have been made. As was already mentioned, the results of this model depend heavily on the parameter ρ (Section 8.3). In Figure 8.16, the $\varepsilon(\omega)$ and $\sigma(\omega)$ values calculated from the "jump-relaxation" model are pre-

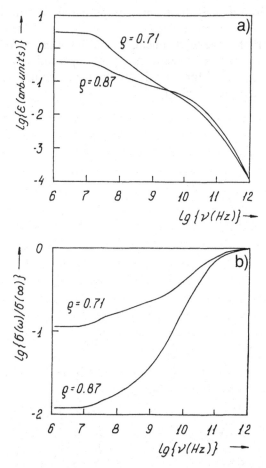

Figure 8.16. Frequency dependence of a) dielectric permittivity and b) conductivity obtained by "jump relaxation" model. The model parameter is ρ (Funke and Hoppe 1990).

sented for $\rho = 0.62$ and 0.87. The other parameters are: $\Delta = 0.9$, $\delta = 0.1$, $\nu_0 = 100$ GHz (for the notations see the reference above). The calculated curves meet the "universal dynamic response" (Equation 8.6) with $n \leq 1$ depending on ρ. Similar behavior was observed in various crystals of the protonic conductors. Of course, the "jump-relaxation" model does not take into account the particular structure of the material and is not very appropriate for TRHSe. Nevertheless, it allows a definitive statement that the fundamental microwave dispersion in perfect ionic conductors can be attributed to the interactions between the mobile ions only.

No distribution of relaxation times or potential barriers is actually needed.

It appears that the final conclusion on the origin of dispersion of dielectric permittivity and conductivity of superionics will be made through the extensive Monte-Carlo study of the hopping conductivity model with interionic coupling.

REFERENCES

Abdulin, R.S., I.N. Penkov, N.M. Nizamutdinov, J. Grigas and I.A. Safin, 1977, *Fiz. Tverd. Tela (Sov. Phys. Solid State)* **19**, 1632.

Afsar, M.N. and G.W. Chantry, 1977, *IEEE Trans. Microwave Theory Tech.* **MTT-25**, 509.

Afsar, M.N. and K.J. Button, 1981, *Int. J. Infrared Millimeter Waves* **2**, 1029.

Afsar, M.N., 1984, *IEEE Trans. Microwave Theory Tech.* **MTT-32**, 1598.

Afsar, M.N., 1985, *IEEE Trans. Microwave Theory Tech.* **MTT-33**, 1410.

Afsar, M.N. and K.J. Button, 1985, *Proc. IEEE* **75**, 131.

Afsar, M.N., J.R. Birch and R.N. Clarke, 1986, *Proc. IEEE* **74**, 183.

Agarwal, O.P. and P. Chand, 1985, *Solid St. Commun.* **54**, 65.

Agrawal, D.K. and C.H. Perry, 1971, *Phys. Rev. B* **4**, 1893.

Aksenov, V.L., A.Y. Didyk and N.M. Plakida, 1984, *Phys. Stat Sol. (b)* **124**, 45.

Albers, J., 1988, *Ferroelectrics* **78**, 3.

Aleksandrov, K.S., A.T. Anistratov, B.V. Beznosikov and N.V. Fedoseeva, 1981, *Phase Transitions in Crystals of Haloid Compounds ABX₃*, Novosibirsk, Nauka (in Russian).

Aliev, R.A., K.R. Allakhverdiev, A.I. Baranov, N.R. Ivanov and R.M. Sardarly, 1984, *Fiz. Tverd. Tela* **26**, 775.

Ambrazevičiene, V., A.A. Volkov, G.V. Kozlov, V.S. Krasikov and E.B. Kriukova, 1983, *Fiz. Tverd. Tela* **25**, 1605.

Andreyev, E.F., V.M. Varikash and L.A. Shuvalov, 1991, *Izv. Akad. Nauk SSSR, Ser. Fiz.* **55**, 572.

Aschmore, J.P. and H.E. Petch, 1975, *Canadian J. Phys.* **53**, 2694.

Auda, H. and R.F. Harrington, 1984, *IEEE Trans. Microwave Theory Techn.*, **MTT-32**, 1328.

Audzijonis, A., J. Grigas and A. Karpus, 1970, *Fiz. Tverd. Tela* **12**, 146.

Audzijonis, A. and A. Karpus, 1978, *Lietuvos Fizikos Rinkinys (Lithuanian Phys. Journ.)* **18**, 127.

Augustyniak, M.A. and S.K. Hoffmann, 1992, *Ferroelectrics* **132**, 129.

Augustyniak, M.A. and J. Grigas, 1994, *Lithuanian Phys. Journ.* **34**, 259.

Averbuch-Pouchot, M.A. and A. Durif, 1984, *Ferroelectrics* **52**, 271.

Avkhutskij, L.M., R.L. Davidovich, L.A. Zemnukhova, P.S. Gordienko, V. Urbonavičius and J. Grigas, 1983, *Phys. Stat. Sol. (b)* **116**, 483.

Banys, J., A. Brilingas, J. Grigas and G. Guseinov, 1987, *Fiz. Tverd. Tela* **29**, 3324.

Banys, J., J. Grigas and G. Guseinov, 1988, *Ferroelectrics* **82**, 3.
Banys, J., J. Grigas and G. Guseinov, 1989, *Lietuvos Fizikos Rinkinys* **29**, 348.
Banys, J., A. Brilingas and J. Grigas, 1990, *Phase Transitions* **20**, 211.
Banys, J., J. Grigas, S. Rudys and I.G. Kosakovskis, 1992a, *Lietuvos Fizikos Rinkinys* **32**, 261.
Banys, J., J. Grigas, V. Valiukenas and K. Wacker, 1992b, *Solid St. Commun.* **82**, 633.
Banys, J., J. Grigas, S. Lapinskas, Ž. Lileikis and Y. Yagi, 1992c, *Phys. Stat. Sol. (a)* **132**, 4071.
Banys, J. and J. Grigas, 1993, *Phys. Stat. Sol. (a)* **136**, 235.
Baranov, A.I., 1987, *Izv. Akad. Nauk SSSR, Ser. Fiz.* **51**, 2146.
Baranov, A.I., R.M. Fedosyuk, N.R. Ivanov, V.A. Sandler, L.A. Shuvalov, J. Grigas and R. Mizeris, 1987, *Ferroelectrics* **72**, 59.
Baranov, A.I., V.P. Khiznichenko and L.A. Shuvalov, 1989, *Ferroelectrics* **100**, 135.
Barker, A.S., 1975, *Phys. Rev. B* **12**, 4071.
Bartzokas, A. and D. Siapkas, 1980, *Ferroelectrics* **25**, 561.
Baryshnikov, S.V., E.V. Bursian and J.G. Girshberg, 1977, *Fiz. Tverd. Tela* **19**, 1163.
Bayliss, P. and W. Nowacki, 1972, *Kristallogr.* **135**, 308.
Beliackas, R., A. Brilingas, J. Grigas and J. Meškauskas, 1975, *Lietuvos Fizikos Rinkinys* **15**, 621.
Beyeler, H.U., P. Brüesch, L. Pietronero, W.R. Schneider, S. Strassler and H.R. Zaller, 1979, in: *Physics of Superionic Conductors*, edited by M.B. Salamon, Berlin, Heidelberg, New York, Springer-Verlag.
Bhartia, P., 1977, *AEÜ*, **31**, 60.
Birch, J.R., 1981, *Proc. SPIE* **289**, 362.
Birch, J.R. and R.N. Clarke, 1982, *Radio Electron. Eng.* **52**, 566.
Birch, J.R., 1983, *Proc. IEE* **130H**, 327.
Birch, J.R., G. Bechtold, F. Krewer and A. Poglitsch, 1983, *NPL Rep. DES 79*.
Birch, J.R., 1987, *Mikrochimica Acta*, III, 105.
Birnbaum, G. and E.R. Cohen, 1970, *J. Chem. Phys.* **53**, 2885.
Black, J., E.M. Conwell, L. Seigle and C.W. Spencer, 1957, *J. Phys. Chem. Solids* **2**, 240.
Blat, D.H. and V.I. Zinenko, 1981, *Fiz. Tverd. Tela* **23**, 3425.
Blinc, R., M. Jamžek-Vilpan, G. Lahajnar and G. Hajdukovič, 1970, *J. Chem. Phys.* **52**, 6407.
Blinc, R. and B. Žekš, 1979, *Soft Modes in Ferroelectrics and Antiferroelectrics*, edited by E.P. Wohlfarth, Amsterdam, North-Holland.
Blinc, R. and V.H. Schmidt, 1984, *Ferroelectrics Letters* **1**, 119.
Blinc, R. and A.P. Levanyuk, editors, 1986, *Incommensurate Phases in Dielectrics*, Elsevier Science Publishers B.V.
Blinc, R., V. Rutar, J. Dolinžek, B. Topič, F. Milia and S. Žumer, 1986, *Ferroelectrics* **66**, 57.
Boned, C. and J. Peyrelasse, 1982, *J. Phys. E: Sci. Instrum.* **15**, 534.
Bouziane, E., M.D. Fontana and G.E. Kugel, 1992, *Ferroelectrics* **125**, 331.
Brandt, A.A., 1963, *Microwave Investigations of Dielectrics*, Moscow, 19 (in Russian)
Brazis, R.S., J.K. Furdyna and J.K. Požela, 1979, *Phys. Stat. Sol. (a)* **11**, 11.
Brilingas, A., J. Grigas and V. Samulionis, 1976a, *Fizika Tverdogo Tela* **18**, 2816.
Brilingas, A., J. Grigas and N.N. Mozgova, 1976b, *Lietuvos Fizikos Rinkinys* **16**, 737.
Brilingas, A., J. Grigas and N.N. Mozgova, 1977a, *Lietuvos Fizikos Rinkinys* **17**, 213.
Brilingas, A., J. Grigas and P. Bendorius, 1977b, *Lietuvos Fizikos Rinkinys* **17**, 501.
Brilingas, A., J. Grigas and A.J. Dmytriv, 1977c, *Fizika Tverdogo Tela* **19**, 3445.
Brilingas, A., N.J. Bucko, J. Grigas, V. Samulionis and J.D. Sheshnitsch, 1978, *Kristall Techn.* **13**, 221.
Brilingas, A., R.L. Davidovich, J. Grigas, S. Lapinskas, M.A. Medkov, V. Samulionis and V. Skritski, 1986, *Phys. Stat. Sol. (a)* **96**, 101.
Brilingas, A., J. Grigas and V. Kalesinskas, 1987, *Izv. Akad. Nauk SSSR, Ser. Fiz.* **15**, 2196.
Brilingas, A., J. Grigas, V. Kalesinskas and R. Mizeris, 1990, *Proc. 6th Intl. School on Microwave Physics and Technics*, 723, Singapore, World Scientific.

Bruce, A.D., 1978, in: *Solitons and Condensed-Matter Physics*, edited by A.R. Bishop and T. Schneider, Berlin, Heidelberg, New York, Tokyo, Springer-Verlag.

Brückner, H.J., H.G. Unruch, G. Fischer and L. Genzel, 1988, *Z. Phys. B* **71**, 225.

Brückner, H.J., 1989, *Z. Phys. B* **75**, 259.

Buerger, M.J. and T. Hann, 1955, *Amer. Min.* **40**, 226.

Burns, G., 1976, *Phys. Rev. B* **13**, 215.

Bussey, H.E., 1967, *Proc. IEEE* **55**, 1046.

Bussman-Holder, A., H. Bilz and P. Vogl, 1983, in: *Dynamical Properties of IV–VI Compounds*, Berlin, Heidelberg, New York, Tokyo, Springer-Verlag.

Bussman-Holder, A. and H. Bilz, 1984, *Ferroelectrics* **54**, 5.

Cach, R. and Z. Czapla, 1990, *Phys. Stat. Sol. (a)* **121**, 299.

Cach, R., S. Dacko and Z. Czapla, 1989, *Phys. Stat. Sol. (a)* **116**, 827.

Carpenter, C.D. and R. Nitsche, 1974, *Mat. Res. Bull.* **9**, 1097.

Carvalko, A.V. and S.R. Salinas, 1978, *J. Phys. Soc. Jpn.* **44**, 238.

Chamberlain, J. and G.W. Chantry, 1973, *High Frequency Dielectric Measurements*, Guildford, UK, Science and Technology Press.

Champlin, K.S. and G.H. Glover, 1966, *IEEE Trans. Microwave Theory Tech.*, **MTT-14**, 397.

Christofferson, G.D. and J.D. McCullough, 1959, *Acta Cryst.* **12**, 14.

Clarke, R.N. and C.B. Rosenberg, 1982, *J. Phys. E: Sci. Instrum* **15**, 9.

Cochran, W., 1960, *Advances Phys.* **9**, 387.

Cole, K.S. and R.H. Cole, 1941, *J. Chem. Phys.* **9**, 341.

Cole, R.H., 1976, *IEEE Trans. Instrum. Meas.* **IM-25**, 371.

Colomban, Ph., editor, 1992, *Proton Conductors*, Cambridge, Cambridge University Press.

Cook, R.J. and C.B. Rosenberg, 1979, *J. Phys. D: App. Phys.* **12**.

Cullen, A.L., 1983, in: *Infrared and Millimetre Waves* **10**, edited by K.J. Button, New York, Academic Press.

Czapla, Z., J. Grigas and R. Sobiestianskas, 1990, in *Ferroelektrizität*, Halle-Wittenberg, 12.

Czapla, Z. and S. Dacko, 1993, *Ferroelectrics* **140**, 271.

Dacko, S., Z. Czapla and R. Sobiestianskas, 1992, in *Ferroelektrizität*, Halle-Wittenberg, 1.

Davidovich, R.L., P.S. Gordienko, J. Grigas, T.A. Kaidalova, V. Urbonavicius and L.A. Zemnukhova, 1984, *Phys. Stat. Sol (a)* **84**, 387.

Davidson, D.V. and R.H. Cole, 1951, *J. Chem. Phys.* **19**, 1484.

Debye, P., 1954, *Polar Molecules*, New York, Dover.

Decreton, M.C. and F.E. Gardiol, 1974, *IEEE Trans. Instrum. Meas.*, **IM-23**, 434.

Deguchi, K., E. Okaue and E. Nakamura, 1982, *J. Phys. Soc. Jpn.* **51**, 349; 3569; 3575.

Deguchi, K., E. Nakamura and K. Hirano, 1986, Ferroelectrics, **67**, 23.

Deguchi, K. and E. Nakamura, 1988, *J. Phys. Soc. Jpn.* **57**, 413.

DiAntonio, P., J. Toulouse, B.E. Vugmeister and S. Pilzer, 1994, *Ferroelectrics Letters* **17**, 115.

Dittmar, G. and H. Schafer, 1974, *Z. Naturforsch.* **29b**, 312.

Dougherty, T.P., G.P. Wiederrecht and K.A. Nelson, 1991, *Ferroelectrics* **120**, 79.

Dvořák, V. and Y. Ishibashi, 1976, *J. Phys. Soc. Jpn.* **41**, 548.

Dvořák, V., 1980, in: *Lecture Notes in Physics*, edited by A. Pekalski and J. Przystawa, Berlin, Heidelberg, New York, Tokyo, Springer-Verlag.

Dvořák, V., 1990, *Ferroelectrics* **104**, 135.

Ech-Chaoui, M., J.L. Miane and A. Daoub, 1993, *Phys. Stat. Sol. (a)* **140**, 567.

Edenharter, A., W. Nowacki and Y. Takeuchi, 1970, *Z. Kristallogr.* **131**, 397.

Elliot, D.S., 1915, *Phys. Rev.* **5**, 53.

Elliot, S.R., 1987, *Adv. Phys.* **36**, 135.

Ermark, F., B. Topič, U. Haeberlen and R. Blinc, 1989, *J. Phys.: Condens. Matter* **1**, 5489.

Fatuzo, E., G. Harbeke, W.J. Merz, R. Nitsche, H. Roetschi and W. Ruppel, 1962, *Phys. Rev.* **127**, 2036.

Fehst, I., M. Paasch, S.L. Hutton, M. Braune, R. Böhmer, A. Loidl, M. Dörfell, T. Narz, S. Haussühl and G.J. McIntyre, 1993, *Ferroelectrics* **138**, 1.

322   MICROWAVE DIELECTRIC SPECTROSCOPY

Fellers, R.G., 1967, Proc. IEEE 55, 1003.
Flocken, J.W., R.A. Guenther, J.R. Hardy and L.L. Boyer, 1992, Ferroelectrics 135, 309.
Fortier, S., M.E. Fraser and R.D. Heyding, 1985, Acta Cryst. 41, 1139.
Fousek, J., 1965, J. Appl. Phys. 36, 588.
Frazer, B.C., D. Demmingsen, W.D. Ellenson and G. Shirane, 1979, Phys. Rev. B 20, 2745.
Freitag, O., H.J. Brückner and H.G. Unruh, 1985, Z. Phys. B 61, 75.
Fridkin, V.M., 1980, Ferroelectric Semiconductors, New York, Consultants Bureau.
Fröhlich, H., 1958, Theory of Dielectrics, Oxford, Oxford University Press.
Funke, K. and R. Hoppe, 1990, Solid State Ionics 40–41, 200.
Gabriel, J. et al., 1984, J. Phys. E: Sci. Instrum. 17, 513.
Gaida, G. and S.S. Stuchly, 1983, IEEE Trans. Instrum. Meas., IM-32, 506.
Gaillard, J., J. Gloux, P. Glouch, B. Lamotte and G. Rius, 1984, Ferroelectrics 54, 81.
Gasanly, N.M., A.F. Gocharov, N.N. Melnik, A.S. Ragimov and V.I. Tagirov, 1983, Phys.
    Stat. Sol. (b) 116, 427.
Gauckler, L.J. and K. Sasaki, 1994, Solid State Ionics (Proc. Int. Workshop Interfaces and
    Ionic Materials), Ringberg.
Gauckler, L.J., K. Sasaki, A. Mitterdorfer, M. Gödickemeier and P. Bohac, 1994, Proc. 1st
    Europ. Solid Oxide Fuel Cell Forum, Lucerne, Switzerland.
Gaumann, A., A. Orliukas and P. Bohac, 1977, Helv. Phys. Acta 50, 773.
Gauss, K.E., H. Hap and G. Rother, 1975, Phys. Stat. Sol. (b) 72, 623.
Gavelis, R., 1977, Lietuvos Fizikos Rinkinys 17, 57.
Gavrilova. N.D., V.M. Varikash, S.V. Rodin and A.N. Galigin, 1982, Fiz. Tverd. Tela 24,
    2183.
Gervais, F., 1984, Ferroelectrics 53, 91.
Gestblom, B. and E. Noreland, 1977, Chem. Phys. Lett. 47, 349.
Gestblom, B. and B. Jonsson, 1980, J. Phys. E: Sci. Instrum. 13, 1067.
Giuntini, I.C., J.V. Zanchetta and F. Henn, 1988, Solid State Ionics 28–30, 142.
Glauber, R.J., 1963, J. Math. Phys. 4, 294.
Gomonnai, A.V., J.M. Vysochansky, M.I. Gurzan and V.Y. Slivka, 1983, Fiz. Tverd. Tela
    23, 1454.
Gödickemeier, M., K. Sasaki, P. Bohac and L.J. Gauckler, 1994, Proc. 6th IEA Workshop
    Advanced Solid Oxide Fuel Cells, Rome.
Gödickemeier, M., M. Michel, A. Orliukas, P. Bohac, K. Sasaki and L. Gauckler, 1994,
    Journ. Mat. Research 9, 1228.
Goncharov, Y.G., G.V. Kozlov, A.A. Volkov, J. Albers and J. Petzelt, 1988, Ferroelectrics
    80, 221.
Grant, E.J., R.J. Steppard and G.P. South, 1978, Dielectric Behaviour of Biological
    Molecules in Solution, Oxford, Oxford University Press.
Grigas, J. and A. Karpus, 1967, Fiz. Tverdogo Tela 9, 2882.
Grigas, J. and V. Shugurov, 1969, Izv. VUZ, Radiofizika 12, 307.
Grigas, J. and J. Meškauskas, 1971, Lietuvos Fizikos Rinkinys 11, 643.
Grigas, J., A. Orliukas and N. Mozgova, 1973, Lietuvos Fizikos Rinkinys 13, 119.
Grigas, J., A. Orliukas and N.N. Mozgova, 1975a, Lietuvos Fizikos Rinkinys 15, 827.
Grigas, J., N.N. Mozgova and A. Orliukas, 1975b, Kristallografyja 20, 1226.
Grigas, J., A. Orliukas, V. Samulionis, A. Brilingas and N.N. Mozgova, 1975c, Lietuvos
    Fizikos Rinkinys 15, 835.
Grigas, J., J. Meškauskas and A. Orliukas, 1976a, Phys. Stat. Sol. (a) 37, K39.
Grigas, J., L.A. Zadoroznaja, V.A. Liachovitskaja and A. Orliukas, 1976b, Lietuvos Fizikos
    Rinkinys 16, 833.
Grigas, J., V. Urbonavičius and A. Orliukas, 1977, Kristall. Techn. 12, 1163.
Grigas, J., 1978, Ferroelectrics 20, 173.
Grigas, J. and R. Beliackas, 1978, Fiz. Tverd. Tela 20, 2123.
Grigas, J., A. Kindurys, J. Meškauskas, A. Orliukas and V. Urbonavičius, 1978a, Kristall
    Technik 13, 683.
Grigas, J., R.L. Davidovich and V. Urbonavičius, 1978b, Fiz. Tverd. Tela 20, 1615.

Grigas, J., R. Davidovich, A. Pečeliunaite and V. Urbonavičius, 1979, *Kristall Techn.* **14**, 877.

Grigas, J., 1980a, *Phase Transitions and Ferroelectricity in Sb₂S₃ Family Crystals*, D.Sc. Thesis, Vilnius University, Lithuania.

Grigas, J., 1980b, *J. Phys. Soc. Jpn.* **49**, Suppl. B, 140.

Grigas, J. and J. Batarunas, 1980, *Fiz. Tverd. Tela* **22**, 707.

Grigas, J., R.L. Davidovich, A. Orliukas, V. Samulionis, V. Skritski and V. Urbonavičius, 1980, *Kristal Techn.* **15**, K73.

Grigas, J., V. Kalesinskas and R. Mizeris, 1982, *Fiz. Tverd. Tela* **24**, 844.

Grigas, J., V. Kalesinskas and I. Stasyuk, 1984a, *Ferroelectrics* **55**, 31.

Grigas, J., I.R. Zachek, V.S. Krasikov, I.V. Kutny, R.R. Levitsky and W. Paprotny, 1984b, *Lietuvos Fizikos Rinkinys* **24**, 33.

Grigas, J., R.R. Levitsky, E.V. Mits, W. Paprotny and I.R. Zachek, 1985, *Ferroelectrics* **64**, 349.

Grigas, J., V. Kalesinskas, S. Lapinskas, M.I. Gurzan, 1988a, *Phase Transitions* **12**, 263.

Grigas, J., V. Kalesinskas, S. Lapinskas and W. Paprotny, 1988b, *Ferroelectrics* **80**, 225.

Grigas, J., I. Zachek, N. Zaitseva, R. Levitsky, R. Mizeris and E. Mits, 1988c, *Lietuvos Fizikos Rinkinys* **28**, 486.

Grigas, J., 1990, in *Ferroelektrizität*, Halle-Wittenberg, 70.

Grigas, J., A. Brilingas and V. Kalesinskas, 1990a, *Ferroelectrics* **107**, 61.

Grigas, J., R. Sobiestianskas, Z. Czapla and E. Narewski, 1990b, *Acta Phys. Polonica* **A78**, 747.

Grigas, J., R. Mizeris and R. Sobiestianskas, 1991, in *Proc. Radio and Microwave Spectroscopy*, Poznan, 177.

Grigas, J. and R. Sobiestianskas, 1994, *Ferroelectrics* **156**, 279.

Guillot Gauthier, S., J.C. Peuzin, M. Olivier and G. Rolland, 1984, *Ferroelectrics* **52**, 293.

Hagiwara, B.T., K. Itoh, E. Nakamura, M. Komukae and Y. Makita, 1984, *Acta Cryst.* **40**, 718.

Happ, H., D. Schuster and U. Döbler, 1985, *Solid State Commun.* **56**, 417.

Hatta, I., 1968, *J. Phys. Soc. Jpn.* **24**, 1043.

Havlin, S. and S. Sompolinsky, 1980, *J. Phys. C* **12**, 3135.

Henkel, W., H.D. Hochheimer, C. Carlone, A. Werner, S. Ves and H.G. Shnering, 1982, *Phys. Rev. B* **26**, 3211.

Hill, R.M. and S.K. Ichiki, 1963, *Phys. Rev.* **132**, 1603.

Hill, R.M. and A.K. Jonscher, 1983, *Contemp. Phys.* **24**, 75.

Höchli, U.T., K. Knorr and L. Loidl, 1990, *Adv. Phys.* **39**, 405.

Horioka, M. and R. Abe, 1979, *Jap. J. Appl. Phys.* **18**, 2065.

Horioka, M. and R. Abe, 1990, *Ferroelectrics* **108**, 267.

Horton, G.K. and A.A. Maradudin, editors, 1975, *Dynamical Properties of Solids, v. I and II*, Amsterdam, North-Holland.

Hoshino, S. and S. Sato, 1963, *Jap. J. Appl. Phys.* **2**, 519.

Hosoya, M. and E. Nakamura, 1970, *Jap. J. Appl. Phys.* **9**, 552.

Hyde, P.J., 1970, *Proc. IEE* **117**, 1891.

Ibuki, S. and S. Yoshimatsu, 1955, *J. Phys. Soc. Jpn.* **10**, 549.

Ichiguchi, T., S. Nishikawa and K. Murase, 1980, *Solid. St. Commun.* **34**, 309.

Ichikawa, M., 1979, *J. Phys. Soc. Jpn.* **47**, 1562.

Ichikawa, M. and K. Motida, 1988, *J. Phys. Soc. Jpn.* **57**, 2217.

Igoshin, I.P., V.V. Gladki and V.A. Kirikov, 1984, *Fiz. Tverd. Tela* **26**, 3688.

Inushima, T., K. Uchinokura, K. Sasahara, E. Matsuura and R. Yoshizaki, 1981, *Ferroelectrics* **38**, 885.

Inushima, T., K. Uchinokura, K. Sasahara and E. Matsuura, 1982, *Phys. Rev. B* **26**, 2525.

Irie, K., 1978, *Ferroelectrics* **21**, 395.

Ise, K. and M. Koshiba, 1986, *IEEE Trans. Microwave Theory Tech.*, **MTT-34**, 103.

Ishii, T., H. Sato and R. Kikuchi, 1986, *Phys. Rev. B* **34**, 8335.

Ishii, T., 1990, *Solid State Ionics* **40–41**, 244.

Itoh, K., H. Matsunaga and E. Nakamura, 1980, *Phys. Stat. Sol.* (b) **97**, 289.
Itoh, K., T. Hagiwara and E. Nakamura, 1983, *J. Phys. Soc. Jpn.* **52**, 2626.
Itoh, K., L.Z. Zeng, E. Nakamura and N. Mishima, 1985, *Ferroelectrics* **63**, 29.
Iwasaki, H., 1969, *J. Phys. Soc. Jpn.* **27**, 513.
Iwata, Y., N. Koyano and I. Shibuya, 1980, *J. Phys. Soc. Jpn.* **49**, 304.
Izatt, J.R. and F. Kremer, 1981, *Appl. Opt.* **20**, 2555.
Jaeger, F.M., 1907, *Proc. Roy. Acad. Sci.* **9**, 808.
Jantsch, W., 1983, in: *Dynamical Properties of IV–VI Compounds*, Berlin, Heidelberg, New York, Tokyo, Springer-Verlag.
Javadov, N.A. and W. Salejda, 1990, *Phys. Stat. Sol.* (b) **158**, 475.
Jones, C.R., J.M. Dutta and H. Dave, 1984, *Int. J. Infrared Millimeter Waves* **5**, 279.
Jonscher, A.K., 1983, *Dielectric Relaxation in Solids*, London, Chelsea Dielectric Press.
Juodviršis, A., V. Kedavičius and P. Pipinis, 1969, *Fizika Tverd. Tela* **11**, 1420.
Kacenelenbaum, B.Z., 1966, *High-frequency Electrodynamics*, Moscow, 238 (in Russian).
Kachalov, N.P., A. Orliukas, I.N. Polandov and J. Grigas, 1975, *Fiz. Tverdogo Tela* **17**, 1790.
Kaczkowski, A. and A. Milewski, 1980, *IEEE Trans. Microwave Theory Tech.* **MTT-28**, 225.
Kalesinskas, V., K. Žičkus and A. Audzijonis, 1981, *Fizika Tverd. Tela* **23**, 1502.
Kalesinskas, V., J. Grigas, A. Audzijonis and K. Žičkus, 1982, *Phase Transitions* **3**, 217.
Kalesinskas, V., J. Grigas, W. Paprotny and V. Spitsyna, 1983a, *Fiz. Tverd Tela* **25**, 421.
Kalesinskas, V., J. Grigas, R. Jankevičius and A. Audzijonis, 1983b, *Phys. Stat. Sol. (b)* **115**, K11.
Kalesinskas, V. and V. Shugurov, 1991, *Lietuvos Fizikos Rinkinys* **31**, 104.
Kanda, E., A. Tamaki and T. Fujimura, 1982, *J. Phys. C: Solid State Phys.* **15**, 3401.
Kanda, E., A. Tamaki, T. Yamakami and T. Fujimura, 1983, *J. Phys. Soc. Jpn.* **52**, 3085.
Karpus, A. and M. Mikalkevičius, 1962, *Lietuvos Fizikos Rinkinys* **2**, 151.
Kavaliauskiene, G., M. Mikalkevičius and V. Rinkevičius, 1973, *Lietuvos Fizikos Rinkinys* **13**, 907.
Keens, A. and H. Happ, 1988, *J. Phys. C: Solid State Phys.* **21**, 1661.
Kent, M. and J. Kohler, 1984, *J. Microwave Power* **19**, 173.
Kikuchi, R., 1966, *Progr. Theor. Phys. Suppl.* **35**, 1.
Kikuchi, A., Y. Oka and E. Sawaguchi, 1967, *J. Phys. Soc. Jpn.* **23**, 337.
Kikuchi, R. and H. Sato, 1971, *J. Chem. Phys.* **53**, 702.
Kiosse, G.A., I.M. Razdobreyev, L.F. Kirpichnikova, 1990a, *Izv. Akad. Nauk SSSR, Ser. Fiz.* **54**, 749.
Kiosse, G.A., S.V. Ursaki, I.N. Gricheshen and K.R. Sbigli, 1990b, *Kristallografyja* **35**, 739.
Kirpichnikova, L.F., L.A. Shuvalov, N.R. Ivanov, B.N. Prasolov and E.F. Andreyev, 1989, *Ferroelectrics* **96**, 313.
Kirpichnikova, L.F., A.A. Urusovskaya, L.A. Shuvalov and V.I. Mozgovoy, 1990, *Ferroelectrics* **111**, 339.
Kliefoth, K., 1972, *Exp. Techn. Phys.* **20**, 119.
Kneubühl, F.K., 1989, *Infrared Phys.* **29** 1925.
Kobayashi, A., Y. Yoshioka, N. Nakamura and H. Chihara, 1988, *Z. Naturforsch.* **43a**, 233.
Kock, E.J. and H. Happ, 1980, *Phys. Stat. Sol.(b)* **97**, 239.
Kohatsu, I. and B.J. Wuensch, 1971, *Acta Cryst.* **27**, 1245.
Kolodziej, H.A., E. Narewski and L. Sobczyk, 1978, *Acta Phys. Polonica A* **53**, 79.
Komukae, M., T. Osaka, Y. Makita, T. Ozaki, K. Itoh and E. Nakamura, 1981, *J. Phys. Soc. Jpn.* **50**, 3187.
Komukae, M. and Y. Makita, 1985, *J. Phys. Soc. Jpn.* **54**, 4359.
Konsin, P., 1978, *Phys. Stat. Sol.* (b) **86**, 57.
Koutsoudakis, A., J. Louizis, A. Bartzokas and D. Siapkas, 1976, *Ferroelectrics* **12**, 131.
Kozlov, G.V., E.V. Kriukova, S.P. Lebedev, J. Grigas, W. Paprotny and Y. Uesu, 1984, *Ferroelectrics* **54**, 321.
Krasikov, V.S., S.V. Ogurtzov and E.F. Uchatkin, 1981, *Fiz. Tverd. Tela* **23**, 3425.

Kraus, J.D. and K.R. Carver, 1973, *Electromagnetics*, New York, McGraw-Hill.

Kremer, F. et al., 1984, in *Infrared and Millimetre Waves* **2**, edited by K.J. Button, New York, Academic Press.

Kriukova, E.V., 1984, *Fiz. Tverd. Tela* **26**, 717.

Kroupa, J., J. Petzelt, G.V. Kozlov and A.A. Volkov, 1987, *Ferroelectrics* **21**, 387.

Kubo, R., 1957, *J. Phys. Soc. Jpn.* **12**, 570.

Kutner, R., K. Binder and K.W. Kehr, 1982, *Phys. Rev. B* **26**, 2967.

Kvedaravičius, S., N. Mykolaitiene and A. Audzijonis, 1993, *Ferroelectrics* **150**, 381.

Lanagan, M.T., J.H. Kim, D.C. Dube, S.J. Jang and R.E. Newnham, 1988, *Ferroelectrics* **82**, 91.

Lapinskas, S., A. Sujeta, V. Urbonavičius, T. Verbitskaya and L. Svetlova, 1990, *Lietuvos Fizikos Rinkinys* **30**, 592.

Laurinavičius, A., P. Malakauskas and J. Požela, 1987, *Int. J. Infrared and Millimeter Waves* **8**, 573.

Laverghetta, T.S., 1984, *Microwave Materials and Fabrication Techniques*, Dedham, Artech House.

Leschenko, M.A., J.M. Poplavko, V.P. Bovtun, I.P. Igoshin and V.A. Jurin, 1989, *Fiz. Tverd. Tela* **31**, 286.

Levitsky, R.R., I.R. Zachek and V.I. Varanitsky, 1979, *J. Ukr. Phys.* **24**, 1486.

Levitsky, R.R., I.R. Zachek and V.I. Varanitsky, 1980, *Fiz. Tverd. Tela* **22**, 2750.

Levitsky, R.R., I.R. Zachek, E.V. Mits, J. Grigas and W. Paprotny, 1986, *Ferroelectrics* **67**, 109.

Levitsky, R.R., I.R. Zachek, I.V. Kutny, J.J. Schur, J. Grigas and R. Mizeris, 1990, *Ferroelectrics* **110**, 85.

Levstik, A., B. Žekš, I. Levstik, H.G. Unruh, G. Luther and H. Roemer, 1983, *Phys. Rev. B* **27**, 5706.

Ligthart, L.P., 1983, *IEEE Trans. Microwave Theory Tech.* **MTT-31**, 249.

Lipskis, K., M. Mikalkevičius, A. Sakalas and G. Juška, 1971, *Fizika Technika Poluprovodnikov* **5**, 694.

Lockwood, D.J., N. Ohno and M.H. Kuok, 1987, *J. Phys. C: Solid State Phys.* **19**, 3751.

Lucat, C., P. Sorbe, J. Portier, J.M. Reau, P. Hagenmüller and J. Grannec, 1977, *Mater. Res. Bull.* **12**, 145.

Lukaszewicz, K. and A. Pietraszko, 1992, *Acta Crystallografica* **C48**, 2069.

Lukaszewicz, K., 1994, private communication.

Lurio, A. and E. Stern, 1960, *J. Appl. Phys.* **31**, 1125.

Luther, G., 1973, *Phys. Stat. Sol. (a)* **20**, 227.

Lynch, A.C., 1974, *IEEE Trans. Instrum. Meas.* **IM-23**, 425.

Lynch, A.C., 1977, *Proc. IEE* **124**, 188.

Macdonald, J.R., *Impedance Spectroscopy*, 1987, John Wiley & Sons.

Maeda, M., 1988, *J. Phys. Soc. Jpn.* **57**, 2162.

Maj, Sz. and J.W. Modelski, 1984, *IEEE Trans. Microwave Theory Tech.* **MTT-5** Dig., 525.

Makarova, I.P., L.A. Shuvalov and V.I. Simonov, 1987, *Ferroelectrics* **79**, 111.

Makita, Y. and M. Sumita, 1971, *J. Phys. Soc. Jpn.* **31**, 792.

Mathias, B.T. and J.P. Remeika, 1957, *Phys. Rev.* **107**, 1727.

Matsubara, T., 1985, *Jap. J. Appl. Phys.* **24** Suppl., 1.

Mayor, M.M., B.M. Koperles, B.A. Savchenko, M.I. Gurzan, O.V. Morozova and N.F. Korda, 1983, *Fiz. Tverd. Tela* **25**, 214.

Mayor, M.M., V.P. Bovtun, J.M. Poplavko, B.M. Koperles and M.I. Gurzan, 1984a, *Fiz. Tverd. Tela* **26**, 659.

Mayor, M.M., B.M. Koperles, Y.M. Vysochansky and M.I. Gurzan, 1984b, *Fiz. Tverd. Tela* **26**, 690.

McKee, D.O. and J.E. McMullan, 1975, *Zeit. Kristall.* **142**, 447.

McMorrow, D., R.A. Cowley, P.D. Hatton and J. Banys, 1990, *J. Phys. Condens. Matter* **2**, 3699.

Merinov, B.V., N.B. Bolotina, A.I. Baranov and L.A. Shuvalov, 1988, *Kristallografyja* **33**, 1387.

Merinov, B.V., N.B. Bolotina, A.I. Baranov and L.A. Shuvalov, 1991, *Kristallografyja* **36**, 639.

Meškauskas, J. and J. Grigas, 1976, *Lietuvos Fizikos Rinkinys* **16**, 729.

Meškauskas, J., V. Samulionis and J. Grigas, 1977, *Fiz. Tverd. Tela* **19**, 1502.

Mikalkevičius, M. and B. Grigas, 1966, *Lietuvos Fizikos Rinkinys* **6**, 543.

Miniewicz, A., J. Lefebvre and R. Jakubas, 1990, *Ferroelectrics* **107**, 183.

Mizaras, R., J. Grigas, V. Valevičius, V. Samulionis and B. Brezina, 1994a, *Ferroelectrics* **158**, 357.

Mizaras, R., V. Valevičius, V. Samulionis, J. Grigas, L.A. Shuvalov and A.J. Baranov, 1994b, *Ferroelectrics* **155**, 201.

Mizeris, R., V. Kalesinskas and J. Grigas, 1985, *Kristallografyja* **30**, 1018.

Mizeris, R., J. Grigas, L.A. Shuvalov and A.I. Baranov, 1987, *Ferroelectrics Letters* **7**, 83.

Mizeris, R., J. Grigas, V. Samulionis, V. Skritski, A.I. Baranov and L.A. Shuvalov, 1988, *Phys. Stat. Sol. (a)* **110**, 429.

Mizeris, R., J. Grigas, R.N. Choudhary and T.V. Narasaiah, 1989, *Phys. Stat. Sol. (a)* **113**, 597.

Mizeris, R., J. Grigas, R.R. Levitsky, J.R. Zachek and S.J. Sorokov, 1990, *Ferroelectrics* **108**, 261.

Mizeris, R., A. Sujeta, V. Urbonavičius, J. Grigas and L.A. Shuvalov, 1991, *Kristallografyja* **36**, 693.

Mizeris, R., J. Grigas, and B. Březina, 1992, *Ferroelectrics* **126**, 133.

Moskvich, Y.N., A.M. Polyakov and A.A. Sukhovskij, 1988, *Ferroelectrics* **81**, 197.

Murch, G.E. and R.I. Thorn, 1977, *Philos. Mag.* **36**, 529.

Murch, G.E., 1982, *Solid State Ionics* **7**, 177.

Nakamura, E., H. Kajikawa and T. Ozaki, 1977, *J. Phys. Soc. Jpn.* **42**, 1472.

Nakamura, E. and H. Kajikawa, 1978, *J. Phys. Soc. Jpn.* **44**, 519.

Nakamura, E., K. Deguchi and T. Nishihara, 1990, *Ferroelectrics* **109**, 57.

Nakamura, E. and K. Deguchi, 1994, *Ferroelectrics* **159**, 167.

Nakamura, K. and S. Kashida, 1993, *J. Phys. Soc. Jpn.* **62**, 19.

Narasaiah, T.V. and R.N.P. Choudhary, 1987, *Phys. Stat. Sol. (a)* **100**, 317.

Negran, T.J., A.M. Glass, C.S. Brickenkamp, R.D. Rosenstein, R.K. Osterheld and R. Susot, 1974, *Ferroelectrics* **6**, 179.

Nelmes, R.J., 1980, *Ferroelectrics* **24**, 237.

Nelmes, R.J. and R.N.P. Choudhary, 1981, *Solid State Commun.* **38**, 321.

Nielsen, E.D., 1969, *IEEE Trans. Microwave Theory Tech.* **MTT-17**, 148.

Niizeki, N. and M.J. Buerger, 1957, *Z. Kristallogr.* **109**, 161.

Nishi, S., H. Kawamura and K. Murase, 1980, *Phys. Stat. Sol. (b)* **97**, 581.

Nishikawa, T., K. Wakino, H. Tanaka and Y. Ishikawa, 1990, *Proc. 20th Europ. Microwave Conference* **1**, Singapore, World Scientific.

Oelgart, G., R. Herrmann and U. Meschter, 1977, *Phys. Stat. Sol. (b)* **83**, 521.

Ogawa, I. and A. Kakimoto, 1978, *Rev. Sci. Instrum.* **49**, 936.

Ohno, N. and D.J. Lockwood, 1989, *Ferroelectrics* **94**, 361.

Oldfield, L.C., J.P. Ide and E.J. Griffin, 1985, *IEEE Trans. Instrum. Meas.* **IM-34**, 198.

Onodera, Y., 1970, *Progr. Theor. Phys.* **44**, 1477.

Orliukas, A. and J. Grigas, 1974, *Lietuvos Fizikos Rinkinys* **14**, 313.

Orliukas, A., J. Grigas and N.N. Mozgova, 1975, *Lietuvos Fizikos Rinkinys* **15**, 637.

Orliukas, A., V. Kalesinskas and V. Valiukenas, 1979, *Lietuvos Fizikos Rinkinys* **19**, 115.

Orliukas, A., V. Valiukenas, J. Grigas, N.I. Bucko, 1980, *Lietuvos Fizikos Rinkinys* **20**, 74.

Osaka, T., M. Sumita and Y. Makita, 1983, *J. Phys. Soc. Jpn.* **52**, 1124.

Ozaki, T., 1980, *J. Phys. Soc. Jpn.* **49**, 234.

Ozeki, Y. and T. Ishikawa, 1986, *J. Phys. Soc. Jpn.* **55**, 3931.

Ozeki, Y. and T. Ishikawa, 1988, *J. Phys. Soc. Jpn.* **57**, 988.

Palik, E.D. and J.K. Furdyna, 1970, *Rep. Progr. Phys.* **33**, 1195.

Pannel, R.M. and B.W. Jervis, 1981, *IEEE Trans. Microwave Theory Tech.* **MTT-29**, 383.

Paprotny, W., J. Grigas, R.R. Levitsky, I.V. Kutny and V.S. Krasikov, 1984a, *Ferroelectrics* **61**, 19.

Paprotny, W., J. Grigas and L. Syrkin, 1984b, *Lietuvos Fizikos Rinkinys* **24**, 72.

Paprotny, W. and J. Grigas, 1985, *Ferroelectrics* **65**, 201.

Parisien, B.R. and S.S. Stuchly, 1979, *IEEE Trans. Instrum. Meas.* **IM-28**, 269.

Parker, T.J., 1987, in *Analytical Spectroscopy Library* **2**, edited by C. Burgess and K.D. Mielenz, Amsterdam, Elsevier.

Pavilonis, A., V. Ivaška and V. Kybartas, 1978, *Lietuvos Fizikos Rinkinys* **18**, 323.

Pawlaczyk, C., R. Jakubas, K. Planta, C. Bruch, J. Stephan and H.G. Unruh, 1992, *Ferroelectrics* **126**, 145.

Penkov, I.N., I.A. Safin and A. Juodviršis, 1970, *Lietuvos Fizikos Rinkinys* **10**, 115.

Pething, R., 1979, *Dielectric and Electronic Properties of Biological Materials*, Chichester, New York, Wiley.

Petrov, V., 1972, *Dielectric Measurements of Ferroelectrics*, Moscow (in Russian).

Petzelt, J., 1969, *Phys. Stat. Sol.* **36**, 321.

Petzelt, J., 1973, *Ferroelectrics* **5**, 219.

Petzelt, J. and J. Grigas, 1973, *Ferroelectrics* **5**, 59.

Petzelt, J., G.V. Kozlov and A.A. Volkov, 1987, *Ferroelectrics* **73**, 101.

Petzelt, J., 1988, *Ferroelectrics* **80**, 221.

Pietraszko, A., K. Lukaszewicz and L.F. Kirpichnikova, 1993, *Polish J. Chem.* **67**, 1877.

Plakida, N.M. and W. Salejda, 1988, *Phys. Stat. Sol.* (b) **148**, 473.

Polynova, T.N. and M.A. Porai-Koshic, 1966, *J. Struct. Chem.* **7**, 146.

Pouget, J.P., S.M. Shapiro and K. Nassau, 1979, *J. Phys. Chem. Sol.* **40**, 267.

Powles, J.G., 1948, *Trans. Faraday Soc.* **44**, 802.

Pratt, G.J. and M.J.A. Smith, 1982, *J. Phys. E: Instrum. Meas.* **15**, 927.

Pykacz, B.H., Z. Czapla, J. Mröz, 1984, *Acta Phys. Pol.* **A66**, 639.

Rachford, F.J. and D.W. Forestor, 1983, *IEEE Trans. Magn.* **MAG-19**, 1883.

Rao, K.R., S.L. Chaplot, V.M. Padmanabhan and P.R. Vijayaraghavan, 1982, *Pramana* **19**, 593.

Reinecke, T.L. and K.L. Ngai, 1977, *Ferroelectrics* **6**, 85.

Ribar, B. and W. Nowacki, 1970, *Acta Cryst.* **26**, 201.

Richter, W., H. Kohler and C.R. Becker, 1977, *Phys. Stat. Sol.* (b) **84**, 619.

Riede, V. and H. Sobota, 1978, *Czechoslovak Journ. of Physics* **B28**, 886.

Rinkevičius, V., M. Mikalkevičius and K. Lipskis, 1967, *Lietuvos Fizikos Rinkinys* **7**, 675.

Ritus, A.I., A.S. Rosslik, J.M. Vysochansky, A.A. Grabar and V.Y. Slivka, 1985, *Fiz. Tverd. Tela* **27**, 2225.

Roemer, H. and G. Luther, 1981, *Ferroelectrics* **38**, 919.

Romain, F. and A. Novak, 1991, *Journ. Molecular Structure* **263**, 69.

Rytz, D., M.B. Klein, B. Bobbs, M. Matloubian and H. Fetterman, 1985, *Jap. Journ. of Appl. Phys.* 24-2 Supplement, 1010.

Rzepecka, M.A. and S.S. Stuchly, 1975, *IEEE Trans. Instrum. Meas.* **IM-24**, 27.

Sahalos, J.N., 1985, *J. Phys. D: Appl. Phys.* **18**, 1415.

Salk, M., K. Wacker, J. Fisher and V. Kramer, 1990, *Thermochimica Acta* **160**, 87.

Samara, G.A., 1975, *Ferroelectrics* **9**, 209.

Samulionis, V., V. Kunigelis and M. Girschovichius, 1971, *JETF* **61**, 1941.

Samulionis, V., A. Orliukas and J. Grigas, 1974, *Fiz. Tverd. Tela* **16**, 208.

Samulionis, V., A. Orliukas and J. Grigas, 1975, *Ultrasound (Lithuanian Journ.)* **7**, 19.

Samulionis, V., V. Valevičius, J. Grigas and J.M. Vysochansky, 1990, *Ferroelectrics* **105**, 397.

Sato, S., 1968, *J. Phys. Soc. Jpn.* **25**, 185.

Sato, H. and R. Kikuchi, 1983, *Phys. Rev. B* **28**, 648.

Sato, H., 1990, *Solid State Ionics* **40–41**, 725.

Scalapino, D.J., Y. Imry and P. Pinkus, 1975, *Phys. Rev. B* **11**, 2042.

Ščavničar, S., 1960, *Z. Kristallogr.* **114**, 85.

Schaack, G., 1990, *Ferroelectrics* **104**, 147.

Sekine, A., M. Sumita, T. Osaka and Y. Makita, 1988, *J. Phys. Soc. Jpn.* **57**, 4004.

Shashikala, M.N., B. Raghunatha Chary, H.L. Bhat and P.S. Narayanan, 1989, *J. Raman Spectrosc.* **20**, 351.

Shin, S., Y. Tezuka, M. Ishigame, K. Deguchi and E. Nakamura, 1990, *Phys. Rev.* **B41**, 10155.

Siapkas, D., 1974, *Ferroelectrics* **7**, 295.

Silverman, B.D., 1964, *Phys. Rev.* **135**, H1596.

Simon, P., F. Gervais and E. Courtens, 1988, *Phys. Rev. B* **37**, 1969.

Singer, H. and I. Peshel, 1980, *Z. Phys. B* **39**, 333.

Slivka, V.Y., J.M. Vysochansky, M.I. Gurzan and D.V. Chepur, 1978, *Fiz. Tverd. Tela* **20**, 3530.

Sobiestianskas, R., J. Grigas and Z. Czapla, 1989, *Ferroelectrics* **100**, 187.

Sobiestianskas, R., J. Grigas and Z. Czapla, 1990, *Acta Phys. Polonica* **A78**, 477.

Sobiestianskas, R., J. Grigas, V. Samulionis and E.F. Andreyev, 1991, *Phase Transitions* **29**, 167.

Sobiestianskas, R., J. Grigas and Z. Czapla, 1992a, *Phase Transitions* **37**, 157.

Sobiestianskas, R., J. Grigas, E.F. Andreyev and V.M. Varikash, 1992b, *Phase Transitions* **40**, 85.

Sobiestianskas, R., Z. Czapla and J. Grigas, 1992c, *Phys. Stat. Sol. (a)* **130**, K69.

Sobiestianskas, R., J. Grigas, Z. Czapla and S. Dacko, 1993, *Phys. Stat. Sol. (a)* **136**, 223.

Sobiestianskas, R., J. Banys, J. Grigas, S. Dacko and Z. Czapla, 1994, *Ferroelectrics Letters* **18**, 39.

Sobiestianskas. R., R. Mizaras, J. Grigas and Z. Czapla, 1995, *Phys. Stat. Sol. (a)* **147**, K49.

Srikrishanan, K. and W. Nowacki, 1975, *Z. Kristallogr.* **141**, 174.

Stadnicka, K., 1994, private communication.

Stasyuk, I.V. and N.M. Kaminskaya, 1974, *Ukr. Phys. Journ.* **19**, 237.

Stasyuk, I.V., J. Grigas and V. Kalesinskas, 1983, Preprint of *Inst. of Theor. Phys.* ITF-83-89R, Kiev.

Steigmeier, E.F., H. Anderset and G. Harbeke, 1975, *Phys. Stat. Sol. (b)* **70**, 705.

Strukov, B.A., S.A. Taraskin and Z.V. Savilova, 1986, *Phys. Stat. Sol. (a)* **95**, 447.

Stuchly, M.A. and S.S. Stuchly, 1980a, *J. Microwave Power* **15**, 19.

Stuchly, M.A. and S.S. Stuchly, 1980b, *IEEE Trans. Instrum. Meas.* **IM-29**, 176.

Sugawara, F. and T. Nakamura, 1972, *J. Phys. Chem. Sol.* **33**, 1665.

Sumita, M., T. Osaka and Y. Makita, 1981, *J. Phys. Soc. Jpn.* **50**, 154.

Sumita, M., T. Osaka and Y. Makita, 1984, *J. Phys. Soc. Jpn.* **53**, 2784.

Suzuki, M. and R. Kubo, 1968, *J. Phys. Soc. Jpn.* **24**, 51.

Takagi, Y., 1979, *J. Phys. Soc. Jpn.* **47**, 567.

Takagi, Y., 1987, *Ferroelectrics* **72**, 67.

Takayama, Y., K. Deguchi and E. Nakamura, 1984, *J. Phys. Soc. Jpn.* **53**, 4121.

Teng, M.K., M. Balkanski and M. Massot, 1972, *Phys. Rev. B* **5**, 1031.

Tominaga, Y., M. Tokunaga and I. Tatuzaki, 1985, *Solid State Commun.* **54**, 979.

Topič, B., R. Blinc and L.A. Shuvalov, 1984, *Phys. Stat. Sol. (a)* **85**, 409.

Torgashev, V.I., L.F. Kirpichnikova, Y.I. Yuzyuk and L.A. Shuvalov, 1990, *Ferroelectrics* **84**, 172.

Toupry, N., H. Poulet and M. Postollec, 1981, *J. Raman Spectr.* **11**, 81.

Tyndall, E.P.T., 1923, *Phys. Rev.* **21**, 1625.

Uchinokura, K., T. Inushima, E. Matsuura and A. Okamoto, 1981, *Ferroelectrics* **38**, 901.

Udovenko, A.A., V.N. Butenko, R.L. Davidovich, V.G. Andrianov, M.J. Antipin and A.I. Janovsky, 1981, *Kristallografyja* **26**, 488.

Urban, S., W. Zajac, R. Jakubas, C.J. Carlile and B. Gabrys, 1992, *Physica B* **180–181**, 1050.

Urbonavičius, V., R.L. Davidovich and J. Grigas, 1982a, *Lietuvos Fizikos Rinkinys* **22**, 81.

Urbonavičius, V., V.E. Shneider, J. Grigas and R.L. Davidovich, 1982b, *Sov. Phys. JETP* **56**, 151.

Vakhrushev, S.B., V.V. Zhdanova, B.E. Krytkovski, N.M. Okuneva, K.R. Allakhverdiev, R.A. Aliev and R.M. Sardarly, 1984, *JETP Lett.* **39**, 291.

Valevičius, V., V. Samulionis, J. Banys and J. Grigas, 1994, *Ferroelectrics* **156**, 315.

Van Germet, M.J.C., 1973, *Philips Res. Repts.* **28**, 530.

Van Loon, R. and R. Finsy, 1975, *J. Phys. D: Appl. Phys.* **8**, 1232.

Vendik, O.G., editor, 1979, *Ferroelectrics at Microwaves*, Moscow, Sov. Radio (in Russian).

Voigt, K.H., 1929, *Z. Phys.* **57**, 154.

Volkov, A.A., J.G. Goncharov, G.V. Kozlov, K.R. Allakhverdiev and R.M. Sardarly, 1983a, *Fiz. Tverd. Tela* **25**, 2061.

Volkov, A.A., G.V. Kozlov, N.I. Afanasjeva, J.M. Vysochansky, A.A. Grabar and V.Y. Slivka, 1983b, *Fiz. Tverd. Tela* **25**, 2575.

Volkov, A.A., J.G. Goncharov, G.V. Kozlov, K.R. Allakhverdiev and R.M. Sardarly, 1984a, *Fiz. Tverd. Tela* **26**, 2196.

Volkov, A.A., J.G. Goncharov, G.V. Kozlov, S.P. Lebedev, A.M. Prohorov, R.A. Aliev and K.R. Allakhverdiev, 1984b, *JETP Lett.* **37**, 517.

Volkov, A.A., J.G. Goncharov, G.V. Kozlov and S.P. Lebedev, 1990, *Submillimeter Dielectric Spectroscopy of Solids*, in *Proc. IOFAN*, Moscow, Nauka (in Russian).

Volkov, A.A., G.V. Kozlov, A.G. Pimenov and A.V. Sinitsky, 1992, *Ferroelectrics* **126**, 157.

Volkov, A.A., B.P. Gorshunov, G. Komandin, W. Fortin, G.E. Kugel, A. Kania and J. Grigas, 1995, *J. Phys. C: Condensed Matter* **7**, 785.

Volkova, L.M. and A.A. Udovenko, 1988, in: *Problems of Crystallochemistry*, Moscow, Nauka (in Russian).

von Hippel, A.R., 1954, *Dielectric Materials and Applications*, New York, MIT Technology Press and Wiley.

Vysochansky, J.M., V.Y. Slivka, J.M. Voroshilov, M.I. Gurzan and D.V. Chepur, 1979, *Fiz. Tverd. Tela* **21**, 2402.

Vysochansky, J.M. and V.Y. Slivka, 1994, *Ferroelectrics of $Sn_2P_2S_6$ Family*, Lvov (in Russian).

Watarai, S. and I. Matsubara, 1984, *J. Phys. Soc. Jpn.* **53**, 3648.

Wieting, T.J. and M. Schlüter, editors, 1979, *Electrons and Phonons in Layered Crystal Structures*, Dordrecht, Boston, London, D. Reidel Publishing Company.

Wolak, J. and Z. Czapla, 1981, *Phys. Stat. Sol.* (a) **67**, K171.

Wong, T., M. Brodwin and R. Dupon, 1981, *Solid State Ionics* **5**, 489.

Yagi, T., 1990, *Ferroelectrics* **109**, 63.

Yasuda, N., S. Fujimoto and T. Asano, 1980, *Phys. Letters A* **76**, 174.

Yokota, S., N. Takanohasshi, T. Osaka and Y. Makita, 1982, *J. Phys. Soc. Jpn.* **51**, 199.

Yoshida, H., M. Endo, T. Kaneko, T. Osaka and Y. Makita, 1984, *J. Phys. Soc. Jpn.* **53**, 910.

Yoshikado, S. and I. Taniguchi, 1989, *IEEE Trans. Microwave Theory Tech.*, **MTT-37**, 984.

Yu, D. et al., 1979, *Inst. and Experimental Tech.* **22**, 611.

Yu, P.K. and A.L. Cullen, 1982, *Proc. Roy. Soc. Lond. A* **380**, 49.

Zachek, I.R. and R.R. Levitsky, 1980, *Theor. and Math. Phys.* **43**, 128.

Zallen, R., 1975, in: *Lattice Dynamics and Intermolecular Forces*, edited by S. Califano, New York, Academic.

Žekš, B., L.C. Schukla and R. Blinc, 1972, *J. Phys.* **C33**, Suppl. 67.

Zemnukhova, L.A., R.L. Davidovich, P.S. Gordienko, J. Grigas, A.N. Kovrianov, S.I. Kuznetsov, T.A. Kaidalova and V. Urbonavicius, 1983, *Phys. Stat. Sol.* (a) **80**, 553.

Zinenko, V.I., 1985, *Ferroelectrics* **63**, 179.

Zubkus, V.E., S. Lapinskas and E.E. Tornau, 1989, *Phys. Stat. Sol.* (b) **156**, 93.

Žumer, S., 1980, *Phys. Rev. B* **21**, 1298.

SUBJECT INDEX

A

absorption edge, 123
activation energy, 148, 304
aikinite, 117, 153
alum, 285
alumina, 294
ammonium hydrogen selenate, 272
amplitron, 4
amplitudon, 195, 199, 206, 209
anharmonicity, 173, 177, 179, 184
anisotropic materials, 31, 61
antimonite (Sb_2S_3), 111, 122
antimony selenate, 112
antimony sulfobromide (SbSBr), 120, 164
antimony sulfoiodide(SbSI), 120, 164
antimony teluride, 114
asymmetry parameter, 146
arsenium selenide, 112
arsenium sulphide, 112
arsenium teluride, 112
approximation
  -cluster, 217, 238, 312
  -mean-field, 163
  -quasi-static, 63
  -random phase, 175
  -sech-, 13
atmospheric windows, 107

ATN, 162
attempt frequency, 292, 310

B

backward wave oscillator, 3, 29, 57, 93
berthierite, 115, 150, 152
betaine compounds, 257
biological substances, 25
birefringence, 150
bismuthinite, 112, 122, 154
bismuth sulphide, 112, 114, 130
bismuth selenide, 114, 142
bismuth teluride, 114, 142
Birnbaum-Cohen relaxation, 15
bonding
  -chemical, 114, 120
  -donor-acceptor, 114
  -ionic-covalent, 114
  -metallic, 114
boundary conditions, 66, 71, 76, 88
bournonite, 116
Brillouin zone, 128, 144, 199

C

capacitor, 31, 40
  -coaxial, 43
  -disk, 35

331

Printed in the United States
by Baker & Taylor Publisher Services